Phytoremediation in Food Safety

Phytoremediation is the process that uses plants to remove pollutants from soils. These pollutants are stored in the edible parts of plants and, if they are consumed above a certain level, they become a health risk for humans and animals. This book is a critical review of phytoremediation, its direct and indirect effects on food products, and the risks posed by this cost-effective technology in food safety. It shows how different plants are suited for phytoremediation, explains the role of toxicants in the environment, and analyzes their effects and risks in the food chain at a global level. It also reviews the extraction methods of toxicants from plants after they are exposed to phytoremediation.

FEATURES:

- Summarizes the phytoremediation technology for effective remediation
- Describes different types of pollutants in soils that render food products useless
- Identifies the role of phytoremediation in the environment and its advantages and disadvantages
- Explains the roles of phytoexclusion and phytostabilization in foods and food safety
- Includes many case studies to describe the extraction protocols in postharvest for food safety

This book is intended for practitioners in public and private companies involved in soil remediation and food production, as well as graduate students and academics, in both developed and developing countries, who are involved in soil and environmental sciences, the food industry, agriculture, and biotechnology.

Phytoremediation in Food Safety

Risks and Prospects

Edited by
Sesan Abiodun Aransiola and Naga Raju Maddela

CRC Press
Taylor & Francis Group
Boca Raton London New York

CRC Press is an imprint of the
Taylor & Francis Group, an **informa** business

Designed cover image: © Shutterstock Templates

First edition published 2025
by CRC Press
2385 NW Executive Center Drive, Suite 320, Boca Raton FL 33431

and by CRC Press
4 Park Square, Milton Park, Abingdon, Oxon, OX14 4RN

CRC Press is an imprint of Taylor & Francis Group, LLC

ISBN: 978-1-032-68374-4 (hbk)
ISBN: 978-1-032-68376-8 (pbk)
ISBN: 978-1-032-68375-1 (ebk)

DOI: 10.1201/9781032683751

Typeset in in Times LT Std
by Apex CoVantage, LLC

Contents

SECTION 2 *The Risks of Phytoremediation in Food Safety*

Foreword

With the increase in anthropogenic activities from agriculture and industry, as well as the ever-increasing global population, contaminants of inorganic, organic, and/or chemical origins find their way into the soil and eventually enter the food cycle, causing acute and/or chronic damage to living organisms, including humans. This calls for a remediation strategy. However, the traditional methods to deal with soil contamination, which include chemical treatment, physical treatment, biological treatment, and heat treatment, are unfortunately not only costly but also have significant side effects on the ecosystem. Therefore, phytoremediation is recommended as more effective, economic, and environmental remediation. But this remediation technology can also cause contaminants to enter the food chain and have serious effects on human health. It is well-known that plants absorb carbon dioxide and release oxygen through photosynthesis while promoting degradation of contaminants through interaction with microorganisms in the rhizosphere and absorbing soil contaminants into their body system. Phytoremediation is the process that uses plants to remove pollutants from the soils. After removal, they are stored in the plants' edible parts and sometimes at more than permissible levels. If the edible parts are consumed, the pollutants will have effects (short/acute or long/chronic) on human beings. There is now a need to look into and review whether phytoremediation is a blessing or a curse, and also, what is to be done to make these food crops safer and globally acceptable for consumption by all.

This book, *Phytoremediation in Food Safety: Risks and Prospects*, is a good collection of independent chapters that presents full insight in the study of phytoremediation technology for the removal of contaminants from the soil and, of course, analyzing its role in food safety.

I therefore have no doubt that the chapters in this book provide adequate information on phytoremediation in food safety and do fill the expected scientific knowledge gaps in this area.

Prof. Ahmed Salihu Dan-Kishiya
Dean, Faculty of Science
University of Abuja, Nigeria
February 17, 2024

Preface

Food and food products are essential commodities globally and the factors affecting its general acceptability need to be critically reviewed. Phytoremediation is one of the most cost-effective remediation options, but as a stand-alone technology, it is often not lucrative enough to make it appealing for farmers, especially in economically vulnerable regions. Phytoremediation is a cost-effective and environmentally friendly technology, but if most of its postharvest protocols are not handled professionally, it could become a threat to food safety. Plants possess the natural ability to accumulate nutrients and any contaminants present in the soil. However, this technology when used for remediation processes accumulates the pollutants from the soils through the root and mostly store these toxicants in the plants' edible parts. When these are not checked or extracted from the plants, they pose a danger to human and animal health if they enter the food chain and are consumed. This book, *Phytoremediation in Food Safety: Risks and Prospects*, was conceived to provide in-depth information on the role of phytoremediation in food safety. Generally speaking, soil is polluted with both organic and inorganic contaminants and there is a need to employ an environmentally friendly technology in its remediation processes. Phytoremediation involves the use of green technology plant to remediate the polluted soils.

The book is divided into three sections. Section 1 is titled "Prospects of Phytoremediation for a Safe Environment and Safe Food," section 2 is "The Risks of Phytoremediation in Food Safety," and section 3 is "Bringing Safe Foods Home through Phytoremediation."

The chapters were contributed by 68 academicians, scientists, and researchers from nine different countries (the United Kingdom, Hong Kong, Nigeria, India, the United States, Australia, Canada, Germany and Malaysia) across the world.

Sesan Abiodun Aransiola, Ph.D.
University of Abuja, Nigeria
National Biotechnology Development Agency (NABDA)
Ogbomoso, Nigeria

Naga Raju Maddela, Ph.D.
Universidad Técnica de Manabí, Portoviejo, Ecuador

About the Editors

Sesan Abiodun Aransiola is a Ph.D. holder in environmental microbiology in the Department of Microbiology, Federal University of Technology, Minna, Nigeria. He obtained his first degree (B.Tech) and master degree (M.Tech) from the same department in 2009 and 2014 respectively. He has demonstrated his research expert in the production of vermicasts from vermicoposting (vermitechnology) of organic wastes to assist plants in the remediation of polluted soil with heavy metals. Currently, he is a Lecturer at the Department of Microbiology, University of Abuja, Nigeria. His area of interest is environmental microbiology with research areas in phytoremediation, vermicomposting, biosorption, and bioremediation of soil contaminated environment. He is an award-winning researcher and has over 70 publications including book chapters and research and review articles of good international repute with high impact factor. He has co-edited scientific books of global interest which include *Microbial Biotechnology for Bioenergy* (Elsevier); *Ecological Interplays in Microbial Enzymology, Vermitechnology: Economic, Environmental and Agricultural Sustainability, Prospects for Soil Regeneration and Its Impact on Environmental Protection* (Springer); *Microbial Biofilms: Applications and Control* and *Marine Greens: Environmental, Agricultural, Industrial and Biomedical Applications* (both CRC Press), and has many more in process. Also, he has worked to the rank of assistant chief scientific officer at Bioresources Development Centre, National Biotechnology Development Agency, Nigeria, where he was involved in the uses of vinasse (a by-product of ethanol) as biofertilizer to reclaim polluted soil for agricultural purposes. He is a member of the Nigerian Society for Microbiology and American Society for Microbiology, among others.

Naga Raju Maddela received his M.Sc. (1996–1998) and Ph.D. (2012) in microbiology from Sri Krishnadevaraya University, Anantapuramu, India. During his doctoral program in the area of environmental microbiology, he investigated the effects of industrial effluents/insecticides on soil micro-organisms and their biological activities. Since 1998 he has worked in the Faculty in Microbiology, teaching undergraduate and postgraduate students. He worked as Prometeo Investigator (fellowship received from SENESCYT) at Universidad Estatal Amazónica, Ecuador during 2013–2015, and he received a postdoctoral fellowship (2016–2018) from Sun Yat-sen University in China. He also received external funding from the China Postdoctoral Science Foundation in 2017 and internal funding from Universidad Técnica de Manabí in 2020. He has participated in national/international conferences and has presented research data in China, Cuba, Ecuador, India, and Singapore. Currently, he is working as a Full Professor at the Facultad de Ciencias de la Salud, Universidad Técnica de Manabí, Portoviejo, Ecuador. He has been actively publishing scientific articles, books (authored and edited) and chapters since 2007. As of now, he has published 70 articles, 12 books, and 40 book chapters.

About the Editors

Contributors

Mustahpa Adams
Department of Microbiology
Federal University of Technology
Minna, Nigeria

Joshua Ibukun Adebomi
Department of Biological and Chemical Engineering
University of Saskatchewan
Saskatoon, Canada

Adedayo O. Adeboye
Department of Food Science and Technology
Osun State University
Osogbo, Nigeria

R. Adeoye
School of Pharmacy and Biomolecular Sciences
John Moores University
Liverpool, UK

Deborah O. Aderibigbe
School of Basic Sciences
Nigeria Maritime University
Warri, Nigeria

Charles Oluwaseun Adetunji
Department of Microbiology
Edo State University
Uzairue, Nigeria

O. A. Adewara
Department of Biological Sciences and Biotechnology
Caleb University
Imota, Nigeria

Mutiu A. Alabi
Medical Biochemistry and Pharmacology Department
Kwara State University
Malete, Nigeria

Iyabode Felicia Areo
School of Health and Life Sciences
Teesside University
Middlesbrough, UK

J. S. Ayedun
Department of Biological Sciences and Biotechnology
Caleb University
Imota, Nigeria

Babafemi Raphael Babaniyi
Bioresources Development Center
National Biotechnology Development Agency
Abuja, Nigeria

Patel Shaifali Ben
Centre for Biotechnology
University College of Engineering Science and Technology
Hyderabad, India

Harsha K. Chandrashekar
Department of Plant Sciences
Manipal Academy of Higher Education
Manipal, India

Unoka Edith Chinyere
Department of Chemical Sciences
Dennis Osadebay University
Asaba, Nigeria

P. Gnana Deepu
Centre for Biotechnology
University College of Engineering Science and Technology
Hyderabad, India

A. I. Domavo
Department of Chemistry and Biochemistry
Caleb University
Imota, Nigeria

Bachha Laxmi Dora
Centre for Biotechnology
University College of Engineering Science and Technology
Hyderabad, India

Parisa Ebrahimbabaie
Department of Environmental Biology
Bethune-Cookman University
Daytona Beach, FL

Chidozie Ekene
Department of Biology and Forensic Science
Admiralty University of Nigeria
Ibusa, Nigeria

Kingsley Erhons Enerijiofi
Department of Biological Sciences
Glorious Vision University
Ogwa, Nigeria

Shine Chikaodi Ewa
Department of Public Health and Tropical Medicine
James Cook University
Townsville, Australia

Kingsley Chijioke Eze
Medical Department, Aquatic Bioresources Training Centre
National Biotechnology Development Agency
Cross River State, Nigeria

Damilola Eunice Fagbamigbe
Department of Chemistry
Federal University of Technology
Akure, Nigeria

Haya Fathima
Manipal School of Life Sciences
Manipal Academy of Higher Education
Manipal, India

Abdur-Rahim A. Giwa
Department of Pure and Applied Chemistry
Ladoke Akintola University of Technology
Ogbomoso, Nigeria

Nisha Govender
Institute of Systems Biology
Universiti Kebangsaan Malaysia
Bangi, Malaysia

Kathuroju Harikrishna
Centre for Biotechnology
University College of Engineering Science and Technology
Hyderabad, India

Jogipeta Harihara
Center for Biotechnology
University College of Engineering Science and Technology
Hyderabad, India

Otobong Anwanabasi Inyang
Information Technology Department, Aquatic Bioresources Training Centre
National Biotechnology Development Agency
Cross River State, Nigeria

C. O. Iyiola
Department of Biological Sciences and Biotechnology
Caleb University
Imota, Nigeria

Srinivasan Kameswaran
Department of Botany
Vikrama Simhapuri University College
Kavali, India

Arya Kaniyassery
Department of Plant Sciences
Manipal Academy of Higher Education
Manipal, India

Ramesh B. Kasetti
North Texas Eye Research Institute
University of North Texas Health Science Center
Fort Worth, TX

B. C. Kotun
Department of Biological Sciences and Biotechnology
Caleb University
Imota, Nigeria

Geethu Krishna Kumar
Manipal School of Life Sciences
Manipal Academy of Higher Education
Manipal, India

Manoj Kumar
Department of Plant Sciences
Manipal Academy of Higher Education
Manipal, India

Mohammed Evuti Mahmud
College of Nursing Science
Bida, Nigeria

Asmau M. Maude
Department of Microbiology
Federal University of Technology
Minna, Nigeria

Bhagyashree Mohakhud
Centre for Biotechnology
University College of Engineering Science and Technology
Hyderabad, India

Nathan Moses
Department of Biology and Forensic Science
Admiralty University of Nigeria
Ibusa, Nigeria

Innocent Ojeba Musa
Department of Microbiology
Skyline University Nigeria
Kano, Nigeria

Saheed Ibrahim Musa
Department of Biology and Forensic Science
Admiralty University of Nigeria
Ibusa, Nigeria

Anjaneyulu Musini
Center for Biotechnology
University College of Engineering Science and Technology
Hyderabad, India

Annamalai Muthusamy
Department of Plant Sciences
Manipal Academy of Higher Education
Manipal, India

Simeon Okiki Naphtali
Obafemi Awolowo University
Department of Microbiology
Ile-Ife, Nigeria

Vivian Nathaniel
Department of Microbiology
Federal University of Technology
Minna, Nigeria

Nnenna Patrick Obasi
Department of Bio-entrepreneurship, Aquatic Bioresources Training Centre
National Biotechnology Development Agency
Cross River State, Nigeria

Comfort Okoji
Department of Biology and Forensic Science
Admiralty University of Nigeria
Ibusa, Nigeria

Peter O. Oladoye
Department of Chemistry and Biochemistry
Florida International University
Miami, FL

Isaac Gbolahan Olamide
Department of Chemistry
Federal University of Technology
Akure, Nigeria

E. A. Onakoya
Department of Chemistry and Biochemistry
Caleb University
Imota, Nigeria

Iboyi Nathaniel Onuche
Department of Biology and Forensic Science
Admiralty University of Nigeria
Ibusa, Nigeria

D. Rama Prasanna
Centre for Bio-Technology
University College of Engineering Science and Technology
Hyderabad, India

Bellamkonda Ramesh
Department of Biochemistry and Molecular Biology
University of Nebraska Medical Center
Omaha, NE

Nikhil Kumar Ramesha
Department of Plant Sciences
Manipal Academy of Higher Education
Manipal, India

Pabbati Ranjit
Centre for Biotechnology
University College of Engineering Science and Technology
Hyderabad, India

Chittepu Obula Reddy
Department of Biotechnology
Chaitanya Bharati Institute of Technology
Hyderabad, India

Kondakindi Venkateswar Reddy
Centre for Biotechnology
University College of Engineering Science and Technology
Hyderabad, India

G. Vidya Sagar Reddy
Department of Biotechnology
Vikrama Simhapuri University
Nellore, India

Job Oloruntoba Samuel
Department of Microbiology
Federal University of Technology
Minna, Nigeria

Srija Sappa
Manipal Academy of Higher Education
Manipal, India

Harshitha Shettigar
Manipal School of Life Sciences
Manipal Academy of Higher Education
Manipal, India

A. A. Soyemi
Department of Biological Sciences and Biotechnology
Caleb University
Imota, Nigeria

Sachin Ashok Thorat
Department of Plant Sciences
Manipal Academy of Higher Education
Manipal, India

Abd'Gafar Tunde Tiamiyu
Department of Mathematics
Chinese University of Hong Kong
Hong Kong

Srivasta Udupa
Department of Plant Sciences
Manipal Academy of Higher Education
Manipal, India

Munachimso Odenakachi Victor-Ekwebelem
Department of Microbiology
Alex Ekwueme Federal University
Ndufu-Alike Ikwo, Nigeria

M. M. Yakiimov
Department of Microbiology
German Research Centre for Biotechnology
Braunschweig, Germany

Acknowledgments

With great honor, we acknowledge the support of the contributors to each chapter for their valuable contribution and timely responses leading to the success of this volume. All contributors are immensely appreciated for their eagerness and strong support for this volume to be ready in the scheduled time; we therefore appreciate their teamwork and partnership. We also thank anonymous reviewers for their constructive criticism, which helped us in improving the quality of this book by inviting the experts to contribute the additional chapters. We greatly acknowledge the Taylor & Francis Group Editorial and Production team for their respected support; without their guidelines, this project would not be finished on time. It is an honor and privilege to work with them all. Finally, yet importantly, we are very much thankful to the management and colleagues at University of Abuja (Nigeria), National Biotechnology Development Agency (Nigeria), and Universidad Técnica de Manabí (Ecuador) for their unrestricted backing and for the establishment of treasured propositions at the time of book proposal and final book preparation.

Sesan Abiodun Aransiola, Ph.D.
University of Abuja, Nigeria
National Biotechnology Development Agency (NABDA)
Ogbomoso, Nigeria

Naga Raju Maddela, Ph.D.
Universidad Técnica de Manabí
Portoviejo, Ecuador

Section 1

Prospects of Phytoremediation for a Safe Environment and Safe Food

1 Phytoremediation Overview and Role in Food Safety

Munachimso Odenakachi Victor-Ekwebelem and Simeon Okiki Naphtali

1.1 INTRODUCTION

The Greek word *phyto*, which means "plant," and the Latin suffix *remedium*, which means "to clean up or mend," are combined to form the term *phytoremediation*. The term covers a broad spectrum of plant-based remedies that use genetically modified or naturally occurring plants to clean up polluted environments (Flathman and Lanza, 1998). It refers to a group of technologies that use plants for waste extraction, containment, and destruction. Using vegetation, soil amendments, and agronomic techniques, phytoremediation – also referred to as green remediation, botano-remediation, agroremediation, or vegetative remediation – is considered an environmentally friendly (green) remediation technique that helps eliminate, contain, or render heavy metals in the soil harmless (Aransiola et al., 2022). It is an emerging method that employs different plants to break down, extract, contain, or immobilize pollutants like metals, herbicides, hydrocarbons, and chlorinated solvents from soil and water (Helmisaari et al., 2007; Abioye et al., 2017). Although the first case of plants being used to remove heavy metals was recorded in 1983 and this concept has been implemented more than 300 years ago on wastewater discharge (Blaylock, 2008), this technology has received unusual attention and limelight very recently. The buildup of heavy metals in soil and water is one of the main issues brought on by inorganic pollutants. Their significant presence in soils has had a major negative influence on both human health and food security. Phytoremediation techniques have been determined to be environmentally benign and safe among the several physicochemical strategies available for remediating heavy metal–polluted locations (Aransiola et al., 2013; Oladoye et al., 2022).

The population, activities, and processes of soil microbes are impacted by the buildup of heavy metals in the soil. The types of heavy metals and their chemical affinity determine how they affect enzymatic activity and alterations in microbial populations (Karaca et al., 2010). For instance, lead (Pb) affects catalases, ureases, invertases, and acid phosphatases (Belyaeva et al., 2005); arsenic (As) affects sulfatase and phosphatase (Speir et al., 1999); and copper (Cu) inhibits b-glucosidase (Geiger et al., 1998). Cadmium (Cd) affects ureases, proteases, and basic phosphatases (Lorenz et al., 2006; Elemba and Ijah, 2016). High chromium(VI) concentration in soil demonstrates different toxicity properties generating detrimental effects in soil microbial cell metabolism (Huang et al., 2009). Human and animal health may be at risk when heavy metal–contaminated food crops are consumed because they can cause heavy metals to bioaccumulate and then grow as they move up the food chain's trophic levels (Singh and Kalamdhad, 2011). The accumulation of heavy metals in the human body can lead to severe health complications, including itai-itai disease caused by chronic exposure to Cd; kidney dysfunction and anemia caused by zinc (Zn); mucosa corrosion, hepatic system failure and central nervous system damage caused by Cu; skin irritation and nervous system complications caused by nickel (Ni); cardiovascular issues; and damage to the nervous system caused by Pb (Kafle et al., 2022). Athalatine, dieldrin, hexachlorobenzene, dichlorodiphenyltrichloroethane (DDT), and 1,2,4-trichlorobenzene are among the pesticides that frequently contaminate soils (Kafle et al., 2022). Due to their toxicity and extended half-life in the soil, these pollutants have a significant negative

DOI: 10.1201/9781032683751-2

impact on the availability of mineral nutrients, soil biota, and soil productivity. Water resources have been tainted by various pollutants, much like soils. The majority of the variables contributing to water contamination are caused by the release of toxic heavy metals through sewage, runoff, and effluents from urbanization and industry (Masindi and Muedi, 2018; Parveen et al., 2020; Saleem et al., 2020b). Water resources are often contaminated by Cu, Ni, Pb, Cd, Cr, uranium (U), and Zn, among other metals (Prasad and Freitas, 2005). Due to heavy metal buildup in the food chain, the heavy metals deteriorate water quality and may cause health risks associated with drinking water (Wongsasuluk et al., 2014). In addition to heavy metals, other organic pollutants that contaminate water resources include petroleum hydrocarbons, explosives, polychlorinated biphenyls (PCBs), polycyclic aromatic hydrocarbons (PAHs), pyrene, and trinitrotoluene (TNT). These pollutants can also come from seepage, erosion, sludge and effluents from industries and municipal waste, as well as runoff from mining and agricultural practices (Kafle et al., 2022). Over time, these pollutants have the potential to accumulate and have detrimental effects on both terrestrial and aquatic ecosystems.

Remediation of water and soil pollution is one of the main issues facing our civilization today in order to maintain ecosystem processes and functioning. Numerous physical, chemical, and biological methods have been used to clean up environmental pollution; however, their applications are restricted because of the risks to ecosystems, labor costs, and safety concerns (Ali et al., 2013; Babaniyi et al., 2023). Phytoremediation is an alternate approach that is becoming more and more accepted and used. It has the potential to be a useful treatment. In order to reduce the effects of heavy metal ingestion from such food crops, this tends to address the usage of ornamental plants as opposed to food crops for phytoremediation.

1.2 ADVANTAGES OF PHYTOREMEDIATION

1. When used appropriately, phytoremediation is a green technology that is both environmentally benign and visually acceptable to the general public (Raskin and Ensley, 2000). It is a green technology that can be used to reduce various organic and inorganic pollutants while also improving the environment's aesthetics (Ghosh and Singh, 2005). For broad areas where other methods would be costly and ineffectual, this green technology is appropriate (Prasad and Freitas, 2005; Aransiola et al., 2013).
2. The technique of phytoremediation can be used to a wide range of organic and inorganic chemicals. Both in situ and ex situ applications of phytoremediation are possible (USEPA, 2000; Raskin and Ensley, 2000). Because in situ applications limit disturbance of the soil and surrounding ecosystem and lessen the spread of contamination from airborne and watery pollutants, they are often taken into consideration.
3. Phytoremediation is reasonably simple to apply and doesn't require costly equipment or highly skilled workers.
4. Phytoremediation's low cost when compared to traditional clean-up procedures is its biggest benefit (USEPA, 2000; Raskin and Ensley, 2000).
5. It is more aesthetically pleasing than traditional methods, making it more accepted by the public.
6. The risk of spreading the contamination is reduced since excavation and transport of polluted media is not necessary.
7. Sites polluted with more than one type of pollutant can be treated at the same time.
8. It helps to prevent leaching and soil erosion that may result from water activity and wind, which in turn helps to reduce the expansion of contaminants to water and air (Ghosh and Singh, 2005).

1.3 DISADVANTAGES AND LIMITATIONS OF PHYTOREMEDIATION

1. It is limited to remediative plants' root depth.
2. Remediation with plants is a time-consuming procedure: cleaning up a hazardous waste site might take years or more, and the pollution might not be completely removed (Rajakaruna et al., 2006).

3. Using non-native, invasive plants may impact biodiversity.
4. The consumption of contaminated plants by wildlife is also of concern. Harvested plant biomass produced from the process of phytoextraction may be classified as hazardous waste by the Resource Conservation and Recovery Act (RCRA), therefore subject to proper handling and disposal.

1.4 PHYTOREMEDIATION AND FOOD SAFETY

Plants absorb pollutants through their roots, most of which are water soluble, and move them through different plant tissues where they can be volatilized, digested, or sequestered (Greenberg et al., 2006; Abhilash, 2007; Doty et al., 2007). According to reports, some 400 plant species from at least 45 plant families hyperaccumulate metals (Ghosh and Singh, 2005). Brassicaceae, Fabaceae, Euphorbiaceae, Asteraceae, Lamiaceae, and Scrophulariaceae are a few of the families (Dushenkov, 2003). Very high bioaccumulation potential for Cd/Zn, Cu, cobalt (Co), selenium (Se), and Ni is present in crops such as *Alpine pennycress* (*Thlaspi caerulescens*), *Ipomea alpine, Haumaniastrum robertii, Astragalus racemosus*, and *Sebertia acuminata* (Lasat, 2002). The first plant species known to accumulate significant amounts of metals in its leaves was *T. caerulescens* (Baumann, 1885). Rascio (1977) conducted follow-up research on *T. caerulescens* and found tolerance and substantial Zn buildup in the plant's shoots. There have been reports of high uptake and resistance to heavy metals in sunflower (*Helianthus annuus*), maize (*Zea mays*), Indian mustard (*Brassica juncea* L.), willow (*Salix viminalis*), and sunflower (Schmidt, 2003). *B. juncea* is one of the plants in the *Brassica* species that merits special attention due to the numerous experiments that have proved its significance to the phytoextraction process of heavy metals from soil. Research has shown that *B. juncea* has a significant propensity to retain Cd, primarily in the shoots, where a Cd level of 1450 μg Cd/g dry wt was discovered.

Furthermore, according to Salt et al. (1998), this plant has a high removal effectiveness for various metals, including Pb (28% reduction) and Se (reduced between 13% and 48%). It also removes zinc from soil more successfully than *T. caerulescens*, a renowned zinc hyperaccumulator. Rather than in the roots, tobacco (*Nicotiana tabacum*) accumulates large levels of Cd in the leaves. It has been discovered that sunflower may remove Pb, U, cesium (Cs), and strontium (Sr) from hydroponic solutions, and Indian mustard roots are useful in removing Cd, Cr, Cu, Ni, Pb, and Zn (Lone et al., 2008). Indian mustard and sunflower plants have increased their uptake of copper, according to Tang et al. (2003).

In Zhao and Huang's 2018 work on Cd phytoremediation, rice was identified as one of the main edible sources that exposes the human population in south China to rising levels of Cd. It builds up in the grains of rice, which are subsequently cooked and eventually eaten. Supported by research by Clemens et al. (2013), who found that food accounts for 90% of the total amount of dietary consumption of Cd in non-smokers, and Song et al. (2017), who found that rice accounts for 56% and 65%, respectively, of the total dietary consumption of Cd in the general Chinese population and the population in south China. This is a concerning finding and figure because Cd is a well-known toxin that affects humans and causes phytotoxicity as well as illnesses including cancer, osteoporosis, and renal failure. Eating rice that has been produced in soil contaminated with Cd over an extended period of time can lead to chronic exposure to Cd, which causes renal tubular osteomalacia, the most well-known Cd-induced illness. The consumption of rice as food should be done so with caution because rice will accumulate a significant amount of Cd in its grains, which is the edible portion of the plant, and because there is no discernible difference between rice that has been exposed to heavy metals and rice that has not, the use of rice for phytoremediation is extremely worrisome.

Numerous academics have looked into jute, an indigenous fiber leafy crop, because of its capacity to remove various heavy metals from the environment. Jute is a good plant for phytoremediation of heavy metals like Cd, but Saleem et al. (2020a) concluded in their review of the plant as a potential candidate for metal phytoremediation that it tends to lead to metal accumulation in the plant, which is a major source of concern as well as a reduction in plant productivity and yield. Concerns increase when jute use is broken down. Jute is the least expensive fiber for mass use, mostly glass fiber, and China imports a lot of it and its products from Bangladesh, India, and other nations.

There are two varieties of jute: *Corchorus olitorius*, also called Tosaa jute, which is native to South Asia, and *Corchorus capsularis*, also known as white jute, which comes from the impoverished peasant population in India (Faruk et al., 2012; Choudhary et al., 2013). Nearly all African nations regularly use *C. olitorius* as a vegetable; its leaves are also grown for their fibrous qualities and are used in soups (Saleem et al., 2019). This is quite concerning since the buildup of heavy metals that are toxic in the jute leaves and shoots indicates that these metals will eventually enter human systems, regardless of how hazardous they are. Natural contaminated soil from the Mymensingh district of Bangladesh (As-contaminated soil) and Bhalukaupazila, Bangladesh (Pb-contaminated soil), respectively, was utilized in studies by Nizam et al. (2019) and Uddin et al. (2019) to develop various jute types. According to their findings, heavy metals were deposited in significant levels by all fibrous crops, and the above-ground portions of the plants exhibited higher concentrations of these harmful pollutants than the below-ground parts. This indicates that the portion of jute that may be harvested and consumed has a higher concentration of these metals (Pb, Cd, Cr, Cu, iron [Fe], and Zn) than the portion that is typically unnecessary.

Bandiera et al. (2016) in their study "Phytoremediation Opportunities with Alimurgic Species in Metal-Contaminated Environments," investigated the food safety of wild edible plants grown on contaminated soil as well as their multifunctional roles in metal uptake for phytoremediation. The responses of a few tap-rooted perennial species, including *Rumex acetosa, Cichorium intybus, Sonchus oleraceus, Taraxacum officinale, Tragopogon porrifolius*, and others, were examined in their cultivation in critical open environments (i.e., landfills, ditch sediments, and the sides of heavily trafficked roads) and in an artificially induced environment containing high levels of soil contaminated with Cd, Co, Cu, Pb, and Zn in a pot-scale trial. The study's objectives were to ascertain whether germination would take place and to demonstrate the food safety of the items derived from these plants. Even though the majority of the plants' germination was unaffected by the rise in tissue metal rate, they came to the conclusion that eating the plants is not advised because the study's findings indicate that the plants' excessive uptake of Pb and Cd in their leaves may surpass EU safety thresholds (Bandiera et al., 2016).

The study's findings indicate that zinc was mostly well translocated and reached a high leaf concentration, particularly in *T. officinale* (~600 mg/kg^{-1} dry weight, DW), one of the plants. Given that this plant's leaves are used as food, its use for phytoremediation should be reduced. The increased Cd translocation also indicated potential uses for phytoextraction, especially with *C. intybus*, where leaf Cd levels reached around 16 mg/kg^{-1} DW. In no-tillage systems, a generally high root retention of Pb and Cu may allow for their phytostabilization in the medium term, along with notable reductions in metal leaching when compared to bare soil. However, this could be concerning if the roots of some of the study's plants are used for food or anything leading to consumption by humans.

Additionally, it has been noted that cereal crops (Table 1.1) such as alfalfa (*Medicago sativa*), sorghum (*Sorghum bicolor*), and maize (*Z. mays* L.) have a high capacity to accumulate heavy metals. If the biomass and metal content of such plants are sufficient to finish remediation in an acceptable amount of time, they can be successfully employed to clean up heavy metal–polluted soils Kafle et al., 2022).

1.5 CONCLUSION

Although cash crops and edible plants may effectively remove heavy metals from the environment, their usage has been restricted due to concerns about potential health risks to humans. As a

TABLE 1.1
Plants That Can Conduct Phytoremediation and Possible Contaminates Remediated

Food Crop/Cash Crop	Possible Contaminants	Compartment of Accumulation
Thlaspi caerulescens	Cd, Zn	Leaves
Maize (*Zea mays*)	Pb, Cd	Grain
Sorghum (*Sorghum bicolor*)	Cd	Grain
Ipomea alpine	Cu	Leaves
Haumaniastrum robertii	Co	Roots
Astragalus racemosus	Se	Roots
Sebertia	Ni	Shoots
Cabbage (*Brassica oleracea*)	Cd	Leaves
Lettuce (*Lactuca sativa*)	Cd	Leaves
Indian mustard	Pb, Cr, Zn, Cd, Ni, Cu	Roots
Rice	Cd	Grain
Jute	Cd	Leaves
Wheat (*Nicotiana glauca*)	Cd, Pb	Grain
Alfalfa (*Medicago sativa*)	Pb	Grain
Arabidopsis thaliana	Cd, Pb	
Pteris vittate	As	Shoots
Warnstorfia fluitans	As	Aerial parts/roots
Puccinellia frigida	B	Shoots
Brassica napus	Zn	Shoots
Helianthus annuus	Cr	Shoots
Phalaris arundinacea	Cr	Aerial parts/roots
Pennisetum annuus	Co	Shoots
Brachiaria decumbens	Cu	Shoots
Pisum sativum	Pb	Shoots

result, a more suitable substitute is required. It has been demonstrated that using ornamentals and beautiful plants is safer and preferable. These ornamental plants are highly resistant to the harmful effects of heavy metals and pose no harm to human health. It has been demonstrated that plants such as Indian mustard, white willow, poplar, Indian grass, and sunflower can remove two or more contaminants within a polluted site while also becoming more aesthetically pleasing and provide no risk to human health.

REFERENCES

Abhilash, P.C. (2007). Phytoremediation: An innovative technique for ecosystem clean up. *Our Earth*. 4: 7–12.

Abioye, O.P., Ijah, U.J.J., and Aransiola S.A. (2017). Phytoremediation of soil contaminants by biodiesel plant Jatropha curcas. In Chapter 4 of K. Bauddh et al. (eds.), *Phytoremediation Potential of Bioenergy Plants, DOI 10.1007/978-981-10-3084-0_4*. Springer Nature Singapore Pte Ltd. https://link.springer.com/book/10.1007%2F978-981-10-3084-0.

Ali, H., Khan, E., and Sajad, M.A. (2013). Phytoremediation of heavy metals—concepts and applications. *Chemosphere* 91: 869–881.

Aransiola, S.A., Ijah, U.J.J., and Abioye, O.P. (2013). Phytoremediation of lead polluted soil by glycine max L. *Appl. Environ. Soil Sci.* Article ID 631619. http://dx.doi.org/10.1155/2013/631619, https://www.hindawi.com/journals/aess/2013/631619/abs/.

Aransiola, S.A., Ijah, U.J.J., Abioye, O.P., and Bala, J.D. (2022). Vermicompost-assisted phytoremediation of toxic trace element-contaminated soil in Madaka, Nigeria using Melissa officinalis L and Sida acuta. *IJEST*. http://dx.doi.org/10.1007/s13762-022-04105-y

Babaniyi, B.R., Ogundele, O.D., Bisi-omotosho, A., Babaniyi, E.E., and Aransiola, S.A. (2023). Remediation approaches in environmental sustainability. In N.R. Maddela, L.K. Winkelstroter Eller, and R. Prasad (eds.),

Microbiology for Cleaner Production and Environmental Sustainability. CRC Press. ISBN: 9781032496061. https://www.routledge.com/Microbiology-for-Cleaner-Production-and-Environmental-Sustainability/Maddela-Eller-Prasad/p/book/9781032496061

Bandiera, M., Dal Cortivo, C., Barion, G., Mosca, G., and Vamerali, T. (2016). Phytoremediation opportunities with alimurgic species in metal-contaminated environments. *Sustainability*. 8(4): 357.

Baumann, A. (1885). Das Verhalten von Zinksatzen gegen Pflanzen und im Boden. *Landwirtsch. Vers.-Statn*. 31: 1–53.

Belyaeva, O., Haynes, R., and Birukova, O. (2005). Barley yield and soil microbial and enzyme activities as affected by contamination of two soils with lead, zinc or copper. *Biol. Fertil. Soils*. 41: 85–94.

Blaylock, M. (2008). *Phytoremediation of Contaminated Soil and Water: Field Demonstration of Phytoremediation of Lead Contaminated Soils*. Lewis Publishers.

Choudhary, S.B., Sharma, H.K., Karmakar, P.G., Kumar, A., Saha, A.R., Hazra, P., and Mahapatra, B.S. (2013). Nutritional profile of cultivated and wild jute ('Corchorus') species. *Aust. J. Crop Sci.* 2013(7): 1973.

Clemens, S., Aarts, M.G., Thomine, S., and Verbruggen, N. (2013). Plant science: The key to preventing slow cadmium poisoning. *Trends Plant Sci*. 18: 92–99.

Doty, S.L., Shang, Q.T., Wilson, A.M., Moore, A.L., Newman, L.A., and Strand, S.E. (2007). Enhanced metabolism of halogenated hydrocarbons in transgenic plants contain mammalian P450 2E1. *Proc Natl Acad Sci USA*. 97: 6287–6291.

Dushenkov, D. (2003). Trends in phytoremediation of radionuclides. *Plant Soil*. 249: 167–175.

Elemba, O.M., and Ijah, U.J.J. (2016). Removal of lead and chromium from soil using biosurfactants produced by bacterial isolates from spent lubricating oil contaminated soil. *Int. J. Adv. Res*. 4(5): 201–211 (IMPACT FACTOR 5.388).

Faruk, O., Bledzki, A.K., Fink, H.-P., and Sain, M. 2012. Biocomposites reinforced with natural fibers: 2000–2010. *Prog. Polym. Sci*. 37: 1552–1596.

Flathman, P.E., and Lanza, G.R. (1998). Phytoremediation: Current views on an emerging green technology. *J. Soil Contam*. 7(4): 415–432.

Geiger, G., Brandl, H., Furrer, G., and Schulin, R. (1998). The effect of copper on the activity of cellulase and β-glucosidase in the presence of montmorillonite or Al-montmorillonite. *Soil. Biol. Biochem.* 30: 1537–1544.

Ghosh, M., and Singh, S.P. (2005). A review on phytoremediation of heavy metals and utilization of it's by products. *Appl. Ecol. Environ. Res*. 3(1): 1–18.

Greenberg, B.M., Hunag, X.D., Gurska, Y., Gerhardt, K.E., Wang, W., and Lampi, M.A. (2006). Successful field tests of a multi-process phytoremediation system for decontamination of persistent petroleum and organic contaminants. Proceedings of the twenty-ninth Artic and Marine Oil Spill Program (AMOP). Technical Seminar Vol. 1. Environment Canada, pp. 389–400

Helmisaari, H.S., Salemaa, M., Derome, J., Kiikkila, O., Uhlig, C., and Nieminen, T.M. (2007). Remediation of heavy metal-contaminated forest soil using recycled organic matter and native woody plants. *J. Environ. Qual*. 36: 1145–1153.

Huang, S.-H., Bing, P., Yang, Z.-H., Chai, L.-Y., and Zhou, L.-C. (2009). Chromium accumulation, microorganism population and enzyme activities in soils around chromium-containing slag heap of steel alloy factory. *Transact. Nonferr. Metal. Soc. China*. 19: 241–248.

Kafle, A., Timilsina, A., Asmita, G., Adhikari, K., Bhattara, A., and Niroj, A. (2022). Phytoremediation: Mechanisms, plant selection and enhancement by natural and synthetic agents. *Environ. Adv.* 8(2022): 100–203.

Karaca, A., Cetin, S.C., Turgay, O.C., and Kizilkaya, R. (2010). *Effects of Heavy Metals on Soil Enzyme Activities. Soil Heavy Metals*. Springer, pp. 237–262.

Lasat, M.M. (2002). Phytoextraction of toxic metals: A review of biological mechanisms. *J. Environ. Qual*. 31: 109–120.

Lone, M.I., Zhen-Li, H., Stoffella, P.J., and Xiao, Y. (2008). Phytoremediation of heavy metal polluted soils and water: Progresses and perspectives. *Journal of Zhejiang University Sci B*. 9: 210–220.

Lorenz, N., Hintemann, T., Kramarewa, T., Katayama, A., Yasuta, T., Marschner, P., and Kandeler, E. (2006). Response of microbial activity and microbial community composition in soils to long-term arsenic and cadmium exposure. *Soil Biol. Biochem*. 38: 1430–1437.

Masindi, V., Muedi, K.L., 2018. Environmental contamination by heavy metals. *J. Heavy Metal*. 10: 115–132.

Nizam, M.U., Zaman, M.W., and Rahman, M.M. (2019). Lead toxicity tolerance of jute, kenaf and Mesta at germination phase. *Bangladesh J. Seed Sci. Technol.* 18: 37–44.

Oladoye, P.O., Olumayowa, M.O., and Olowe, M.D.A. (2022). Phytoremediation technology and food security impacts of heavy metal contaminated soils: A review of literature. *J. Chemosphere*. 288(Pt 2): 132555.

Parveen, A., Saleem, M.H., Kamran, M., Haider, M.Z., Chen, J.-T., Malik, Z., Rana, M.S., Hassan, A., Hur, G., Javed, M.T., 2020. Effect of citric acid on growth, ecophysiology, chloroplast ultrastructure, and phytoremediation potential of jute (Corchorus capsularis L.) seedlings exposed to copper stress. *Biomolecules*. 10: 592.

Prasad, M., and Freitas, H. (2005). *Metal-tolerant plants: Biodiversity prospecting for phytoremediation technology*. Trace Elements in the Environment. CRC Press, pp. 501–524.

Rajakaruna, N., Tompkins, K.M., and Pavicevic, P.G. (2006). Phytoremediation: An affordable green technology for the clean-up of metal-contaminated sites in Sri Lanka. *Ceylon J. Sci. (Biological Sciences)*. 35: 25–39.

Rascio, W. (1977). Metal accumulation by some plants growing on Zn mine deposits. *OIKOS*. 29: 250–253.

Raskin, I., and Ensley, B.D. (2000). *Phytoremediation of Toxic Metals: Using Plants to Clean Up the Environment*. John Wiley & Sons, Inc. Publishing, p. 304.

Saleem, M.H., Ali, S., Kamran, M., Iqbal, N., Azeem, M., Tariq Javed, M., . . . Rizwan, M. (2020a). Ethylenediaminetetraacetic acid (EDTA) mitigates the toxic effect of excessive copper concentrations on growth, gaseous exchange and chloroplast ultrastructure of Corchorus capsularis L. and improves copper accumulation capabilities. *Plants*. 9: 756.

Saleem, M.H., Ali, S., Rehman, M., Hasanuzzaman, M., Rizwan, M., Irshad, S., . . . Qari, S.H. (2020b). Jute: A potential candidate for phytoremediation of metals – A review. *Plants*. 9(2): 258.

Saleem, M.H., Fahad, S., Khan, S.U., Din, M., Ullah, A., Sabagh, A.E., Hossain, A., Llanes, A., and Liu, L. (2019). Copper-induced oxidative stress, initiation of antioxidants and phytoremediation potential of flax (Linum usitatissimum L.) seedlings grown under the mixing of two different soils of China. *Environ. Sci. Pollut. Res*. 8(2): 545.

Salt, D.E., Smith, R.D., and Raskin, I. (1998). Phytoremediation. *Ann. Rev. Plant Physiol.* 49: 643–668. https://doi.org/10.1146/annurev.arplant.49.1.643.

Schmidt, U. (2003). Enhancing phytoextraction: The effects of chemical soil manipulation on mobility, plant accumulation, and leaching of heavy metals. *J. Environ. Qual.* 32: 1939–1954.

Singh, J., and Kalamdhad, A.S. (2011). Effects of heavy metals on soil, plants, human health and aquatic life. *Int. Res. Chem. Environ.* 1: 15–21.

Song, Y., Wang, Y., Mao, W., Sui, H., Yong, L., Yang, D., Jiang, D., Zhang, L., and Gong, Y. (2017). Dietary cadmium exposure assessment among the Chinese population. *PLoS One*. 12: e0177978.

Speir, T., Kettles, H., Parshotam, A., Searle, P., and Vlaar, L. (1999). Simple kinetic approach to determine the toxicity of As [V] to soil biological properties. *Soil Biol. Biochem.* 31: 705–713.

Tang, S., Xi, L., Zheng, J., and Li, H. (2003). Response to elevated CO_2 of Indian mustard and sunflower growing on copper contaminated soil. *Bull. Environ. Contam. Toxicol.* 71: 988–997.

Uddin, M.N., Mokhlesur Rahman, M., and Kim, J.E. (2019). Phytoremediation potential of kenaf (Hibiscus cannabinus L.), Mesta (Hibiscus sabdariffa L.), and Jute (Corchorus capsularis L.) in Arsenic-contaminated Soil. *Korean J. Environ. Agric*. 35: 111–120.

United States Environmental Protection Agency (USEPA). (2000). *Introduction to Phytoremediation*. USEPA.

Vidali, M. (2001). Bioremediation. An overview. *Pure Appl. Chem.* 73(7): 1163–1172.

Wongsasuluk, P., Chotpantarat, S., Siriwong, W., and Robson, M. (2014). Heavy metal contamination and human health risk assessment in drinking water from shallow groundwater wells in an agricultural area in Ubon Ratchathani province. *Thai. Environ. Geo. Health*. 36: 169–182.

Zhao, F.J. and Huang, X.Y. (2018). Cadmium phytoremediation: Call rice CAL1. *Molecular Plant*. 11(5): 640–642.

2 Phytoremediation and its Role in Climate Change for a Safe Environment

Pabbati Ranjit, Patel Shaifali Ben, D. Rama Prasanna, and Kondakindi Venkateswar Reddy

2.1 INTRODUCTION

Phytoremediation is the use of plants to eliminate contaminants or pollutants to ensure a safer and healthier environment. Plants are generally used as primary sources to fulfill domestic needs, industries, and so on. However, these days to balance the healthier conditions of the environment, plants are being used as custodians (Aransiola et al. 2019). Pollutants are generated from industrial waste factories, combustion, and accumulation of toxic compounds and effluents. As we know, plants take up CO_2 and release O_2; recent studies show that they absorb or consume gaseous materials released from industries. Conventional techniques of remediation may cost \$50–\$500 per ton and specific techniques may exceed \$1000 per ton. However, nowadays vegetative plant-based processes have been receiving greater attention. As plants grow in soil and water, their roots and shoots will develop by obtaining nutrients from the region where they are growing (i.e. soil and water). Even minerals and elements associated with plant growth are present in the source of growth. According to the studies, a few plants can grow even in contaminated sites by utilizing the pollutants as a source of nutrition (Farraji et al. 2016; Abioye et al. 2021). In ancient times, soil used to have higher fertility and less contaminated water bodies, fresh air, and abundant natural resources were present. Industrialization and anthropogenic movements, misuse of resources, application of hazardous chemicals as fertilizers and pesticides in agriculture, the liberation of toxic gases from chimneys of factories, release of untreated industrial effluent, and so on (Cunningham et al. 1996) harmed the healthy and safe environment. Because of the above activities, there has been a drastic negative impact on climate change which has arisen from specific pollutants; phytoremediation is the green practice of assuring a pollutant-free environment. Many in situ and ex situ methods were developed by employing phytoremediation, which can provide a safer environment. It has emerged as a non-invasive, cost-effective method for eliminating contaminants. Phytoremediation mechanisms include rhizofiltration, phytoextraction, phytofiltration, phytostabilization, phytovolatilization, phytoimmobilization, and phytodegradation. As emerging green technology, phytoremediation will remove pollutants from the environment.

But it is laborious and challenging, and changes cannot be observed immediately (Arthur et al. 2005).

2.2 REASONS FOR CLIMATE CHANGE

Climate change is a blueprint that issues a threat alarm regarding the environment. For instance, changes subjected to uncertain rainfall result in severe floods or drought, changes in temperature, too hot or too cold conditions, and increase or decrease in the moisture content. Also, deforestation, combustion of fossil fuels, and animal husbandry are other reasons that are influencing climate change. Natural gases are present already in the atmosphere, but the above all add a boundless amount of greenhouse gases, thus resulting in global warming. The following factors are the major reasons for climate change (Ferreira and Kroll 2023).

DOI: 10.1201/9781032683751-3

2.2.1 Greenhouse Gases

The greenhouse effect is the motivation for climate change. Even naturally there is an effect of greenhouse gases, but anthropogenic activities are expanding a few gas concentrations in the atmosphere, particularly methane, CO_2, fluorinated gases, and nitrous oxide. The immense donor of global warming is CO_2 liberated by anthropogenic activities. The National Oceanic and Atmospheric Administration (NOAA) indicates that per every 1 million molecules there are 424 CO_2 molecules in the atmosphere. Carbon dioxide and nitrous oxide both have a longer shelf life and are called long-lived greenhouse gases. Methane is another powerful greenhouse gas compared to CO_2 and has a shorter shelf life in the atmosphere. Fluorinated gases are found in the atmosphere, including chlorofluorocarbons liberated from air conditioners, aerosol sprays, refrigerators, and insulation (Mac Dowell et al. 2017).

2.2.2 Petroleum Industry

There is an abundant usage of petroleum products and petroleum, and it is a crucial element in several industries and the production of various materials. Because of petroleum's necessity, it has key concerns overall. Despite its key role, it has a greater negative impact on climate change due to its composition resulting in issues like steam intrusion, petrol spills, air pollution, acid rain, liberation of greenhouse gases, and so on. Numerous petroleum-based products and oils have multiple levels of toxic effects. A variety of variables including soluble content, weathering, and chemical properties like tenacity, influence, and the level of toxicity. Increased weathering often lowers toxicity levels. Also, tightly soluble compounds are generally more dangerous than less soluble ones. On the whole, oils with benzene rings and longer carbon chains are hazardous. Benzene is the petroleum product with the highest degree of toxicity; others include toluene, xylene BETX, and methylbenzene. Motor oil and rough oil are the least harmful compounds (Ashish et al. 2024).

2.2.3 Vapor Intrusion

Vapors or gases that are emitted through different matters are categorized as volatile organic compounds (VOCs). The explosive potentiality of petroleum hydrocarbons, which includes gasoline, jet fuel, and diesel, can leak into underground storage tanks and harm the welfare of individuals when inhaled (Consequences for potential vapor intrusion hazards 2012).

2.2.4 Oil Spills

Oil spillage is a form of pollution that takes place as liquid petroleum hydrocarbons are discharged into the environment. It tends to be the most common in marine scenarios. The following may be connected to oil spills: tankers, rail cars, crude oil spills from pipelines, drilling plants, wells, and offshore platforms; gasoline and diesel are refined petroleum products. The Gulf War oil spills, the Deepwater Horizon oil spill, and the Lakeview Gusher are examples of significant oil spills. Numerous variables which include the sort of oil spilled, shorelines, beaches, and water temperature will all impact the cleaning and restoration procedure of oil spills.

2.2.5 Exhaust Emissions

Release from the petroleum industry originates at the extraction process and extends throughout all phases of the oil processing chain. Following the extraction stage, the gas gives off not just carbon dioxide and methane but additionally pollutants (e.g., nitrous oxide and aerosols); carbon monoxide and methanol are by-products. When solvents of oil burn char, it is usually an incomplete combustion, resulting in a chemical reaction that creates by-products that are not carbon dioxide and water.

2.2.6 Waste Oil

Oil regarded as waste has impurities from application and is also a deteriorate by-product. Waste oils include brake, motor-driven, hydraulic, synthetic, and crankcase. Waste oil along with natural petroleum encounters several identical constraints. When oil from motor vehicles or cars spills, it contains benzene-associated carcinogens. Leakages or rainwater runoff enters the water bodies and contaminates them.

2.3 CLIMATE CHANGE AND ITS IMPACT ON THE ENVIRONMENT

Several ties exist in numerous fields of society that are being impacted by climate change. Production of food and wellness of the population may be defined amid drought. Flooding holds the possibility of wrecking infrastructure and may induce the spreading of diseases. Challenges regarding individual well-being may hinder an increase in death rate, affect the supply of food, and may negatively impact productivity in the workforce. The repercussions related to climate change are apparent in each aspect of the worldwide community. Also, the consequences of climate change may not be consistent across the country and the globe; they could differ between communities and residents of the same place or town.

People can experience the ripples of climate change around the earth's extremes. Various modifications including the following have been observed by NOAA (Pilon-Smits and Elizabeth 2005):

- Rise in the sea level.
- Shrinking of glaciers.
- Rise in the amount of carbon dioxide in the environment from 25% to 45%.
- Increase in the global temperature by about 1.8°F.

As the major source of nutrition is food, its cultivation, harvesting, and supply is significantly affected by climate change. Particularly, agricultural strategies and procedures may be adjustable and convertible by scientists and farmers, or they might turn to novel techniques, even though navigating particular modifications is going to be tricky.

Workers on farms are susceptible to ailments associated with heat, which include dehydration, heat exhaustion, and cardiac arrest. Even animals struggle with heat-associated stress and warmer temperatures. Ecosystems have experienced a lot of changes primarily as a consequence of the effects of climate change. Hurricanes and tropical cyclones can wipe out coral reefs, and higher waters might choke corals. The bleaching of corals is the consequence of climate change. Even the ocean is experiencing transition. Nearly 20% of the CO_2 generated in the atmosphere due to the combustion of fuels is captured by the ocean. As a result, the pH of the ocean is decreasing gradually (turning acidic), and this hurts aquatic life. Coupled with the thawing of glaciers and ice sheets, thermal expansion has raised sea levels, boosting the potential risk of erosion along with wave surges in areas with coastlines (Lindsey and Dahlmal 2024).

2.4 CLIMATE CHANGE AND SAFE ENVIRONMENT THROUGH PHYTOREMEDIATION

The term *phytoremediation* is derived from the Greek prefix *phyto*, which means "plant," and the Latin suffix *remedium*, which means "to clean or restore." The term describes a wide range of plant-based solutions that employ naturally occurring plants or genetically modified plants to purify the contaminated location. This developing technique is eco-friendly, less expensive, and natural; it uses different types of plants and their natural abilities to break down, remove, confine, or immobilize pollutants in water, soil, and air.

Though the main aim of phytoremediation is to purify air, water, and soil, it can also have unexpected advantages that help to reduce climate change and create a safer environment. The following sections explain how phytoremediation helps to ease the effects of climate change.

2.4.1 Carbon Sequestration

A major part of reducing climate change is carbon sequestration through phytoremediation. Plants use photosynthesis to take up carbon dioxide from the atmosphere (Govindaraju et al. 2021), which they then store in the soil and biomass of the plants during the phytoremediation process. By removing carbon dioxide, one of the greenhouse gases, from the atmosphere, this activity effectively lowers the concentration of the gas. Phytoremediation is a process that absorbs carbon dioxide and other greenhouse gases from the environment and stores them for future use. This procedure helps to lessen the greenhouse effect, which is necessary to maintain stable world temperatures and lessen the effects of climate change. Additionally, the carbon that is stored in soil and plant biomass supports biodiversity and increases the resilience of natural habitats, both of which are benefits of ecosystem health. Reducing the influence of human activity on the global climate system helps not only the environment but also the broader efforts to prevent climate change.

Therefore, by immediately lowering the atmospheric concentration of carbon dioxide and promoting the stability and health of ecosystems, phytoremediation's carbon sequestration greatly aids in the mitigation of climate change.

2.4.2 Soil Health Improvement

By eliminating or breaking down contaminants, phytoremediation methods like phytoextraction and rhizosphere bioremediation can enhance the quality of soil (Eapen et al. 2005). Healthy soils encourage microbial activity and plant growth, improving carbon storage and building resistance to pressures brought on by climate change.

2.4.2.1 Phytoextraction

The process of removing contaminants from soil and moving them to plant parts above ground for additional degradation is known as phytoextraction. Therefore, both the phyllosphere and the rhizosphere are engaged in phytoextraction (Gul et al. 2019). According to Favas et al. (2014), this process is also known as phytoaccumulation, phytoabsorption, or phytosequestration. This technique makes use of hyperaccumulator plants (which can concentrate 100 times more metals than non-accumulating plants) for absorbing the contaminants from the polluted soils and collecting them in the phyllosphere.

High biomass production, high pollutant translocation factor into the surface biomass, high detoxification, contaminant tolerance, and ease of harvesting are the selection criteria for plants used in this strategy (Etim 2012; Aransiola et al. 2013). Following the harvest of these plants from the contaminated places, incineration is the last step in the cleanup of contaminants (Pandey and Bajpai 2019). The advantage of phytoextraction over traditional methods for pollution remediation includes the permanent removal of contaminants from locations at a lower cost. The effectiveness of the phytoremediation process mainly relies on the movement and the availability of the heavy metals in the root zone (Mousavi et al. 2021; Morikawa et al. 2003; Krishnasamy et al. 2021).

2.4.2.2 Water Quality Management

Polluted water bodies can be cleaned up with the use of phytoremediation techniques like rhizofiltration. Aquatic ecosystems depend on clean water supplies to survive, and a balanced carbon cycle is necessary to control climate change.

2.4.2.3 Rhizofiltration

The process of rhizofiltration uses plants to absorb and precipitate contaminants from an aqueous environment by way of the rhizosphere. In a hydroponic system, where the roots or other rhizosphere organs absorb pollutants, plants are frequently grown. Similar to the phytoextraction approach, the plants were picked whenever they reached the saturation limit of the pollutants. This technique is frequently applied to wastewater, surface water, and groundwater remediation. This method's plant selection requirements include large adsorption surface area, high root biomass, high accumulation capacity, and tolerance to pollutants (Favas et al. 2014). Rhizofiltration advantages include the fact that it doesn't require the transmission of pollutants to the shoots and that it may be applied to both terrestrial and aquatic plants.

Different forms of phytoremediation, such as rhizodegradation, phytovolatilization, phytostabilization, and phytodegradation, address environmental sustainability and climate change in different ways.

2.4.2.3.1 Phytovolatilization

Transport of pollutants into the phyllosphere of plants via the epidermis by diffusion over the cuticle and open stomata is known as phytovolatilization (Vadim et al. 2003). These pollutants undergo further degradation to become volatile components that are subsequently discharged into the atmosphere through transpiration from the stomata (Favas et al. 2014). Volatile substances that have evaporated may either remain in the environment as air pollutants or decompose due to the action of hydroxyl radicals (Karaghool and Haneen 2022). There is a process of phytovolatilization for several pollutants, both inorganic and organic pollutants. pollutants are taken up by plants through the phyllosphere and converted into volatile molecules during the process of phytovolatilization. Eventually, stomata allowed these broken-down volatile chemicals to escape into the atmosphere (Kumar et al. 2023).

2.4.2.3.2 Phytodegradation

Phytodegradation is also known as phytotransformation as the pollutants are converted by plants and integrated into their tissues in the phyllosphere. In the process of phytodegradation, organic contaminants such as polycyclic aromatic hydrocarbons (PAHs), total petroleum hydrocarbons (TPHs), and polychlorinated biphenyls (PCBs), and inorganic contaminants such as sulfur and nitrogen oxides from the atmosphere, are absorbed by plants and converted into less hazardous, simpler forms (Aransiola et al. 2019; Zhang et al. 2020). Both the phyllosphere and rhizosphere contain the products that were produced by phytotransformation. Particular plant enzymes are essential to the phytotransformation process, including laccases, peroxidase, dehalogenases, and nitroreductases (Favas et al. 2014). The various mechanisms of phytoremediation are given in Figure 2.1.

Phytoremediation employs plants to mitigate environmental pollution by various mechanisms. First, plants absorb contaminants through their roots, which can then be either enzymes that can directly break down pollutants within their tissues. Third, some plants enhance microbial activity in the soil surrounding their roots, leading to accelerated degradation of contaminants. Fourth, plants can act as filters, physically trapping pollutants on their surfaces or within their tissues. Fifth, through a process known as rhizofiltration, plants can remove contaminants from water as it passes through their root systems.

2.5 PHYTOREMEDIATION OF VARIOUS POLLUTANTS

Phytoremediation of various air pollutants includes the remediation of particulate matter, volatile organic and inorganic compounds, gaseous pollutants, and polyaromatic hydrocarbons.

2.5.1 Phytoremediation of Particulate Matter

Plants absorb particulate matter (PM) through their branches and foliage, storing it in the phyllosphere. Afterward, the adsorbed PM is immobilized and phytostabilized within the plant wax layers. Plant-specific traits including pubescence, wax content, surface roughness, thickness, ultrastructure,

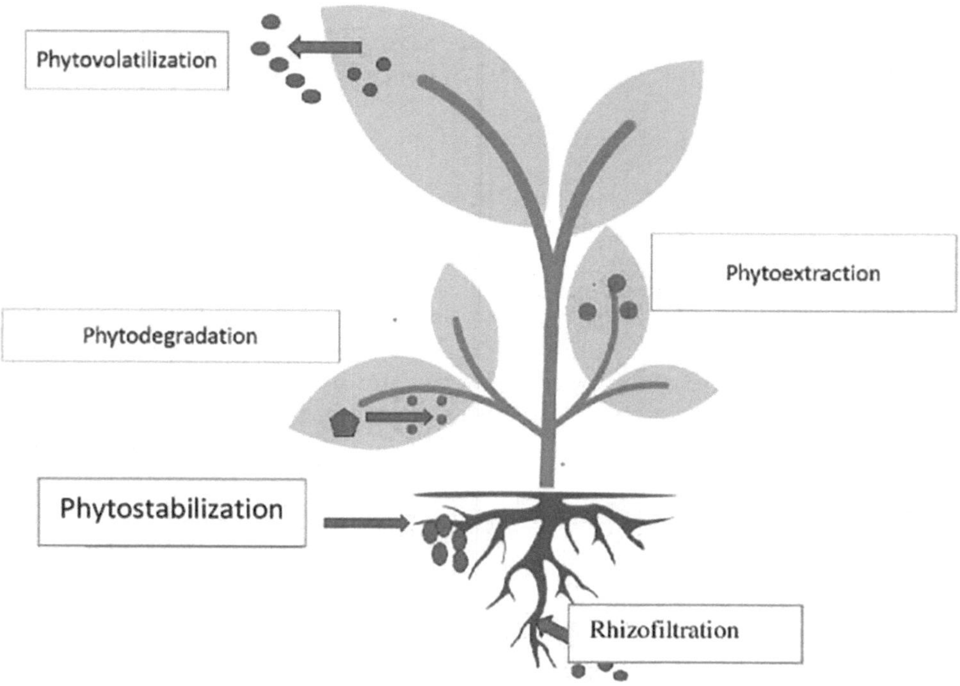

FIGURE 2.1 The various mechanisms of phytoremediation.

leaf size, and structure can all affect the ability of PM to filter. In addition, variations in wind, precipitation, the amount and makeup of PM in the atmosphere, and other climatic factors affect the ability of plants to scavenge PM. Herbaceous vegetation and trees are efficient PM remediators, as demonstrated by (Weber et al. (2014). Plant microbes stimulate plant growth, increase biomass, and improve the plants' ability to remediate PM because they increase the surface area available for the adsorption of pollutants. Additionally, by solubilizing nutrients for plants, bacteria also function as biofertilizers. As a result, plants have a higher tolerance for environmental challenges brought on by PM. Additionally, bacteria can degrade and detoxify substances, which improves the effectiveness of phytoremediation in cleaning up PM.

2.5.2 Volatile Organic Compounds

VOCs such as formaldehyde, xylene, toluene, benzene, and ethylbenzene can be successfully removed from the environment using phytoremediation (Agarwal et al. 2019). Plants use low-molecular-weight VOCs as a defense against virus, animals, and other environmental stresses (Kelly and Fussell 2015). VOCs are typically absorbed by plants through their leaf stomata; however, a small percentage of plants also absorb VOCs through their cuticles (Doty et al. 2007). Phloem then transports the absorbed VOCs to specific plant organs. VOCs are broken down, stored, or eliminated by a variety of processes in the plant system, and they diffuse from the trunks, stems, roots, and leaves of the plants into the atmosphere (Farr 2021; Toome et al. 2010). The characteristics of VOCs affect the phytoremediation efficacy of plants. Specifically, hydrophilic VOCs are absorbed by leaf stomata, while lipophilic VOCs are absorbed through the cuticle (Frontczak et al. 2012). VOCs are also deposited in the rhizosphere of plants and soil as a result of runoff and leaf fall. These chemical substances are metabolized by microbes found in plants into less harmful forms such as carbon dioxide, water, and cellular biomass. Low levels of formaldehyde are detoxified by the spider plant (*Chlorophytum comosum* L.) into amino acids, carbohydrates, lipids, and components of cell walls. Formaldehyde is transformed into serine and phosphatidylcholine by the soybean plant (*Glycine max* L.). Furthermore, by inducing an immune response, microorganisms can alter the VOC emissions from plants (Mikkonen et al. 2018).

2.5.3 Gaseous Pollutants

Oxides of sulfur, carbon, and nitrogen, along with ozone, are examples of gaseous air pollutants (Kim et al. 2016). Carbon is metabolized by plants either by reduction into amino acids or by oxidation into CO_2. It has also been shown that plant canopy can absorb and act as a sink for NO_2. Ammonia (NH_4) in the soil and air can be absorbed by plants (Roupsard et al. 2013). Phytoremediation is an effective way to reduce pollutants like carbon dioxide (CO_2), sulfur dioxide (SO_2), nitrogen dioxide (NO_2), and ozone (O_3). Plants absorb carbon dioxide from the atmosphere through the process of photosynthesis, storing it for later use or converting it to humus (Frontczak et al. 2012). Carbon sequestration is the mechanism through which plants store carbon dioxide for a long time (Hu et al. 2019).

2.5.4 Removal of Dyes

Water containing toxic substances damages the aquatic ecosystem by preventing light from penetrating, which prevents aquatic plants from conducting photosynthesis (Abbass et al. 2016), (Aljeboree and Aseel M 2015). It has been demonstrated through experimental research that the *Salvinia molesta* plant is more effective at eliminating harmful dyes, and as a result, it is currently a viable strategy utilized by a variety of companies (Al-Baldawi et al. 2020). Colored dyes can be eliminated using a variety of chemical, physical, and biological techniques

(Kandisa et al. 2016). Adsorption, filtration, plants, fungi, bacteria, and algae are examples of physical and biological techniques, whereas ozone, and the Fenton process are components of the oxidation process (Holkar et al. 2016). Adsorption treatments and plant-based green technology are popular methods for eliminating dyes. Phytoremediation is primarily a plant-based approach in which dyes from various sources are removed through plants. Plant parts that are effective at eliminating dyes include leaves, peels, and other sections. Macrophytes are an excellent way to get rid of chemicals such as hazardous sulfonated anthraquinones and acid orange 7. Numerous ferns and other plant species, such as *Zinnia angustifolia*, *Aster amellus*, and *Glandularia pulchella*, have been successful in eliminating dyes (Khandare et al. 2015).

2.5.5 Removal of Heavy Metals

Heavy metals pose a major risk to human health as well as the environment. While several methods have been developed to eliminate heavy metals, their cost is higher than that of removing hazardous metals from businesses using plants. In developing nations, heavy metal pollution is becoming a bigger issue. Many investigations on the removal of heavy metals using less costly and environmentally acceptable (plant-derived) adsorbents have been conducted (Joseph et al. 2019). The buildup of heavy metals in the air, soil, and water as a result of human activities like mining, urbanization, and industrialization has become a global problem. It is strongly recommended to use decorative plant species because of their many roles in the ecosystem (Asgari Lajayer et al. 2019). Without proper treatment, industrial wastewater discharges contaminate land and water. Heavy metals found in these effluents are currently being removed by plant-based methods called phytoremediation. The pollution of wastewater with heavy metals poses a serious risk to aquatic life. A long-term, environmentally benign, and reasonably priced approach is phytoremediation. Aquatic plants are highly effective in removing heavy metal pollutants. A few other plants, including duckweed (*Lemna minor*), have a high metal accumulator efficiency (Ali et al. 2020). Adsorption of heavy metals by inexpensive, environmentally benign adsorbents such as plants is the favored method for treating wastewater. The best natural lignin residues for use as bioadsorbents are those that have been separated using black liquor (Naseer et al. 2019). Biosorption is a technique that has been extensively studied through multiple research studies in the past and is successful in removing heavy metals and other hazardous compounds found in the environment. The utilization of plant-based waste for the removal of metals from industrial wastewater has garnered substantial attention due to its cost-effectiveness and notably higher removal rate, which can be attributed to several functional groups. Lead, cadmium, and mercury can be eliminated from the body using coconut waste and black oak bark (Alalwan et al. 2020). The trunk fiber waste of various date palms has demonstrated outstanding performance in removing heavy metals, but due to its availability and affordability, large-scale projects including this waste are still understudied (Shafiq et al. 2018). Many industries employ plants to repair metal-affected habitats, remove heavy metals from polluted soil, and stop continued negative environmental effects on living things (Tewes et al. 2018). Using green plants over different conventional techniques is more effective in removing heavy metals from soil and wastewater (Thakur et al. 2016). The pollutant removal and their mechanism are given in Table 2.1.

TABLE 2.1
Pollutant Removal and Its Mechanisms

S. No.	Type of Process Involved	Type of Mechanism Involved	Type of Pollutant	Types of Plants Used in Phytoremediation Technique
1.	Phytoextraction	Works by utilizing plants to absorb and accumulate contaminants from the soil through their roots, which are then harvested and disposed of, thus effectively decontaminating the soil	Metals including lead, cadmium, zinc, copper, etc.	Indian mustard, sunflower, and some hyperaccumulators like *Alyssum*, and *Sedum* species
2.	Phytovolatilization	Plants absorb contaminants from soil or water and release them into the atmosphere as volatile compounds through their leaves	Metals including mercury, selenium, and arsenic Organic compounds (e.g., chlorinated solvents)	Willow trees, poplars, and certain grass species
3.	Phytodegradation	Plants break down the contaminants in the soil through various metabolic processes, including enzymatic reactions and microbial interactions within the root zone	Organic pollutants like hydrocarbons, pesticides, chlorinated solvents, and explosives	Plant species include hybrid poplars, willow trees, and certain grasses and legumes
4.	Phytostabilization	Immobilizes and contains contaminants in the soil, preventing them from spreading and minimizing their potential for harm to the environment and human health	Inorganic contaminants including heavy metals (e.g., lead, cadmium, and arsenic) and metalloids (e.g., selenium)	Grasses such as perennial ryegrass, fescues, and certain clover species. Some woody species like willow trees and poplars
5.	Rhizofiltration	Plant roots absorb and filter contaminants, effectively reducing the concentration of pollutants	Heavy metals and metalloids, including contaminants like lead, cadmium, arsenic and selenium	Water hyacinth, cattails, various types of water lilies, and some species of willow trees

2.6 CONCLUSION

In addition to providing a way to deal with environmental contamination, phytoremediation is essential for reducing climate change and fostering a more secure and sustainable environment for present and future generations. There is much greater potential for the removal of hazardous contaminants from the environment through phytoremediation. This method of phytoremediation is easy to use, more affordable, environmentally friendly, and sustainable. Through the processes of phytoextraction, phytostabilization, rhizofiltration, phytovolatilization, phytodegradation, and phytodesalination, the plants and microorganisms used in phytoremediation eliminate contaminants. But it is important to consider the kind of plant and microbe that will help with certain pollution in a particular location. According to the literature review, compared to other remediation options, phytoremediation techniques are 5 to 13 times more cost-effective, have a greater success rate, are well received, and are environmentally friendly.

REFERENCES

"Consequences for potential vapour intrusion hazards" The Ministry of Health in Hawaii. Checked back on 8 December 2012.

Abbass, Aamir, Adnan M. Al-Amer, Tahar Laoui, Mohammed J. Al-Marri, Mustafa S. Nasser, Majeda Khraisheh, and Muataz Ali Atieh. "Heavy metal removal from aqueous solution by advanced carbon nanotubes: Critical review of adsorption applications." *Separation and Purification Technology* 157 (2016): 141–161. https://doi.org/10.1016/j.seppur.2015.11.039

Abioye, O. P., U. J. J. Ijah, S. A. Aransiola, S. H. Auta, and M. I. Ojeba. "Bioremediation of toxic pesticides in soil using microbial products." In R. Prasad, S. C. Nayak, R. N. Kharwar, and N. K. Dubey (eds.), *Mycoremediation and Environmental Sustainability. Fungal Biology*. Springer, Cham, 2021. https://doi.org/10.1007/978-3-030-54422-5_1.

Agarwal, Prerita, Mangaldeep Sarkar, Binayak Chakraborty, and Tirthankar Banerjee. "Phytoremediation of air pollutants: Prospects and challenges." *Phytomanagement of Polluted Sites* (2019): 221–241.

Alalwan, Hayder A., Mohammed A. Kadhom, and Alaa H. Alminshid. "Removal of heavy metals from wastewater using agricultural byproducts." *Journal of Water Supply: Research and Technology – AQUA* 69, no. 2 (2020): 99–112. https://doi.org/10.2166/aqua.2020.133

Al-Baldawi, Israa Abdulwahab, Siti Rozaimah Sheikh Abdullah, Asia Fadhile Almansoory, Nur'Izzati Ismail, Hassimi Abu Hasan, and Nurina Anuar. "Role of Salvinia molesta in biodecolorization of methyl orange dye from water." *Scientific reports* 10, no. 1 (2020): 13980.

Ali, Shafaqat, Zohaib Abbas, Muhammad Rizwan, Ihsan Elahi Zaheer, İlkay Yavaş, Aydın Ünay, Mohamed M. Abdel-Daim, May Bin-Jumah, Mirza Hasanuzzaman, and Dimitris Kalderis. "Application of floating aquatic plants in phytoremediation of heavy metals polluted water: A review." *Sustainability* 12, no. 5 (2020): 1927. https://doi.org/10.3390/su12051927

Aljeboree, Aseel M. "Adsorption of methylene blue dye by using modified Fe/Attapulgite clay." *Research Journal of Pharmaceutical Biological and Chemical Sciences* 6, no. 4 (2015): 778–788.

Aransiola, S.A., U. J. J. Ijah, and O. P. Abioye. (2013). "Phytoremediation of lead polluted soil by *Glycine max* L." *Applied and Environmental Soil Science* 2013, Article ID 631619. doi: 10.1155/2013/631619. https://www.hindawi.com/journals/aess/2013/631619/abs/.

Aransiola, S.A., U. J. J. Ijah, O. P. Abioye, and J. D. Bala. "Microbial-aided phytoremediation of heavy metals contaminated soil: A review." *European Journal of Biological Research* 9, no. 2 (2019): 104–125. http://www.journals.tmkarpinski.com/index.php/ejbr/article/view/157.

Arthur, Ellen L., Pamela J. Rice, Patricia J. Rice, Todd A. Anderson, Sadika M. Baladi, Keri LD Henderson, and Joel R. Coats. "Phytoremediation – an overview." *Critical Reviews in Plant Sciences* 24, no. 2 (2005): 109–122.

Asgari Lajayer, Behnam, Nader Khadem Moghadam, Mohammad Reza Maghsoodi, Mansour Ghorbanpour, and Khalil Kariman. "Phytoextraction of heavy metals from contaminated soil, water and atmosphere using ornamental plants: Mechanisms and efficiency improvement strategies." *Environmental Science and Pollution Research* 26 (2019): 8468–8484. https://doi.org/10.1007/s11356-019-04241-y

Ashish, Patre Rajat, Rohan Vashistha, Rakesh Singh, and Krishnan Thakur. "Petroleum industry and its effect on climate." In *AIP Conference Proceedings*, vol. 2978, no. 1. AIP Publishing, 2024.

Cunningham, Scott D., and David W. Ow. "Promises and prospects of phytoremediation." *Plant Physiology* 110, no. 3 (1996): 715.

Doty, Sharon L., C. Andrew James, Allison L. Moore, Azra Vajzovic, Glenda L. Singleton, Caiping Ma, Zareen Khan et al. "Enhanced phytoremediation of volatile environmental pollutants with transgenic trees." *Proceedings of the National Academy of Sciences* 104, no. 43 (2007): 16816–16821. https://doi.org/10.1073/pnas.0703276104

Eapen, Susan, and S. F. D'Souza. "Prospects of genetic engineering of plants for phytoremediation of toxic metals." *Biotechnology advances* 23, no. 2 (2005): 97–114. https://doi.org/10.1016/j.biotechadv.2004.10.001

Etim, E. E. "Phytoremediation and its mechanisms: A review." *International Journal of Environment and Bioenergy* 2, no. 3 (2012): 120–136.

Farr, Charlotte. *Indoor Air Quality: Causes, Controls and Consequences*. Lancaster University Press, 2021.

Farraji, Hossein, Nastaein Q. Zaman, R. Tajuddin, and Hamed Faraji. "Advantages and disadvantages of phytoremediation: A concise review." *International Journal of Environmental Science* 2 (2016): 69–75.

Favas, Paulo J. C., João Pratas, Mayank Varun, Rohan D'Souza, and Manoj S. Paul. "Phytoremediation of soils contaminated with metals and metalloids at mining areas: Potential of native flora." *Environmental Risk Assessment of Soil Contamination* 3 (2014): 485–516.

Ferreira, Fernanda and Jesse Kroll (2023). https://climate.mit.edu/ask-mit/how-can-such-small-amount-carbon-dioxide-atmosphere-only-around-420-parts-million-cause-so#:~:text=In%20June%202023%2C%20the%20National,parts%20per%20million%20(ppm).&text=This%20means%20that%20of%20every,atmosphere%2C%20424%20are%20CO2

Frontczak, Monika, Rune Vinther Andersen, and Pawel Wargocki. "Questionnaire survey on factors influencing comfort with indoor environmental quality in Danish housing." *Building and Environment* 50 (2012): 56–64.

Govindaraju, M., J. Fowmitha Banu, S. Senthamil Selvi, and Malti Goel. "CO 2 Sequestration through phytoremediation techniques with special emphasis on urban forestry to mitigate climate change impact." *Climate Change and Green Chemistry of CO2 Sequestration* (2021): 263–271.

Gul, Iram, Maria Manzoor, Imran Hashmi, Muhammad Faraz Bhatti, Jean Kallerhoff, and Muhammad Arshad. "Plant uptake and leaching potential upon application of amendments in soils spiked with heavy metals (Cd and Pb)." *Journal of Environmental Management* 249 (2019): 109408. https://doi.org/10.1016/j.jenvman.2019.109408

Holkar, Chandrakant R., Ananda J. Jadhav, Dipak V. Pinjari, Naresh M. Mahamuni, and Aniruddha B. Pandit. "A critical review on textile wastewater treatments: Possible approaches." *Journal of Environmental Management* 182 (2016): 351–366. https://doi.org/10.1016/j.jenvman.2016.07.090

Hu, Hao, Qi Zhou, Xiang Li, Wei Lou, Cheng Du, Qing Teng, Dongmei Zhang, Hongyu Liu, Yuanyuan Zhong, and Chunping Yang. "Phytoremediation of anaerobically digested swine wastewater contaminated by oxytetracycline via Lemna aequinoctialis: Nutrient removal, growth characteristics and degradation pathways." *Bioresource Technology* 291 (2019): 121853. https://doi.org/10.1016/j.biortech.2019.121853

Joseph, Lesley, Byung-Moon Jun, Joseph R. V Flora, Chang Min Park, and Yeomin Yoon. "Removal of heavy metals from water sources in the developing world using low-cost materials: A review." *Chemosphere* 229 (2019): 142–159. https://doi.org/10.1016/j.chemosphere.2019.04.198

Kandisa, Ravi Vital, K.V. Narayana Saibaba, Khasim Beebi Shaik, and R. Gopinath. "Dye removal by adsorption: A review." *Journal of bioremediation and Biodegradation* 7, no. 6 (2016). https://climate.ec.europa.eu/climate-change/causes-climate-change_en

Karaghool, Haneen A. Kh. "The Employment of Endophytic Bacteria for Phytodegradation of Pyridine." In *IOP Conference Series: Earth and Environmental Science*, vol. 961, no. 1, p. 012021. IOP Publishing, 2022. doi: 10.1088/1755–1315/961/1/012021

Kelly, Frank J., and Julia C. Fussell. "Air pollution and public health: Emerging hazards and improved understanding of risk." *Environmental Geochemistry and Health* 37 (2015): 631–649.

Khandare, Rahul V., and Sanjay P. Govindwar. "Phytoremediation of textile dyes and effluents: Current scenario and future prospects." *Biotechnology Advances* 33, no. 8 (2015): 1697–1714. https://doi.org/10.1016/j.biotechadv.2015.09.003

Kim, K. J., H. J. Kim, M. Khalekuzzaman, E. H. Yoo, H. H. Jung, and H. S. Jang. "Removal ratio of gaseous toluene and xylene transported from air to root zone via the stem by indoor plants." *Environmental Science and Pollution Research* 23 (2016): 6149–6158. https://doi.org/10.1007/s11356-016-6065-y

Krishnasamy, Selvanayaki, Ramkumar Lakshmanan, and Mythili Ravichandran. "Phytoremediatiation of metal and metalloid pollutants from farmland: An in-situ soil conservation." In *Biodegradation Technology of Organic and Inorganic Pollutants*. IntechOpen, 2021.

Kumar, Kundan, Akshay Shinde, Varad Aeron, Aanchal Verma, and Naseera Sayed Arif. "Genetic engineering of plants for phytoremediation: Advances and challenges." *Journal of Plant Biochemistry and Biotechnology* 32, no. 1 (2023): 12–30. https://doi.org/10.1007/s13562-022-00776-3

Mac Dowell, Niall, Paul S. Fennell, Nilay Shah, and Geoffrey C. Maitland. "The role of CO2 capture and utilization in mitigating climate change." *Nature Climate Change* 7, no. 4 (2017): 243–249.

Mikkonen, Anu, Tao Li, Mari Vesala, Jatta Saarenheimo, Viivi Ahonen, Sirpa Kärenlampi, James D. Blande, Marja Tiirola, and Arja Tervahauta. "Biofiltration of airborne VOC s with green wall systems – Microbial and chemical dynamics." *Indoor Air* 28, no. 5 (2018): 697–707. https://doi.org/10.1111/ina.12473

Morikawa, Hiromichi, and Özgür Cem Erkin. "Basic processes in phytoremediation and some applications to air pollution control." *Chemosphere* 52, no. 9 (2003): 1553–1558. https://doi.org/10.1016/S0045-6535(03)00495-8

Mousavi, Afsaneh, Latifeh Pourakbar, Sina Siavash Moghaddam, and Jelena Popović-Djordjević. "The effect of the exogenous application of EDTA and maleic acid on tolerance, phenolic compounds, and cadmium phytoremediation by okra (Abelmoschus esculentus L.) exposed to Cd stress." *Journal of Environmental Chemical Engineering* 9, no. 4 (2021): 105456. https://doi.org/10.1016/j.jece.2021.105456

Naseer, Ayesha, Anum Jamshaid, Almas Hamid, Nawshad Muhammad, Moinuddin Ghauri, Jibran Iqbal, Sikander Rafiq, Shahzad Khuram, and Noor Samad Shah. "Lignin and lignin based materials for the removal of heavy metals from waste water-an overview." *Zeitschrift für Physikalische Chemie* 233, no. 3 (2019): 315–345. https://doi.org/10.1515/zpch-2018-1209

Pandey, Vimal Chandra, and Omesh Bajpai. "Phytoremediation: From theory toward practice." In *Phytomanagement of Polluted Sites*, pp. 1–49. Elsevier, 2019. https://doi.org/10.1016/B978-0-12-813912-7.00001-6

Pilon-Smits, Elizabeth. "Phytoremediation." *Annual Review of Plant Biology* 56, no. 1 (April 29, 2005): 15–39. doi:10.1146/annurev.arplant.56.032604.144214. ISSN 1543-5008. PMID 15862088.

Rebecca Lindsey and Luann Dahlmal (2024). https://www.nrdc.org/stories/what-are-effects-climate-change#environment.

Roupsard, Pierre, Muriel Amielh, Didier Maro, Alexis Coppalle, Hubert Branger, Olivier Connan, P. Laguionie, D. Hébert, and M. Talbaut. "Measurement in a wind tunnel of dry deposition velocities of submicron aerosol with associated turbulence onto rough and smooth urban surfaces." *Journal of Aerosol Science* 55 (2013): 12–24. https://doi.org/10.1016/j.jaerosci.2012.07.006

Shafiq, M., A. A. Alazba, and M. T. Amin. "Removal of heavy metals from wastewater using date palm as a biosorbent: A comparative review." *Sains Malaysiana* 47, no. 1 (2018): 35–49 http://dx.doi.org/10.17576/jsm-2018-4701-05

Tewes, Lisa Johanna, Clemens Stolpe, Aylin Kerim, Ute Krämer, and Caroline Müller. "Metal hyperaccumulation in the Brassicaceae species Arabidopsis halleri reduces camalexin induction after fungal pathogen attack." *Environmental and Experimental Botany* 153 (2018): 120–126. https://doi.org/10.1016/j.envexpbot.2018.05.015

Thakur, Sveta, Lakhveer Singh, Zularisam Ab Wahid, Muhammad Faisal Siddiqui, Samson Mekbib Atnaw, and Mohd Fadhil Md Din. "Plant-driven removal of heavy metals from soil: Uptake, translocation, tolerance mechanism, challenges, and future perspectives." *Environmental Monitoring and Assessment* 188 (2016): 1–11.

Toome, Merje, Pille Randjärv, Lucian Copolovici, Ülo Niinemets, Katrin Heinsoo, Anne Luik, and Steffen M. Noe. "Leaf rust induced volatile organic compounds signalling in willow during the infection." *Planta* 232 (2010): 235–243.

Vadim, Sergey N. Shabala, Katherine B. Coutts, Mark A. Tester, and Julia M. Davies. "Free oxygen radicals regulate plasma membrane Ca2+-and K+-permeable channels in plant root cells." *Journal of Cell Science* 116, no. 1 (2003): 81–88.

Weber, Frauke, Ingo Kowarik, and Ina Säumel. "Herbaceous plants as filters: Immobilization of particulates along urban street corridors." *Environmental Pollution* 186 (2014): 234–240.

Zhang, Ying, Paul J. Beggs, Alice McGushin, Hilary Bambrick, Stefan Trueck, Ivan C. Hanigan, Geoffrey G. Morgan et al. "The 2020 special report of the MJA–Lancet Countdown on health and climate change: Lessons learnt from Australia's 'Black Summer'." *Medical Journal of Australia* 213, no. 11 (2020): 490–492. https://doi.org/10.5694/mja2.50869 52.

3 Organic and Inorganic Pollutant Uptake Mechanisms by Plants

Munachimso Odenakachi Victor-Ekwebelem, Simeon Okiki Naphtali, and M. M. Yakiimov

3.1 INTRODUCTION

Severe strains on the environment and natural resources have resulted from the rapid increase in the world's population and the corresponding demands for food, water, energy, timber, and other ecosystem services (Kamran et al., 2021; Saleem et al., 2020c; Zaheer et al., 2020). Heavy metals (HMs) and metalloids, radionuclides, organic compounds, agrochemicals, and oil spills are among the anthropogenic factors leading to significant environmental degradation (Abioye et al., 2017). These include agricultural intensification practices, fast urbanization, and industrialization. Soils have become contaminated due to mining operations, the release of industrial and residential effluents, excessive use of pesticides and fertilizers, and irrigation of contaminated water (Aransiola et al., 2022; Victor-Ekwebelem et al., 2020). Numerous soil characteristics, such as pH, electrical conductivity, and cation exchange capacity, are disturbed as a result of mining (Saleem et al., 2020a). As a result, HMs like cadmium (Cd), zinc (Zn), lead (Pb), Cu, and nickel (Ni) are released. Thus, in an effort to maximize the advantages of each mechanism through plant selection, this chapter compiles data and conclusions on processes that plants use. High amounts of organic pollutants, including hydrocarbons, aromatic compounds, phenols, organochlorinated compounds, and pesticides, as well as elements like chromium (Cr), iron (Fe), and manganese (Mn) have been discovered in soil and sediments (Liu et al., 2017b; Song et al., 2017). The population, activities, and processes of soil microbes are impacted by HM accumulation in the soil (Huang et al., 2009; Singh & Kalamdhad, 2011 Aransiola et al., 2019).

The effects of HMs on changes in microbial population and enzymatic activities are dependent upon the type of HMs and their chemical affinity (Karaca et al., 2010; Aransiola et al., 2023). For example, Cd affects ureases, proteases, and basic phosphatases (Lorenz et al., 2006); arsenic (As) affects sulfatase and phosphatase; copper (Cu) restricts β-glucosidase (Kafle et al., 2022); and lead (Pb) affects catalases, ureases, invertase, and acid phosphatase (Belyaeva et al., 2005; Aransiola et al., 2013). High chromium(VI) concentration in soil exhibits distinct toxicity characteristics causing negative effects in soil microbial cell metabolism (Huang et al., 2009). Bioaccumulation and biomagnification of HMs can arise from ingestion of food crops contaminated with HMs. These can threaten both animal and human health (Singh & Kalamdhad, 2011). HMs accumulation in human body can cause serious complications in human health, such as itai-itai disease due to chronic Cd exposure (Pan et al., 2010); mucosa corrosion, hepatic system failure, and central nervous system damage due to Cu (Singh & Kalamdhad, 2011); abortion and breakdown of nervous and skin rashes due to mercury (Hg); kidney dysfunction and anemia due to Zn; skin irritation and nervous system complications due to Ni (Argun et al., 2007; Druruibe et al., 2007); cardiovascular problems and damage to nervous system due to Pb (Kazemipour et al., 2008; Singh & Kalamdhad, 2011); cancer due to Cr (Zohdi et al., 2012); and carcinogenic and immune system dysfunction due to As (Druruibe et al., 2007).

DOI: 10.1201/9781032683751-4

Among different organic contaminants, chlorinated solvents, petroleum hydrocarbons, explosives, polychlorinated biphenyls (PCBs), polycyclic aromatic hydrocarbons (PAHs), pyrene, and trinitrotoluene (TNT) are major soil contaminants. Pesticides that often contaminate soils include atrazine, dieldrin, hexachlorobenzene, dichlorodiphenyltrichloroethane (DDT), and 1,2,4-trichlorobenzene. These pollutants seriously affect the soil biota, mineral nutrient availability, and soil productivity because of their persistence in soil for long period of time and their toxicity. Like soils, water resources have also been contaminated by different pollutants. Urbanization and industrialization activities release toxic HMs through sewage runoff and effluent, contributing to the largest share of factors for water contamination (Masindi & Muedi, 2018; Parveen et al., 2020; Saleem et al., 2020a). Over 80% of the world's wastewater reaches the environment without treatment, causing nearly 58% of health issues like diarrhea, which is a major cause of early child death (Connor et al., 2017). Among metals, Cu, Ni, Pb, Cd, Cr, uranium (U), and Zn frequently contaminate water resources (Prasad & Freitas, 2005). The HMs degrade the quality of water and can lead to waterborne health hazards through HM accumulation in the food chain (Wongsasuluk et al., 2014). Apart from HMs, organic pollutants contaminate water resources by erosion, seepage, sludge, and effluents from municipal waste and industries, as well as runoff from agricultural practices and mining (Kafle et al., 2022). These contaminants can accumulate over time and negatively affect both aquatic and terrestrial ecosystems. Hence there is a need to understand their methods and mechanisms so that the right approach will be used. This chapter attempts to bring together reports and findings on mechanisms that plants utilize and how to maximize the benefits of each mechanism, on the strengths and limitations of the phytoremediation technology, and on ways to further enhance phytoremediation of pollutants including HMs and organic pollutants in soil and water.

3.2 MECHANISM OF PHYTOREMEDIATION

The essential physiological processes that higher plants go through form the basis of phytoremediation. Among these systems are those of respiration, photosynthesis, metabolism, and mineral feeding. Apart from taking in both organic and inorganic elements, roots can also bind and stabilize materials on their outside surfaces and communicate with rhizosphere bacteria. Plants extend roots in water, sediments, and soils; plant cells and tissues have the ability to move, store, modify, and accumulate ingested substances; and aerial portions of plants may exchange gases with the atmosphere, allowing molecules to be absorbed or released. It would be more appropriate to refer to the wide range of uses of plants for pollution management and control – even in situations where they are not destroyed or eliminated – as "phytotechnology." There are seven distinct strategies for performing phytoremediation; which one is best suited for a successful remediation depends on the kind of pollution, the properties of the soil, and the material's bioavailability. The several forms of phytoremediation are phytoextraction, phytoimmobilization, phytostabilization, phytodegradation, and rhizodegradation (Figure 3.1). To remediate soil and water, different plants employ different strategies, or combinations of them, depending on the types, forms, and source of the pollutant. For example, rhizofiltration, phytodegradation, phytovolatilization, phytodegradation, and phytodegradation can all be used to remediate contaminated groundwater, whereas rhizofiltration, phytodegradation, and phytodegradation can be used to treat wastewater and surface water contamination. Remediation of contaminated soil, sediments, or sludges is accomplished by phytoextraction, phytodegradation, phytostabilization, rhizodegradation, or phytovolatilization. The ideal plant for phytoremediation must possess certain key characteristics like ease of cultivation, high biomass production rate, tolerance to the deleterious effects of metals and contaminants, high absorption capacity, and lack of herbivore attraction (Adesodun et al., 2010; Kafle et al., 2022; Shabani & Sayadi, 2012; Aransiola et al., 2019). Assistance is required to improve the phytoremediation capacity of certain plant species because they do not always function adequately. These enhancement techniques include *Arbuscular mycorrhiza*, ethylenediaminetetraacetic acid (EDTA), transgenic plants, charcoal, and endophytic bacteria (Paz-Ferreiro et al., 2014; Shahid et al., 2014; Afzal et al.,

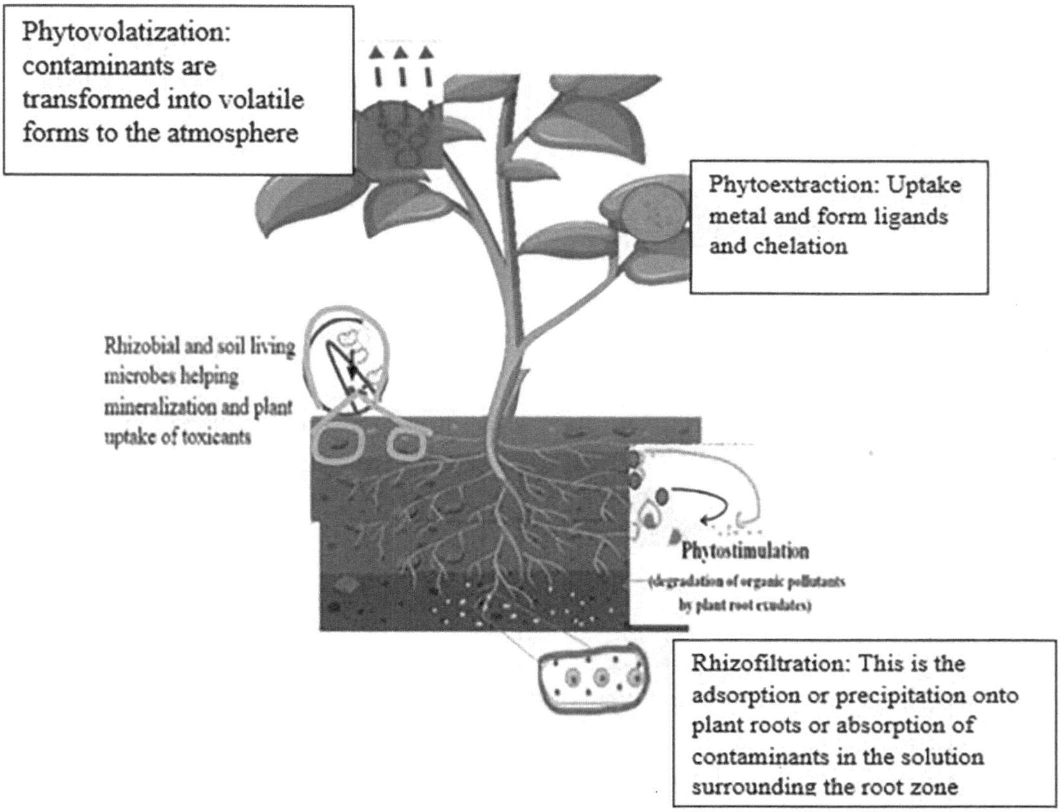

FIGURE 3.1 Some mechanisms used by plants to remediate inorganic pollutants.

2014; Elemba & Ijah, 2016). These compounds support the phytoremediation process by increasing the availability of pollutants, reducing their toxicity, or promoting plant development.

Phytoextraction, also known as "phytoaccumulation," describes the process by which plant roots absorb and relocate metal pollutants from the soil into the above-ground sections of the plants. The main application of phytoextraction is in the remediation of polluted soils (United States Environmental Protection Agency [USEPA], 2000). This method involves using plants to absorb, concentrate, and precipitate hazardous metals from polluted soils into the above-ground biomass (leaves, shoots, etc.) in order to remove the pollution from the soil. A hyperaccumulator is a plant that can accumulate 100 times as much metal as a typical non-accumulating plant (UNEP, undated). The majority of plants that take up and absorb abnormally high concentrations of metals favor metals like Ni, Zn, and Cu, making these the greatest candidates for phytoextraction removal. Additionally, the technique is typically restricted to metals and other inorganic substances found in sediment or soil (Saleh, 2012; EPA, 2000). Phytoextraction techniques are used by plants such as maize (Zamani et al., 2018), lettuce, and tobacco to phytoremediate pollutants like Pb, Ni, and Cd (Table 3.1). Phytomass must be disposed off properly.

Rhizofiltration is primarily used to remediate extracted groundwater, surface water, and wastewater with low contaminant concentrations. It is the adsorption or precipitation onto plant roots or absorption of contaminants in the solution surrounding the root zone. Rhizofiltration is typically exploited in groundwater (either in situ or extracted), surface water, or wastewater for removal of metals or other inorganic compounds (EPA, 2000). To achieve rhizofiltration plants like, water hyacinth, water velvet, and duckweed (Table 3.2) can be used for As, Cd, Cu, Ni, Zn, and Cr, which

TABLE 3.1
Plants That Utilize Phytoextraction during Phytoremediation and Possible Contaminates Remediated

Common Name	Scientific Name	Contaminants Easily Remediated
Corn	*Zea mays*	Pb, Ti
Lettuce	*Lactuca sativa*	Ni, Co, and Fe
Tobacco	*Nicotiana tabacum*	Cd
Rattlebush	*Sesbania drummondii*	Pb
Chinese brake fern	*Pteris vittata*	Arsenic (As)
Indian mustard	*Brassica juncea*	Pb
Common cocklebur	*Xanthium strumarium*	Cd, Pb, Ni, Zn
Vygies	*Mesembryanthemum criniflorum*	Pb
Perennial ryegrass	*Lolium perenne*	Ni, Co, and Fe
Black nightshade	*Solanum nigrum*	Cd, Pb
Garden geranium	*Pelargonium hortorum*	Pb, Cd
Hance	*Sedum alfredii*	Pb

TABLE 3.2
Plants That Use the Rhizofiltration Mechanism for Phytoremediation

Common Name	Scientific Name	Contaminants
Carolina mosquitofern/water velvet	*Azolla caroliniana*	As
Water hyacinth	*Eichhornia crassipes*	Fe, Cr, Cu, Cd, Zn, Ni, As
Water-starwort	*Callitriche lusitanica*	As
Duckweed	*Lemna minor*	U
Threadleaf crowfoot	*Ranunculus trichophyllus*	As

are primarily retained within the roots (USEPA, 2000). Rhizofiltration is similar to phytoextraction, but the plants are used primarily to address contaminated groundwater rather than soil. The plants to be used for cleanup are raised in greenhouses with their roots in water rather than in soil. To acclimatize the plants, once a large root system has been developed, contaminated water is collected from a waste site and brought to the plants where it is substituted for their water source. The plants are then planted in the contaminated area where the roots take up the water and the contaminants along with it. As the roots become saturated with contaminants, they are harvested.

The advantages associated with rhizofiltration are the ability to use both terrestrial and aquatic plants for either in situ or ex situ applications. Another advantage is that contaminants do not have to be translocated to the shoots. Thus species other than hyperaccumulators may be used. Terrestrial plants are preferred because they have a fibrous and much longer root system, increasing the amount of root area (Raskin & Ensley, 2000). Rhizofiltration is a cost-effective, environmentally friendly, self-sustaining method of contaminant removal. Disadvantages include continuous adjustment of pH in influent solution to obtain optimum metals uptake and slow-growing root systems.

Phytovolatilization involves using plants to absorb pollutants from the soil, changing them into volatile forms, and then moving them into the atmosphere via transpiration (Figure 3.1). According to the EPA (2000), phytovolatilization also entails the uptake of toxins into the plant's body, but

TABLE 3.3
Plants That Utilize Phytovolatilization Mechanism during Phytoremediation

Common Name	Scientific Name	Possible Contaminants
Perennial reed grass	*Phragmites australis*	OCs
Mustard	*Brassica juncea*	Se
Broccoli	*Brassica oleracea* var. *italica*	Se
Rabbitsfoot grass	*Polypogon monspeliensis*	As
Parrot's feather	*Myriophyllum brasiliensis*	Se

after that, the contaminant, a volatile version of it, or a volatile breakdown product is transpired with water vapor from the leaves. Diffusion of pollutants from the stems or other plant parts that the contaminant passes through before reaching the leaves is another possible aspect of phytovolatilization (Kafle et al., 2022). Contaminants in soil, sediment, or water can cause phytovolatilization. Mercury is the primary metal contaminant for which this process has been used. Additionally, it has been discovered to happen with volatile inorganic chemicals like As and selenium (Se) and volatile organic substances like trichloroethene (EPA, 2000). This approach has the benefit of potentially converting the pollutant, mercuric ion, into elemental Hg, which is less hazardous. The drawback of this is that precipitation is likely to recycle Hg discharged into the atmosphere, redistributing it into lakes and oceans and so continuing the process of anaerobic bacteria producing methylmercury (USEPA, 2000). phytovolatilization techniques are used by plants such as perennial reed grass and rabbitsfoot grass to phytoremediate pollutants like organochlorines (OCs), As, and Se (Table 3.3).

Phytostabilization is also referred to as in-place inactivation. It is primarily used for the remediation of soil, sediment, and sludge (USEPA, 2000). It is the use of certain plant species to immobilize contaminants in the soil and groundwater through absorption and accumulation by roots, adsorption onto roots, or precipitation within the root zone of plants (rhizosphere). This process reduces the mobility of the contaminant and prevents migration to the groundwater; it also reduces bioavailability of metal into the food chain. This technique can also be used to re-establish vegetation cover at sites where natural vegetation fails to survive due to high metals concentrations in surface soils or physical disturbances to surface materials. Metal-tolerant species are used to restore vegetation at contaminated sites, thereby decreasing the potential migration of pollutants through wind erosion and transport of exposed surface soils and leaching of soil contamination to groundwater. Phytostabilization can occur through sorption, precipitation, complexation, or metal valence reduction. It is useful for the treatment of Pb, As, Cd, Cr, Cu, and Zn. Phytostabilization takes advantage of the changes that the presence of the plant induces in soil chemistry and environment. These changes in soil chemistry may induce adsorption of contaminants onto the plant roots or soil or cause metals precipitation onto the plant root. Advantages associated with this technology include that the disposal of hazardous material/biomass is not required, and it is very effective when rapid immobilization is needed to preserve groundwater and surface water (Zhang et al., 2009). The presence of plants also reduces soil erosion and decreases the amount of water available in the system (USEPA, 2000). However, this clean-up technology has several major disadvantages, including contaminant remaining in soil, application of extensive fertilization or soil amendments, mandatory monitoring, and contaminant stabilization primarily due to the soil amendments. Willow is an important phytoremediation plant (Table 3.4) due to its huge biomass and its ability to use various phytoremediation mechanism. Phytostabilization can be favored by the addition of about 20 mg/L of citric acid.

Phytodegradation is also referred to as phytotransformation. It involves the degradation of complex organic molecules into simple molecules or the incorporation of these molecules into plant tissues (Kafle et al., 2022). When the phytodegradation mechanism is at work, contaminants are broken down after they have been taken up by the plant. As with phytoextraction and phytovolatilization,

TABLE 3.4
Plants That Utilize Phytostabilization/Phytoimmobilization Processes

Plant Name	Scientific Name	Pollutants
Cowpea	*Vigna unguiculata*	Pb, Zn
Willow	*Salix* sp.	Cd
Holy rope/hemp-agrimony	*Eupatorium cannabinum*	Citric acid (CA)
Sunflower	*Helianthus annuus*	Cu, Zn, Pb, Hg, As, Cd, Ni
Coastal mallow/seashore mallow	*Kosteletzkya pentacarpos*	Cd, Zn
Water-starwort	*Callitriche brutia*	As
Black nightshade	*Solanum nigrum*	Cu, Zn, Cd

Sources: González et al. (2019) Zhou et al. (2019) Yang et al. (2019) Li et al. (2019).

plant uptake generally occurs only when the contaminants' solubility and hydrophobicity fall into a certain acceptable range. Phytodegradation has been observed to remediate some organic contaminants, such as chlorinated solvents, herbicides, and munitions, and it can address contaminants in soil, sediment, or groundwater (EPA, 2000).

Rhizodegradation is also referred to as phytostimulation. Rhizodegradation refers to the breakdown of contaminants within the plant root zone, or rhizosphere. It is believed to be carried out by bacteria or other microorganisms whose numbers typically flourish in the rhizosphere. Studies have documented up to 100 times as many microorganisms in rhizosphere soil as in soil outside the rhizosphere (USEPA, 2000). Microorganisms may be so prevalent in the rhizosphere because the plant exudes sugars, amino acids, enzymes, and other compounds that can stimulate bacterial growth. The roots also provide additional surface area for microbes to grow on and a pathway for oxygen transfer from the environment. The localized nature of rhizodegradation means that it is primarily useful in contaminated soil, and it has been investigated and found to have at least some successes in treating a wide variety of mostly organic chemicals, including petroleum hydrocarbons, polycyclic aromatic hydrocarbons (PAHs), chlorinated solvents, pesticides, polychlorinated biphenyls (PCBs), benzene, toluene, ethylbenzene, and xylenes (EPA, 2000). It can also be seen as plant-assisted bioremediation, the stimulation of microbial and fungal degradation by release of exudates/enzymes into the root zone (rhizosphere; Zhuang et al., 2005).

3.3 CONCLUSION

With current intensive agriculture practices and industrialization, pollution of natural resources like land and water with HMs, organic pollutants, radionuclides, pesticides, and fertilizers has become a major concern. Phytoremediation is a cost-effective and environmentally friendly technique that utilizes plants to immobilize, uptake, reduce toxicity, stabilize, or degrade the compounds that are released into the environment from different sources. Studies have shown that HMs, organic contaminants, radionuclides, antibiotics, and pesticides can be remediated using plants. Ideal plant species for phytoremediation purpose should be hardy in nature, able to produce high biomass, tolerant to toxic effects of metals and contaminants, easy to cultivate, have high absorption capacity, and be non-attractive for herbivory. Though phytoremediation has been practiced for decades, it is still an emerging technology. This chapter summarizes existing information and synthesizes the recent findings on plant species suitable for use in phytoremediation through utilizing different mechanisms, aids that can enhance the efficiency of phytoremediation processes. Diverse plants remediate different pollutants at different rates through one or multiple mechanisms. The limitations of phytoremediation can be overcome by using several aids including natural and chemical amendments, genetic engineering, and natural microbial stimulation. Given the low-cost of phytoremediation compared to conventional technology and sustainability associated with plants and use of renewable energy, phytoremediation can be a reliable solution for a sustainable and economical remediation of soil and water from the organic and inorganic pollutants.

REFERENCES

Abioye, O.P., Ijah, U.J.J., Aransiola, S.A., 2017. Phytoremediation of soil contaminants by biodiesel plant Jatropha curcas. In Chapter 4 of K. Bauddh et al. (eds.), *Phytoremediation Potential of Bioenergy Plants, DOI 10.1007/978-981-10-3084-0_4*. Springer Nature Singapore Pte Ltd. https://link.springer.com/book/10.1007%2F978-981-10-3084-0.

Adesodun, J.K., Atayese, M.O., Agbaje, T., Osadiaye, B.A., Mafe, O., Soretire, A.A., 2010. Phytoremediation potentials of sunflowers (Tithonia diversifolia and Helianthus annuus) for metals in soils contaminated with zinc and lead nitrates. *Water Air Soil Pollut.* 207, 195–201.

Afzal, M., Khan, Q.M., Sessitsch, A., 2014. Endophytic bacteria: Prospects and applications for the phytoremediation of organic pollutants. *Chemosphere* 117, 232–242.

Aransiola, S.A., Afolabi, F., Joseph, F., Maddela, N.R., 2023. Soil Enzymes: Distribution, interactions and influencing factors. In P. Jeschke, E.B. Starikov (eds.), *Agricultural Biocatalysis: Enzymes in Agriculture and Industry* (1st ed.). Jenny Stanford Publishing, pp. 303–333. ISBN 9789814968478. https://www.routledge.com/Agricultural-Biocatalysis-Enzymes-in-Agriculture-and-Industry/Jeschke-Starikov/p/book/9789814968478

Aransiola, S.A., Ijah, U.J.J., Abioye, O.P., 2013. Phytoremediation of lead polluted soil by *Glycine max* L. *Appl. Environ. Soil Sci.* 2013, Article ID 631619. doi: 10.1155/2013/631619. https://www.hindawi.com/journals/aess/2013/631619/abs/.

Aransiola, S.A., Ijah, U.J.J., Abioye, O.P., Bala, J.D., 2019. Microbial-aided phytoremediation of heavy metals contaminated soil: A review. *Eur. J. Biol. Res.* 9(2), 104–125. http://www.journals.tmkarpinski.com/index.php/ejbr/article/view/157.

Aransiola, S.A., Ijah, U.J.J., Abioye, O.P., Bala, J.D., 2022. Vermicompost-assisted phytoremediation of toxic trace element-contaminated soil in Madaka, Nigeria using *Melissa officinalis* L and *Sida acuta*. *IJEST*. http://dx.doi.org/10.1007/s13762-022-04105-y

Argun, M.E., Dursun, S., Ozdemir, C., Karatas, M., 2007. Heavy metal adsorption by modified oak sawdust: Thermodynamics and kinetics. *J. Hazard. Mater*. 141, 77–85.

Belyaeva, O., Haynes, R., Birukova, O., 2005. Barley yield and soil microbial and enzyme activities as affected by contamination of two soils with lead, zinc or copper. *Biol. Fertil. Soils*. 41, 85–94.

Connor, R., Renata, A., Ortigara, C., Koncagül, E., Uhlenbrook, S., Lamizana-Diallo, B.M., Zadeh, S.M., Qadir, M., Kjellén, M., Sjödin, J., 2017. The United Nations world water development report 2017. Wastewater: The untapped resource. *Unit. Nat. World Water Dev. Report.* deFontenoy, Paris, France.

Druruibe, J.O., Ogwuegbu, M., Egwurugwu, J., 2007. Heavy metal pollution and human biotoxic effects. *Int. J. Phys. Sci.* 2, 112–118.

Elemba, O.M., Ijah, U.J.J., 2016. Removal of lead and chromium from soil using biosurfactants produced by bacterial isolates from spent lubricating oil contaminated soil. *Int J Adv Res*. 4(5), 201–211 (IMPACT FACTOR 5.388).

EPA, 2000. *A Citizen's Guide to Phytoremediation*. EPA 542-F-98-011. United States Environmental Protection Agency, p. 6. http//www.bugsatwork.com/XYCLONYX/EPA_GUIDES/PHYTO.PD

González, H., Fernández-Fuego, D., Bertrand, A., González, A., 2019. Effect of pH and citric acid on the growth, arsenic accumulation, and phytochelatin synthesis in Eupatorium cannabinum L., a promising plant for phytostabilization. *Environ. Sci. Pollut. Res.* 26, 26242–26253.

Huang, S.-H., Bing, P., Yang, Z.-H., Chai, L.-Y., Zhou, L.-C., 2009. Chromium accumulation, microorganism population and enzyme activities in soils around chromium-containing slag heap of steel alloy factory. *Transact. Nonferr. Metal. Soc. China*. 19, 241–248.

Kafle, A., Timilsina, A., Asmita, G., Adhikari, K., Bhattara, A., Niroj, A., 2022. Phytoremediation: Mechanisms, plant selection and enhancement by natural and synthetic agents *Environ. Adv.* 8(2022), 100–203.

Kamran, M., Danish, M., Saleem, M.H., Malik, Z., Parveen, A., Abbasi, G.H., Jamil, M., Ali, S., Afzal, S., Riaz, M., 2021. Application of abscisic acid and 6-benzylamino-purine modulated morpho-physiological and antioxidative defense responses of tomato (Solanum lycopersicum L.) by minimizing cobalt uptake. *Chemosphere* 263, 128169.

Karaca, A., Cetin, S.C., Turgay, O.C., Kizilkaya, R., 2010. *Effects of Heavy Metals on Soil Enzyme Activities. Soil Heavy Metals*. Springer, pp. 237–262.

Kazemipour, M., Ansari, M., Tajrobehkar, S., Majdzadeh, M., Kermani, H.R., 2008. Removal of lead, cadmium, zinc, and copper from industrial wastewater by carbon developed from walnut, hazelnut, almond, pistachio shell, and apricot stone. *J. Hazard. Mater*. 150, 322–327.

Li, X., Zhang, X., Wang, X., Cui, Z., 2019. Phytoremediation of multi-metal contaminated mine tailings with Solanum nigrum L. and biochar/attapulgite amendments. *Ecotoxicol. Environ. Saf.* 180, 517–525.

Liu, Y., Zeng, G., Zhong, H., Wang, Z., Liu, Z., Cheng, M., Liu, G., Yang, X., Liu, S., 2017b. Effect of rhamnolipid solubilization on hexadecane bioavailability: Enhancement or reduction? *J. Hazard. Mater*. 322, 394–401.

Lorenz, N., Hintemann, T., Kramarewa, T., Katayama, A., Yasuta, T., Marschner, P., Kandeler, E., 2006. Response of microbial activity and microbial community composition in soils to long-term arsenic and cadmium exposure. *Soil Biol. Biochem*. 38, 1430–1437.

Masindi, V., Muedi, K.L., 2018. Environmental contamination by heavy metals. *Heavy Metal.* 10, 115–132.

Pan, J., Plant, J.A., Voulvoulis, N. Oates, C.J., Ihlenfeld, C. 2010. Cadmium levels in Europe: implications for human health. *Environ. Geochem. Health*. 32, 1–12. https://doi.org/10.1007/s10653-009-9273-2.

Parveen, A., Saleem, M.H., Kamran, M., Haider, M.Z., Chen, J.-T., Malik, Z., Rana, M.S., Hassan, A., Hur, G., Javed, M.T., 2020. Effect of citric acid on growth, ecophysiology, chloroplast ultrastructure, and phytoremediation potential of jute (Corchorus capsularis L.) seedlings exposed to copper stress. *Biomolecules*. 10, 592.

Paz-Ferreiro, J., Lu, H., Fu, S., Méndez, A., Gascó, G., 2014. Use of phytoremediation and biochar to remediate heavy metal polluted soils: A review. *Solid Earth* 5, 65.

Prasad, M., Freitas, H., 2005. *Metal-tolerant Plants: Biodiversity Prospecting for Phytoremediation Technology*. Trace Elements in the Environment. CRC Press, pp. 501–524.

Saleem, M.H., Ali, S., Kamran, M., Iqbal, N., Azeem, M., Tariq Javed, M., Ali, Q., Zulqurnain Haider, M., Irshad, S., Rizwan, M., 2020a. Ethylenediaminetetraacetic acid (EDTA) mitigates the toxic effect of excessive copper concentrations on growth, gaseous exchange and chloroplast ultrastructure of Corchorus capsularis L. and improves copper accumulation capabilities. *Plants*. 9, 756.

Saleem, M.H., Ali, S., Rehman, M., Rana, M.S., Rizwan, M., Kamran, M., Imran, M., Riaz, M., Soliman, M.H., Elkelish, A., 2020c. Influence of phosphorus on copper phytoextraction via modulating cellular organelles in two jute (Corchorus capsularis L.) varieties grown in a copper mining soil of Hubei Province, China. *Chemosphere* 248, 126032.

Saleh, H.M., 2012. Water hyacinth for phytoremediation of radioactive waste simulate contaminated with cesium and cobalt radionuclides. *Nucl. Eng. Des*. 242, 425–432. https://doi.org/10.1016/j.nucengdes.2011.10.023.

Shabani, N., Sayadi, M., 2012. Evaluation of heavy metals accumulation by two emergent macrophytes from the polluted soil: An experimental study. *Environment*. 32, 91–98.

Shahid, M., Austruy, A., Echevarria, G., Arshad, M., Sanaullah, M., Aslam, M., Nadeem, M., Nasim, W., Dumat, C., 2014. EDTA-enhanced phytoremediation of heavy metals: A review. *Soil and Sedim. Contam.* 23, 389–416.

Singh, J., Kalamdhad, A.S., 2011. Effects of heavy metals on soil, plants, human health and aquatic life. *Int. Res. Chem. Environ*. 1, 15–21.

Song, B., Zeng, G., Gong, J., Zhang, P., Deng, J., Deng, C., Yan, J., Xu, P., Lai, C., Zhang, C., 2017. Effect of multi-walled carbon nanotubes on phytotoxicity of sediments contaminated by phenanthrene and cadmium. *Chemosphere*. 172, 449–458.

UNEP, Undated. Phytoremediation: An environmentally sound technology for pollution prevention, control and remediation. An Introductory Guide to Decision-Makers. Newsletter and Technical Publications Freshwater Management Series No. 2, United Nations Environment Programme Division of Technology, Industry, and Economics.

United States Environmental Protection Agency (USEPA). (2000). Introduction to phytoremediation. EPA 600/R-99/107, U.S. Environmental Protection Agency, Office of Research and Development Cincinnati, OH.

Victor-Ekwebelem, M.O., Ijah, J.J., Abioye, O.P., Alkali, Y.B., 2020. Physicochemical and microbiological properties of soil in Ahoko: A Suspected Petroleum Bearing Site (SPBS). *Sci. Res. J. (SCIRJ)*. VIII(V), 15–31.

Wongsasuluk, P., Chotpantarat, S., Siriwong, W., Robson, M., 2014. Heavy metal contamination and human health risk assessment in drinking water from shallow groundwater wells in an agricultural area in Ubon Ratchathani province. *Thai. Environ. Geochem. Health*. 36, 169–182.

Yang, W., Yang, Y., Ding, Z., Yang, X., Zhao, F., Zhu, Z., 2019. Uptake and accumulation of cadmium in flooded versus non-flooded Salix genotypes: Implications for phytoremediation. *Ecol. Eng.* 136, 79–88.

Zaheer, I.E., Ali, S., Saleem, M.H., Imran, M., Alnusairi, G.S., Alharbi, B.M., Riaz, M., Abbas, Z., Rizwan, M., Soliman, M.H., 2020. Role of iron–lysine on morpho-physiological traits and combating chromium toxicity in rapeseed (Brassica napus L.) plants irrigated with different levels of tannery wastewater. *Plant Physiol. Biochem.* 155, 70–84.

Zamani, J., Hajabbasi, M.A., Mosaddeghi, M.R., Soleimani, M., Shirvani, M., Schulin, R., 2018. Experimentation on degradation of petroleum in contaminated soils in the root zone of maize (Zea Mays L.) inoculated with Piriformospora indica. *Soil and Sedim. Contam.* 27, 13–30.

Zhang, H., Zheng, L.C., Yi, X.Y., 2009. Remediation of soil co-contaminated with pyrene and cadmium by growing maize (Zea mays L.). *Int. J. Environ. Sci. Tech.* 6, 249–258.

Zhou, M., Ghnaya, T., Dailly, H., Cui, G., Vanpee, B., Han, R., Lutts, S., 2019. The cytokinin trans-zeatine riboside increased resistance to heavy metals in the halophyte plant species Kosteletzkya pentacarpos in the absence but not in the presence of NaCl. *Chemosphere.* 233, 954–965.

Zhuang, P., Ye, Z.H., Lan, C.Y., Xie, Z.W., Hsu, W.S., 2005. Chemically assisted phytoextraction of heavy metal contaminated soils using three plant species. *Plant Soil.* 276, 153–162.

Zohdi, H., Emami, M., Shahverdi, H.R., 2012. Galvanic corrosion behavior of dental alloys. In *Environmental and Industrial Corrosion-Practical and Theoretical Aspects*. IntechOpen.

4 Soil Regeneration through Phytoremediation for Safe Foods

Innocent Ojeba Musa, Job Oloruntoba Samuel, Mustahpa Adams, Vivian Nathaniel, Asmau M. Maude, Mohammed Evuti Mahmud, and Abd'Gafar Tunde Tiamiyu

4.1 INTRODUCTION

Soil contamination poses a significant risk to both global food safety and environmental integrity. The need for effective and long-term remediation techniques is crucial due to increasing concerns about polluted soils and their impact on human health and ecosystems (Farooq et al., 2019). Industrialization and urbanization have led to a significant increase in the presence of heavy metals in the environment, causing global concerns (Suman et al., 2018). Heavy metals are metallic chemical elements that possess high densities, atomic weights, and atomic numbers. Cadmium (Cd), mercury (Hg), lead (Pb), arsenic (As), zinc (Zn), copper (Cu), nickel (Ni), and chromium (Cr) are common heavy metals/metalloids. These elements can originate from natural or human activities, such as the water produced from the oil and gas industries (Pichtel, 2016), the application of phosphate fertilizers in agriculture (Hamzah et al., 2016; Rafique and Tariq, 2016), sewage sludge, and metal mining and smelting (Chen et al., 2016).

Heavy metals are non-biodegradable and persist in soil for an extended period, causing long-term environmental harm (Suman et al., 2018). Heavy metals are categorized as either essential or non-essential depending on their biological importance. Excessive presence of heavy metals, including Cu, iron (Fe), manganese (Mn), Ni, and Zn, can be toxic, although they are necessary for physiological and biochemical processes throughout the plant life cycle (Ashraf et al., 2019). Non-essential heavy metals, including Pb, Cd, As, and Hg, are highly toxic to plants and lack any known physiological function (Fasani et al., 2018). Pesticides have the potential to harm the environment, interfere with physiological and biochemical processes in crops, and decrease agricultural productivity (Ashraf et al., 2019). Pesticides can enter the food chain through crops and accumulate in the human body, presenting a substantial risk to human health (Rehman et al., 2017; Abioye et al., 2017).

In order to mitigate the contamination of terrestrial, atmospheric, and aquatic environments by heavy metals and improve polluted land, the implementation of remediation methods is necessary (Gerhardt et al., 2017; Hasan et al., 2019). Several remediation techniques have been developed for soil contaminated by heavy metals. The procedures used for remediation of contaminated soil can be classified as mechanical or physiochemical methods. Examples of these methods include soil burning, excavation and landfill, soil washing, solidification, and electric field application (DalCorso et al., 2019). Physicochemical approaches have limitations including high cost, inefficiency at low contaminant concentrations, irreversible changes to soil properties, degradation of the soil ecosystem, and the introduction of secondary pollutants. Therefore, it is crucial to develop cost-effective, efficient, and environmentally acceptable remediation technology for reclaiming soil contaminated with heavy metals (DalCorso et al., 2019).

 DOI: 10.1201/9781032683751-5

Phytoremediation is a plant-based technique used to extract and eliminate elemental contaminants from soil or reduce their bioavailability (Aransiola et al., 2013; Jacob et al., 2018). Plants have the ability to uptake ionic substances from the soil through their roots, even when present in low concentrations. Plants establish rhizosphere ecosystems by extending their root systems into the soil matrix. This allows them to absorb heavy metals and regulate their bioavailability, thereby restoring damaged soil and maintaining soil fertility (DalCorso et al., 2019). Phytoremediation offers numerous benefits including the following:

1. It is economically feasible. Phytoremediation is an autotrophic system that relies on solar energy, resulting in a straightforward operation and low costs for installation and maintenance.
2. Environmentally friendly practices can effectively minimize pollutant exposure to the environment and ecosystem.
3. The method is applicable on a large scale and can be easily disposed of.
4. It prevents erosion and the leaching of heavy metals by stabilizing them, thereby reducing the risk of contaminant spread.
5. It can enhance soil fertility by releasing various organic substances into the soil (Thakur et al., 2016; Aransiola et al., 2022).

This chapter explores phytoremediation, an approach that harnesses the natural abilities of plants to restore soil quality, eliminate contaminants, and ensure the cultivation of safe and uncontaminated crops.

4.2 PHYTOREMEDIATION TECHNIQUES

Phytoremediation encompasses a range of methodologies that harness the distinctive capabilities of plants to mitigate many forms of soil contaminants. Every strategy possesses its own distinct mechanism for the elimination of pollution.

4.2.1 Phytoextraction: Extracting Heavy Metals from Soil

Phytoextraction involves the utilization of plants to take up pollutants from soil or water and subsequently transport and accumulate these toxins in their above-ground biomass (Jacob et al., 2018). Phytoextraction is a prominent phytoremediation method for extracting heavy metals and metalloids from contaminated soil. It is a permanent method for removing heavy metals from damaged soil, unlike phytostabilization where plants temporarily retain the heavy metals below ground. According to Zakari et al. (2021), phytoextraction is more suitable for commercial purposes.

Phytoextraction of heavy metals involves several steps, including mobilization of heavy metals in the rhizosphere, uptake of heavy metals by plant roots, translocation of heavy metal ions from roots to aerial parts of the plant, and sequestration and compartmentation of heavy metal ions in plant tissues (Gavrilescu, 2022). Phytoextraction efficiency is influenced by plant selection, plant performance, heavy metal bioavailability, soil variables, and rhizosphere variables. Strategies to enhance the efficiency of phytoextraction are developed and discussed further in the following sections.

Selecting the right plant species is crucial for achieving optimal phytoextraction. Plant species suitable for phytoextraction should possess the following characteristics: (1) high tolerance to heavy metal toxicity; (2) high extraction ability with high levels of heavy metal accumulation in aboveground parts; (3) fast growth with high biomass production; (4) abundant shoots and extensive root systems; (5) good adaptation to the prevailing environment, a strong ability to grow in poor soils, and easy cultivation and harvest; and (6) high resistance to pathogens and pests, and repellant to herbivores to avoid heavy metals entering the food chain.

Metal accumulation and above-ground biomass are key factors in determining the ability of a plant species to perform phytoextraction. Two plant selection strategies are being employed: (1) the utilization of hyperaccumulator plants that have a greater capacity to accumulate heavy metals in above-ground parts, and (2) the utilization of plants with high above-ground biomass production, which may have lower metal-accumulating capacities but exhibit overall heavy metal accumulation comparable to hyperaccumulators (Masoudi et al., 2020; Bhat et al., 2022).

Hyperaccumulators are plant species that can accumulate high levels of heavy metals in their above-ground parts without showing any signs of phytotoxicity. In the same conditions, naturally occurring heavy metal hyperaccumulators can accumulate metals at levels 100 times higher than non-hyperaccumulating species (Bhat et al., 2022). The definition of hyperaccumulator must adhere to the following criteria: (1) a shoot-to-root ratio of heavy metal concentration greater than 1, indicating efficient transportation of metals from roots to shoots (Ahmad et al., 2021); (2) a shoot-to-soil ratio of heavy metal concentration greater than 1, indicating a high capacity to uptake heavy metals from soil (Ahmad et al., 2021); and (3) the shoot contains metal concentrations that surpass the following thresholds: 10 mg/kg for Hg; 100 mg/kg for selenium (Se) and Cd; 1000 mg/kg for Pb, Cr, Co, Ni, and Cu; and 10,000 mg/kg for Mn and Zn (Huang et al., 2021).

The identification of effective hyperaccumulators is a crucial approach for efficient phytoremediation of heavy metals. Over 450 plant species belonging to at least 45 angiosperm families have been recognized as metal hyperaccumulators (Suman et al., 2018). These species encompass a wide range of growth forms, including annual herbs, perennial shrubs, and trees. Some of the families that contain metal hyperaccumulators include Brassicaceae, Fabaceae, Euphorbiaceae, Asteraceae, Lamiaceae, and Scrophulariaceae. Certain species, like *Sedum alfredii*, have the ability to accumulate multiple elements including Cd, Pb, and Zn (Zhao et al., 2023).

4.2.2 Rhizofiltration: Soil and Water Purification through Roots

Phytofiltration involves the use of plant roots (rhizofiltration), seedlings (blastofiltration), or shoots (caulofiltration) to remove contaminants from polluted surface water or wastewater (Clay and Pichtel, 2019). Heavy metals can be deposited onto the root surface or absorbed by the roots during the process of rhizofiltration. Root exudates can modify the pH of the rhizosphere, leading to the precipitation of heavy metals onto plant roots. Consequently, this process reduces the transportation of heavy metals to subterranean water sources (Javed et al., 2019).

Plants utilized for rhizofiltration are initially cultivated hydroponically in clean water to promote robust root development. Subsequently, the clean water is substituted with contaminated water to facilitate the plants' acclimation process. After acclimation, the plants are transferred to the contaminated site for the purpose of removing heavy metals. The roots are discarded after being soaked (Lapan et al., 2019). Ideal plants for rhizofiltration should possess thick roots, high biomass production, and resistance to heavy metals. Rhizofiltration can be achieved using both terrestrial and aquatic plants. Aquatic species, such as hyacinth, mosquito fern, duckweed, cattail, and poplar, are commonly used in wetland water remediation due to their ability to accumulate heavy metals, their tolerance to adverse conditions, and their fast growth and biomass production (Lapan et al., 2019). Terrestrial plants, such as Indian mustard (*Brassica juncea*) and sunflower (*Helianthus annuus*), possess longer and more hairy root systems in comparison to aquatic plants. They exhibit significant potential for accumulating heavy metals through rhizofiltration (Rezania et al., 2016; Dhanwal et al., 2017).

4.2.3 Phytostabilization: Immobilizing Contaminants in Soil

Phytostabilization involves employing metal-tolerant plant species to immobilize heavy metals in the soil, thereby reducing their bioavailability. This process effectively prevents the migration of metals into the ecosystem and mitigates the risk of metals entering the food chain (Radziemska et al., 2021). Phytostabilization can occur through heavy metal precipitation or reduction in metal valence in the rhizosphere, absorption and sequestration within root tissues, or adsorption onto root cell walls (Gerhardt et al., 2017). Plant growth aids in the preservation of soil health in environments contaminated with heavy metals. The presence of established plant cover has multiple benefits, including stabilizing underground heavy metals and reducing their leaching into groundwater. Additionally, it prevents wind dispersion of soil particles containing heavy metals (Radziemska et al., 2020). Phytostabilization offers an advantage over phytoextraction as it eliminates the need for disposal of hazardous biomass (Gerhardt et al., 2017).

The selection of appropriate plant species is crucial for the process of phytostabilization. Plants need to possess resistance to heavy metal environments to fulfill the requirement for highly efficient phytostabilization. Thick plant roots are essential for immobilizing heavy metals, stabilizing soil structure, and preventing soil erosion. Plants require high biomass production and rapid development to successfully establish vegetation in a specified location within a given time frame. Additionally, it is important that the plant cover is easily manageable in the field (Li et al., 2021). Several plant species meeting the aforementioned criteria have been identified and employed for the purpose of phytostabilizing soils contaminated with heavy metals (Burges et al., 2018).

Organic and inorganic amendments can enhance the efficiency of phytostabilization in polluted soil. Soil amendments can alter the pH and redox status of the soil, leading to changes in metal speciation, reduced solubility of heavy metals, and decreased bioavailability (Burges et al., 2018). Amendments have the potential to enhance soil organic matter content, vital nutrients, physicochemical properties, and biological properties, thereby facilitating plant colonization and improving water holding capacity (Visconti et al., 2020).

Rhizosphere microorganisms, including bacteria and mycorrhiza, contribute to phytostabilization. Microbes have the ability to enhance the immobilization of heavy metals through mechanisms such as adsorption onto their cell walls, production of chelators, and promotion of precipitation processes (Burges et al., 2018). In addition, mycorrhizal fungi can enhance the size and reach of plant roots, thereby aiding in phytostabilization. They also serve as a protective barrier, preventing the transfer of heavy metal ions from roots to shoots (Radziemska et al., 2021).

4.2.4 Phytodegradation: Breaking Down Organic Pollutants

Phytodegradation is an environmentally friendly and innovative approach for addressing the prevalent issue of organic pollutants in contaminated soils, including hydrocarbons and pesticides. Organic pollutants pose significant risks to human health and ecosystems due to their long-lasting nature and potential toxicity (Kaur et al., 2020). Phytodegradation is a process that harnesses the metabolic capabilities of plants and their symbiotic relationships with microorganisms to facilitate the decomposition of complex substances, leading to the improvement of soil quality and health (Hassan and Jalil, 2022).

Phytodegradation, as explained by Aadil et al. (2023), relies on the plant's capacity to produce specialized enzymes, including peroxidases and dehydrogenases, which play a crucial role in initiating the breakdown process. Plants respond to organic contaminants by activating their enzymatic machinery, which initiates a series of chemical reactions to transform these toxins into less harmful metabolites. Enzymatic breakdown processes, such as oxidation, reduction, and hydrolysis, can convert pollutants into forms that are more easily degraded by microorganisms or assimilated by plants (Hassan and Jalil, 2022). The rhizosphere, which refers to the area surrounding plant roots, plays a crucial role in enhancing the effectiveness of phytodegradation through the interaction between plants and microorganisms. Microbes in this zone enhance plant functionality through the provision of supplementary enzymes and metabolic pathways, thereby augmenting the degradation of organic pollutants. The dynamic interaction accelerates degradation rates and expands the scope of chemicals that can be remediated (Shivaraju et al., 2020).

Phytodegradation surpasses conventional cleanup methods in multiple aspects. Phytodegradation, unlike mechanical or chemical methods, is a non-invasive process that enhances soil health and fertility without causing disruptions to ecosystems and soil structure (Qi et al., 2022). Phytodegradation by-products are often less toxic and more readily assimilated by the environment, thereby reducing the risk of secondary contamination (Qi et al., 2022).

Phytodegradation is a soil remediation technique that utilizes the natural abilities of plants and their microbial partners to achieve holistic and sustainable outcomes (Hassan and Jalil, 2022). This method reduces organic pollutants, promotes ecosystem regeneration, and enhances environmental resilience. Phytodegradation shows potential for transforming contaminated areas into productive landscapes through plant-microbe interactions and improved degradation processes (Aadil et al., 2023).

4.2.5 Phytovolatilization: Transforming Contaminants into Gas

Phytovolatilization is a phytoremediation technique where plants absorb pollutants from the soil, convert them into less harmful volatile forms, and release them into the atmosphere through transpiration via the leaves or foliage. Lebrun et al. (2021) demonstrated the efficacy of this method in detoxifying both organic contaminants and heavy metals, including As, Se, and Hg. Members of the Brassicaceae family, including *B. juncea*, have been found effective in volatilizing selenium (Bhalla et al., 2022). Selenocysteine and selenomethionine (seMet), which are organic selenoamino acids, are initially derived from inorganic Se. SeMet undergoes biomethylation to produce dimethylselenide, a volatile compound that poses less risk when dispersed in the atmosphere compared to inorganic selenium (Chen et al., 2022). Mercury exists in its elemental form as a liquid at standard temperature and readily undergoes volatilization. Due to its high reactivity, Hg is primarily found in the environment as the divalent cation Hg^{2+}. Methylmercury undergoes transformation into ionic Hg after being absorbed by plant roots or leaves. This ionic form of Hg is subsequently converted into a less dangerous elemental form and released into the atmosphere (Lee et al., 2020).

Phytovolatilization is advantageous in comparison to other phytoremediation methods as it effectively eliminates heavy metal (metalloid) pollutants by converting them into gaseous molecules, without the need for plant harvesting and disposal. Phytovolatilization, while a remedial method, does not completely eliminate contaminants, resulting in the persistence of pollutants in the environment (Lee et al., 2020). The transportation of pollutants from the soil to the atmosphere results in the release of hazardous volatile chemicals, thereby contaminating the air. Additionally, precipitation has the potential to redeposit these substances back into the soil. A risk assessment is necessary before field application (Visconti et al., 2020).

4.3 PLANT SPECIES FOR PHYTOREMEDIATION

Phytoremediation allows for the strategic use of a wide variety of plant species, each of which can be tailored to address a particular type of contamination.

4.3.1 Hyperaccumulators: Nature's Metal Cleaners

Hyperaccumulators are plants that thrive in soils contaminated with metals and have the ability to absorb and accumulate substantial amounts of metals within their tissues (Aziz and Mujeeb, 2022). Metal-tolerant plants, commonly found in naturally metal-rich environments, possess specialized mechanisms enabling them to efficiently absorb and distribute metals within their structures without causing harm (Sytar et al., 2021). Hyperaccumulators are of great interest in the field of phytoremediation due to their ability to remediate contaminated soils and potentially recover valuable metals (Aziz and Mujeeb, 2022).

Thlaspi caerulescens, a Zn-tolerant species, exemplifies hyperaccumulation. Skuza et al. (2022) found that this plant species exhibits significantly higher Zn accumulation in its tissues compared to non-hyperaccumulator plants. *Alyssum* species have been noted for their Ni accumulation capacity. Hyperaccumulator plants have the ability to remove metals from the soil and store them in their leaves, stems, and roots, effectively functioning as natural "metal cleaners" (Shrivastava et al., 2019).

Hyperaccumulators have both commercial and ecological significance. Hyperaccumulator plants can be harvested and processed to extract the metals they have accumulated (Shrivastava et al., 2019). Phytomining is a metal recovery technique that involves extracting valuable materials from polluted soils. After harvesting, metals can be obtained from plant biomass using conventional refining methods. This method offers a sustainable approach to mining and also helps in the restoration of damaged soils (Sytar et al., 2021).

The identification and utilization of hyperaccumulator plants have expanded the potential applications of phytoremediation. These plants can be strategically cultivated in contaminated regions to effectively sequester metals and ultimately restore soil quality. Phytomining has garnered attention from researchers and industry due to its potential as an environmentally friendly and socially responsible alternative to conventional mining operations (Aziz and Mujeeb, 2022).

4.3.2 Native Plants: Tailoring Remediation to Local Ecosystems

Utilizing native plant species in phytoremediation endeavors is a prudent and environmentally responsible approach that aligns with the principles of sustainability and ecosystem rehabilitation. Native plants that are well suited to the soil compositions, climatic conditions, and ecological interactions of their local habitats offer numerous advantages when employed in phytoremediation procedures (Mensah et al., 2021).

The adaptability of native plants is a result of their long-standing coevolution with local environmental factors. These plants possess physiological, morphological, and biochemical traits that enable them to flourish in their natural environments. Transplanting native species to contaminated areas for phytoremediation allows them to utilize their adaptations more effectively in response to the stressors posed by contaminants. This adaptation enhances plant survival, growth, and their capacity to absorb and sequester soil pollutants (Castle et al., 2019; Shen et al., 2023).

Native plants are highly valuable for phytoremediation due to their capacity to contribute to ecosystem restoration. These programs aim to restore degraded natural environments by reintroducing plant species from the native plant community (Shen et al., 2023). Native plants play a crucial role in restoring a well-functioning ecosystem by offering nourishment, refuge, and reproductive habitats for indigenous fauna (Valliere et al., 2020).

Additionally, the incorporation of native plants in phytoremediation systems mitigates the risk of introducing invasive species. Non-indigenous plant species possess the ability to outcompete native flora and disrupt established ecological processes, thereby exacerbating the degradation of ecosystems. In contrast, native plants are already integrated into the local ecological network, reducing the likelihood of unforeseen ecological disturbances and facilitating the restoration of natural ecological processes (Fletcher et al., 2022). The utilization of native plant species in remediation efforts leads to heightened community engagement and public support. Local residents often develop a strong affinity for native plants due to cultural and aesthetic considerations. The utilization of familiar plant species in restoration initiatives fosters a feeling of ownership and pride, thereby promoting active involvement and a stronger dedication to environmental conservation (Shen et al., 2023).

4.3.3 Crop Plants: Synergizing Remediation with Agriculture

The integration of crop plants into phytoremediation systems shows potential for addressing soil contamination and enhancing agricultural productivity (Grifoni et al., 2021). This approach utilizes the inherent abilities of certain crop species not only to flourish in polluted soils but also to enhance soil fertility for sustainable food production in the long run. The symbiotic relationship between phytoremediation and agriculture offers a mutually beneficial outcome by integrating environmental restoration and the need to sustainably feed the expanding global population (Liu et al., 2022).

Crop plants chosen for this dual-purpose approach possess traits that render them appropriate for both phytoremediation and agricultural productivity. These plants have exhibited the capacity to endure and accumulate pollutants without negatively affecting their growth or the quality of their yield. Maize (corn) and sunflowers are significant crop species that have been effectively employed in this synergistic technique. Gondal and Tayyiba (2022) have identified these plants as having robust growth, extensive root systems, and the ability to accumulate metals, which makes them suitable for soil remediation purposes.

Integrating phytoremediation with agriculture offers a key advantage: the restoration of soil health. Crop plants absorb toxins from the soil, thereby reducing pollution levels and mitigating the risk of contaminants entering the food chain. In addition, cultivating these plants enhances soil structure, nitrogen cycling, and microbial activity, thereby creating a favorable environment for optimal crop growth (Fu et al., 2022).

This method also offers significant economic benefits. Crop phytoremediation enables farmers to engage in soil remediation while maintaining their agricultural livelihoods. The safety of harvested crops for consumption or non-food purposes depends on the level of contamination. In addition, the financial benefits to the community's economy can be derived from both cleanup efforts and the sale of crops (Fu et al., 2022; Liu et al., 2022).

Nevertheless, there are challenges associated with the integration of phytoremediation and agriculture. Choosing suitable crop species for effective remediation of specific pollutants while ensuring economic viability necessitates thorough assessment. Additionally, it is crucial to conduct a comprehensive assessment of the potential impacts on food safety and the environment in order to ensure that the agricultural products adhere to regulatory standards (Afridi et al., 2022).

4.4 FACTORS INFLUENCING PHYTOREMEDIATION

Phytoremediation outcomes are strongly influenced by factors such as soil composition, contaminant types and levels, climate conditions, plant-microbe interactions, and the effectiveness of monitoring and assessment methods. Understanding and optimizing these factors are essential for successful and sustainable soil remediation.

4.4.1 Soil Composition and Its Role in Remediation

Soil composition is crucial for the success of phytoremediation. Soil parameters such as texture, pH, and organic matter content have significant effects on plant growth, pollutant absorption, and the overall outcome of restoration efforts (Sun et al., 2021).

Soil texture refers to the proportions of sand, silt, and clay particles within the soil matrix. Soil texture impacts the soil's ability to retain water, facilitate drainage, and provide adequate aeration. Sandy soils have high drainage capacity but low water and nutrient retention, while clayey soils have good water retention but tend to become compacted, limiting root penetration. Loam soils, which consist of a balanced combination of sand, silt, and clay, are commonly considered to be ideal for phytoremediation due to their ability to retain water effectively and provide adequate aeration and root growth conditions (Akimbekov et al., 2021).

Soil pH is a crucial factor that determines the acidity or alkalinity of soil. It plays a significant role in regulating nutrient availability and influencing plant health. Plants generally prefer pH levels ranging from slightly acidic to neutral, typically around 6 to 7. Extreme pH levels can hinder nutrient absorption and decelerate plant growth. The addition of lime or sulfur is often necessary to raise soil pH and create optimal conditions for the growth of phytoremediation plants (Neina, 2019).

Soil organic matter comprises decomposed plant and animal material and plays a crucial role in soil structure, water retention, and nutrient availability. Sufficient organic matter enhances soil microbial activity, thereby promoting phytoremediation through the facilitation of pollutant degradation and nutrient cycling. The addition of organic matter, such as compost or organic mulch, can enhance soil quality and promote plant health (Xiang et al., 2022).

Enhancing soil characteristics is crucial for optimizing the effectiveness of phytoremediation. The addition of organic matter to sandy soils with low water retention can enhance moisture retention and nutrient availability, thereby creating a more favorable environment for plant growth and pollutant absorption. Enhancing aeration in compacted clayey soils can be achieved by implementing practices such as soil tilling or incorporating organic matter. These measures have the potential to promote root penetration and subsequently aid in pollution elimination (DalCorso et al., 2019). Soil composition significantly affects phytoremediation and is crucial for developing remediation plans tailored to specific sites. Tailoring soil amendments and management practices to address the specific challenges posed by different soil types is crucial for the success of phytoremediation programs. The role of soil composition in phytoremediation underscores the importance of considering the intricate relationship among plants, contaminants, and the soil matrix for achieving sustainable environmental restoration (Abdullah et al., 2020).

4.4.2 Types and Levels of Contaminants: Navigating Complexity

The effectiveness of phytoremediation is closely tied to the type and amount of pollutants present in the soil. These variables impact the choice of phytoremediation methods and the selection of suitable plant species. The presence of multiple pollutants in mixed environments underscores the importance of meticulous planning and thoughtful consideration in phytoremediation efforts (Zhu et al., 2023). Phytoremediation is an effective method for targeting heavy metals, organic pollutants, and radioactive compounds. Different phytoremediation strategies are suited to specific types of pollutants (Liu et al., 2021). Phytoextraction is effective for removing heavy metals from soil, while phytodegradation is more effective for removing organic pollutants like petroleum hydrocarbons. A thorough comprehension of the chemical properties and behavior of pollutants is essential in determining the most suitable approach (Liu et al., 2021).

The concentration of pollutants in soil significantly affects the viability and effectiveness of phytoremediation. Phytoremediation is a viable and cost-efficient approach for addressing low to moderate levels of pollution. Excessive levels of contaminants can exceed plants' capacity to absorb and detoxify pollutants, necessitating more rigorous pre-treatment or soil amendment methods (Parveen et al., 2022). Contaminated sites often exhibit multiple contaminants, leading to complex challenges. Contaminant interactions can have synergistic, antagonistic, or neutral effects, which can modify the bioavailability and toxicity of the contaminants. Selecting plant species and strategies capable of simultaneously tolerating multiple toxins necessitates thorough investigation. This may involve the selection of hyperaccumulator species or the combination of multiple phytoremediation strategies (Shen et al., 2022). Phytoremediation is a viable method for remediating various contaminated environments due to its adaptability and versatility. Success depends on a comprehensive understanding of the pollutants and their interactions in the soil. Phytoremediation programs can effectively and sustainably restore soil quality, ensuring safe and productive land use in the future. This is achieved by analyzing the types and quantities of pollutants present (Constable et al., 2019).

4.4.3 Climate and Environmental Variables in Remediation

Climate and environmental factors greatly influence the effectiveness of phytoremediation, as they directly impact plant performance, microbial activity, and pollutant behavior in contaminated soils. Customizing phytoremediation programs to the specific climatic conditions of a site is imperative for attaining effective and sustainable remediation outcomes (Grifoni et al., 2022). Plant development and adaptation are essential processes for the survival and success of plants in various environments. These processes involve the growth and differentiation of plant cells, tissues, and organs, as well as the ability of plants to adjust. Temperature, precipitation, and sunlight are key factors in influencing plant growth and development. Selecting plant species that are suitable for the local climate is crucial for ensuring their successful establishment and long-term survival. Drought-resistant plants are desirable in arid regions, while cold-tolerant species are necessary in temperate climates. Plant adaptability to climate enhances their capacity to absorb toxins, thereby contributing to the broader phytoremediation endeavors (Ravaoarinorotsihoarana et al., 2023). Climate significantly influences soil microbial activity, which plays a crucial role in the degradation of pollutants. Temperature and moisture affect microbial populations and metabolic rates. Warmer temperatures enhance microbial activity, leading to accelerated degradation of pollutants. Microbial activity can be diminished in cold or moist environments. Climate plays a crucial role in creating a conducive environment for microbial involvement in phytoremediation (Allouzi et al., 2021). Climatic factors can affect the movement of pollutants in the soil. Enhanced precipitation or irrigation can facilitate the infiltration of toxins into the soil, thereby potentially impacting groundwater. Drought conditions can lead to increased concentration of pollutants in the root zone. The comprehension of the relationship between climate and pollution transport is essential for mitigating risks and optimizing remediation efforts (Li et al., 2021). Climate factors affect the choice of suitable plant species: indigenous plants that have acclimated to the specific climatic conditions are more inclined to thrive and effectively restore the area. In addition, it is important to synchronize planting and harvesting activities with favorable weather conditions in order to optimize plant growth and enhance pollution absorption (Li et al., 2021). Climate change must be taken into account when planning phytoremediation initiatives. Climate factors such as temperature fluctuations, precipitation variations, and extreme weather events have implications for the suitability of specific plant species and necessitate adjustments in remediation strategies (Grifoni et al., 2022; Ravaoarinorotsihoarana et al., 2023).

4.4.4 Plant-Microbe Interactions: Below-Ground Collaborations

The rhizosphere is a small area of soil that is directly influenced by plant roots. It is a dynamic environment where complex interactions between plants and microorganisms play a significant role in the effectiveness of phytoremediation. Below-ground collaborations play a crucial role in pollutant degradation, nutrient cycling, and soil structure enhancement, making them a key focus in optimizing phytoremediation performance (Honvault et al., 2021). Microbial collaborations occur within the rhizosphere, where a diverse population of microorganisms, such as bacteria, fungus, and archaea, engage in symbiotic interactions with plants. Partnerships can be either mutual (i.e., benefiting both parties) or commensal (i.e., benefiting one party while negatively affecting the other). Microbes play a crucial role in phytoremediation by aiding in the degradation of pollutants, acquisition of nutrients, and suppression of diseases (Odelade and Babalola, 2019). Microbes possess significant capabilities in the degradation of pollutants. These organisms possess enzymes and metabolic pathways capable of breaking down diverse toxins, such as organic pollutants and heavy metals. Root exudates are multifaceted compounds released by plants, serving as both energy sources and signaling molecules for microorganisms. Exudates facilitate the growth of bacteria that degrade pollutants, thereby enhancing the conversion of pollutants into less harmful forms (Kishore et al., 2022). Microbes are crucial in the process of nutrient cycling within the rhizosphere. Microorganisms break down organic matter and release essential nutrients such as nitrogen, phosphorus, and potassium, facilitating their absorption by plants. Microbial-mediated nutrient recycling enhances plant growth and increases their resistance to phytoremediation stresses, including high metal concentrations (Enebe and Babalola, 2021). Microbes play a crucial role in enhancing soil structure through the production of extracellular compounds, including polysaccharides and glomalin. These compounds facilitate the binding of soil particles, leading to improved soil aggregation. According to Redmile-Gordon et al. (2020), this enhances rhizosphere aeration, root penetration, and water retention, creating a more conducive environment for plant growth and pollutant absorption. It is crucial to comprehend and foster subterranean interactions in order to maximize the effectiveness of phytoremediation. Bioaugmentation and rhizoremediation are two techniques that can be employed to address pollution. Bioaugmentation involves introducing specific microorganisms that are capable of degrading pollutants into the rhizosphere. Rhizoremediation, on the other hand, combines plant-based and microbial strategies to remediate polluted environments. In addition, the selection of plant species that produce exudates promoting the growth of beneficial bacteria can enhance these interactions (Afridi et al., 2022).

4.5 CONCLUSION

Phytoremediation is a promising and eco-friendly approach for addressing soil contamination problems. This holistic technique utilizes plants, their microbial partners, and environmental conditions to remove pollutants, improve soil health, and restore ecosystems. To optimize the effectiveness of this strategy, it is crucial to carefully choose plant species and phytoremediation techniques that are well suited to the particular pollutant and environmental conditions. Phytoremediation is a potentially cost-effective, community-engaged, and environmentally friendly approach for addressing contamination in various global locations. Advancements in research on plant-microbe interactions and remediation processes have enhanced our comprehension of phytoremediation's potential to transform polluted landscapes into sustainable ecosystems. Implementing this innovative approach can lead to improved soil quality, enhanced ecological conditions, and a more favorable outlook for both human and biological communities. However, it is important to acknowledge the limitations of the approach, conduct continuous research, and adapt methods in accordance with evolving environmental constraints. Phytoremediation offers a potential solution to the escalating pollution issues and the necessity for sustainable remedies. It holds promise in improving soil quality, enhancing biodiversity, and fostering global resilience. The utilization of this technology in environmental remediation endeavors holds the potential to create a healthier and more sustainable planet for future generations.

REFERENCES

Aadil, M., Kousar, T., Hussain, M., Somaily, H. H., Abdulla, A. K., Muhammad, E. R., & Ansari, M. Z. (2023). Generation of mesoporous npn (ZnO–CuO–CeO2) heterojunction for highly efficient photodegradation of micro-organic pollutants. *Ceramics International*, *49*(3), 4846–4854.

Abdullah, S. R. S., Al-Baldawi, I. A., Almansoory, A. F., Purwanti, I. F., Al-Sbani, N. H., & Sharuddin, S. S. N. (2020). Plant-assisted remediation of hydrocarbons in water and soil: Application, mechanisms, challenges and opportunities. *Chemosphere*, *247*, 125932.

Abioye, O.P., Ijah, U.J.J., & Aransiola, S.A. (2017). Phytoremediation of soil contaminants by biodiesel plant Jatropha curcas. In Chapter 4 of K. Bauddh et al. (eds.), *Phytoremediation Potential of Bioenergy Plants, DOI 10.1007/978-981-10-3084-0_4*. Springer Nature Singapore Pte Ltd. https://link.springer.com/book/10.1007%2F978-981-10-3084-0.

Afridi, M. S., Ali, S., Salam, A., César Terra, W., Hafeez, A., Sumaira, A., & Karunakaran, R. (2022). Plant microbiome engineering: Hopes or hypes. *Biology*, *11*(12), 1782.

Ahmad, I., Malik, S. A., Saeed, S., Rehman, A. U., & Munir, T. M. (2021). Phytoextraction of heavy metals by various vegetable crops cultivated on different textured soils irrigated with City Wastewater. *Soil Systems*, *5*(2), 35.

Akimbekov, N. S., Digel, I., Tastambek, K. T., Sherelkhan, D. K., Jussupova, D. B., & Altynbay, N. P. (2021). Low-rank coal as a source of humic substances for soil amendment and fertility management. *Agriculture*, *11*(12), 1261.

Allouzi, M. M. A., Tang, D. Y. Y., Chew, K. W., Rinklebe, J., Bolan, N., Allouzi, S. M. A., & Show, P. L. (2021). Micro (nano) plastic pollution: The ecological influence on soil-plant system and human health. *Science of the Total Environment*, *788*, 147815.

Aransiola, S. A., Ijah, U. J. J., & Abioye, O. P. (2013). Phytoremediation of lead polluted soil by glycine max L. *Applied and Environmental Soil Science*, Article ID 631619. doi: 10.1155/2013/631619. https://www.hindawi.com/journals/aess/2013/631619/abs/.

Aransiola, S. A., Ijah, U. J. J., Abioye, O. P., & Bala, J. D. (2022). Vermicompost-assisted phytoremediation of toxic trace element-contaminated soil in Madaka, Nigeria using Melissa officinalis L and Sida acuta. *International Journal of Environmental Science and Technology*. http://dx.doi.org/10.1007/s13762-022-04105-y

Ashraf, S., Ali, Q., Zahir, Z. A., Ashraf, S., & Asghar, H. N. (2019). Phytoremediation: Environmentally sustainable way for reclamation of heavy metal polluted soils. *Ecotoxicology and Environmental Safety*, *174*, 714–727.

Aziz, I., & Mujeeb, A. (2022). Halophytes for phytoremediation of hazardous metal (loid) s: A terse review on metal tolerance, bio-indication and hyperaccumulation. *Journal of Hazardous Materials*, *424*, 127309.

Bhalla, G., Bhalla, B., Kumar, V., & Sharma, A. (2022). Bioremediation and phytoremediation of pesticides residues from contaminated water: A novel approach. In *Pesticides Remediation Technologies from Water and Wastewater* (pp. 339–363). Elsevier.

Bhat, S. A., Bashir, O., Haq, S. A. U., Amin, T., Rafiq, A., Ali, M., & Sher, F. (2022). Phytoremediation of heavy metals in soil and water: An eco-friendly, sustainable and multidisciplinary approach. *Chemosphere*, *303*, 134788.

Burges, A., Alkorta, I., Epelde, L., & Garbisu, C. (2018). From phytoremediation of soil contaminants to phytomanagement of ecosystem services in metal contaminated sites. *International Journal of Phytoremediation*, *20*(4), 384–397.

Castle, S., Rejmánková, E., Foley, J., & Parmenter, S. (2019). Hydrologic alterations impact plant litter decay rate and ecosystem resilience in Mojave wetlands. *Restoration Ecology*, *27*(5), 1094–1104.

Chen, B., Stein, A. F., Castell, N., Gonzalez-Castanedo, Y., De La Campa, A. S., & De La Rosa, J. D. (2016). Modeling and evaluation of urban pollution events of atmospheric heavy metals from a large Cu-smelter. *Science of the Total Environment*, *539*, 17–25.

Chen, W., Hou, X., Mao, X., Jiao, S., Wei, L., Wang, Y., . . . Jiang, G. (2022). Biotic and abiotic transformation pathways of a short-chain chlorinated paraffin congener, 1, 2, 5, 6, 9, 10-C10H16Cl6, in a rice seedling hydroponic exposure system. *Environmental Science & Technology*, *56*(13), 9486–9496.

Clay, L., & Pichtel, J. (2019). Treatment of simulated oil and gas produced water via pilot-scale rhizofiltration and constructed wetlands. *International Journal of Environmental Research*, *13*, 185–198.

Constable, D. J., Jiménez-González, C., & Matlin, S. A. (2019). Navigating complexity using systems thinking in chemistry, with implications for chemistry education. *Journal of Chemical Education*, *96*(12), 2689–2699.

DalCorso, G., Fasani, E., Manara, A., Visioli, G., & Furini, A. (2019). Heavy metal pollutions: State of the art and innovation in phytoremediation. *International Journal of Molecular Sciences*, *20*(14), 3412.

Dhanwal, P., Kumar, A., Dudeja, S., Chhokar, V., & Beniwal, V. (2017). Recent advances in phytoremediation technology. *Advances in Environmental Biotechnology*, 227–241.

Enebe, M. C., & Babalola, O. O. (2021). Soil fertilization affects the abundance and distribution of carbon and nitrogen cycling genes in the maize rhizosphere. *AMB Express*, *11*(1), 1–10.

Farooq, M., Rehman, A., & Pisante, M. (2019). Sustainable agriculture and food security. *Innovations in Sustainable Agriculture*, 3–24.

Fasani, E., Manara, A., Martini, F., Furini, A., & DalCorso, G. (2018). The potential of genetic engineering of plants for the remediation of soils contaminated with heavy metals. *Plant, Cell & Environment*, *41*(5), 1201–1232.

Fletcher, J., Willby, N., Oliver, D. M., & Quilliam, R. S. (2022). Resource recovery and freshwater ecosystem restoration – prospecting for phytoremediation potential in wild macrophyte stands. *Resources, Environment and Sustainability*, *7*, 100050.

Fu, X., Zheng, Z., Sha, Z., Cao, H., Yuan, Q., Yu, H., & Li, Q. (2022). Biorefining waste into nanobiotechnologies can revolutionize sustainable agriculture. *Trends in Biotechnology*, *40*(12), 1503–1518. doi: 10.1016/j.tibtech.2022.09.013. PMID: 36270903.

Gavrilescu, M. (2022). Enhancing phytoremediation of soils polluted with heavy metals. *Current Opinion in Biotechnology*, *74*, 21–31.

Gerhardt, K. E., Gerwing, P. D., & Greenberg, B. M. (2017). Opinion: Taking phytoremediation from proven technology to accepted practice. *Plant Science*, *256*, 170–185.

Gondal, A. H., & Tayyiba, L. (2022). Prospects of using nanotechnology in agricultural growth, environment and industrial food products. *Reviews in Agricultural Science*, *10*, 68–81.

Grifoni, M., Franchi, E., Fusini, D., Vocciante, M., Barbafieri, M., Pedron, F., & Petruzzelli, G. (2022). Soil remediation: Towards a resilient and adaptive approach to deal with the ever-changing environmental challenges. *Environments*, *9*(2), 18.

Grifoni, M., Pedron, F., Barbafieri, M., Rosellini, I., Petruzzelli, G., & Franchi, E. (2021). Sustainable valorization of biomass: From assisted phytoremediation to green energy production. In *Handbook of assisted and amendment: Enhanced sustainable remediation technology* (pp. 29–51). Wiley.

Hamzah, A., Hapsari, R. I., & Wisnubroto, E. I. (2016). Phytoremediation of cadmium-contaminated agricultural land using indigenous plants. *International Journal of Environmental & Agriculture Research*, *2*(1), 8–14.

Hasan, M. M., Uddin, M. N., Ara-Sharmeen, I. F., Alharby, H., Alzahrani, Y., Hakeem, K. R., & Zhang, L. (2019). Assisting phytoremediation of heavy metals using chemical amendments. *Plants*, *8*(9), 295.

Hassan, N. S., & Jalil, A. A. (2022). A review on self-modification of zirconium dioxide nanocatalysts with enhanced visible-light-driven photodegradation of organic pollutants. *Journal of Hazardous Materials*, *423*, 126996.

Honvault, N., Houben, D., Firmin, S., Meglouli, H., Laruelle, F., Fontaine, J., & Faucon, M. P. (2021). Interactions between below-ground traits and rhizosheath fungal and bacterial communities for phosphorus acquisition. *Functional Ecology*, *35*(7), 1603–1619.

Huang, R., Cui, X., Luo, X., Mao, P., Zhuang, P., Li, Y., & Li, Z. (2021). Effects of plant growth regulator and chelating agent on the phytoextraction of heavy metals by Pfaffia glomerata and on the soil microbial community. *Environmental Pollution*, *283*, 117159.

Jacob, J. M., Karthik, C., Saratale, R. G., Kumar, S. S., Prabakar, D., Kadirvelu, K., & Pugazhendhi, A. (2018). Biological approaches to tackle heavy metal pollution: A survey of literature. *Journal of Environmental Management*, *217*, 56–70.

Javed, M. T., Tanwir, K., Akram, M. S., Shahid, M., Niazi, N. K., & Lindberg, S. (2019). Phytoremediation of cadmium-polluted water/sediment by aquatic macrophytes: Role of plant-induced pH changes. In *Cadmium toxicity and tolerance in plants* (pp. 495–529). Academic Press.

Kaur, K., Badru, R., Singh, P. P., & Kaushal, S. (2020). Photodegradation of organic pollutants using heterojunctions: A review. *Journal of Environmental Chemical Engineering*, *8*(2), 103666.

Kishore, S., Malik, S., Shah, M. P., Bora, J., Chaudhary, V., Kumar, L., & Ranjan, A. (2022). A comprehensive review on removal of pollutants from wastewater through microbial nanobiotechnology-based solutions. *Biotechnology and Genetic Engineering Reviews*, 1–26.

Lapan, O. V., Mikhyeyev, A. N., & Madzhd, S. M. (2019). Development of a new method of rhizofiltration purification of water objects of Zn (II) and Cd (II). *Journal of water Chemistry and Technology*, *41*, 52–56.

Lebrun, M., Michel, C., Joulian, C., Morabito, D., & Bourgerie, S. (2021). Rehabilitation of mine soils by phytostabilization: Does soil inoculation with microbial consortia stimulate Agrostis growth and metal (loid) immobilization?. *Science of the Total Environment*, *791*, 148400.

Lee, B. X. Y., Hadibarata, T., & Yuniarto, A. (2020). Phytoremediation mechanisms in air pollution control: A review. *Water, Air, & Soil Pollution*, *231*(8), 437.

Li, J., Chen, Q., Li, Q., Zhao, C., & Feng, Y. (2021). Influence of plants and environmental variables on the diversity of soil microbial communities in the Yellow River Delta Wetland, China. *Chemosphere*, *274*, 129967.

Liu, J., Zhao, L., Liu, Q., Li, J., Qiao, Z., Sun, P., & Yang, Y. (2021). A critical review on soil washing during soil remediation for heavy metals and organic pollutants. *International Journal of Environmental Science and Technology*, 1–24.

Liu, K., Fang, L., Li, F., Hou, D., Liu, C., Song, Y., & Dou, F. (2022). Sustainability assessment and carbon budget of chemical stabilization based multi-objective remediation of Cd contaminated paddy field. *Science of the Total Environment*, *819*, 152022.

Masoudi, F., Shirvani, M., Shariatmadari, H., & Sabzalian, M. R. (2020). Performance of new biodegradable chelants in enhancing phytoextraction of heavy metals from a contaminated calcareous soil. *Journal of Environmental Health Science and Engineering*, *18*, 655–664.

Mensah, A. K., Marschner, B., Antoniadis, V., Stemn, E., Shaheen, S. M., & Rinklebe, J. (2021). Human health risk via soil ingestion of potentially toxic elements and remediation potential of native plants near an abandoned mine spoil in Ghana. *Science of the Total Environment*, *798*, 149272.

Neina, D. (2019). The role of soil pH in plant nutrition and soil remediation. *Applied and Environmental Soil Science*, *2019*, 1–9.

Odelade, K. A., & Babalola, O. O. (2019). Bacteria, fungi and archaea domains in rhizospheric soil and their effects in enhancing agricultural productivity. *International Journal of Environmental Research and Public Health*, *16*(20), 3873.

Parveen, S., Bhat, I. U., Khanam, Z., Rak, A. E., Yusoff, H. M., & Akhter, M. S. (2022). Phytoremediation: In situ alternative for pollutant removal from contaminated natural media: A brief review. *Biointerface Research in Applied Chemistry*. doi: 10.33263/BRIAC124.49454960. ISSN: 20695837.

Pichtel, J. (2016). Oil and gas production wastewater: Soil contamination and pollution prevention. *Applied and Environmental Soil Science*. doi: 10.1155/2016/2707989.

Qi, K., Ye, Y., Wei, B., Li, M., Lun, Y., Xie, X., & Xie, H. (2022). N-CQDs from reed straw enriching charge over BiO2– x/BiOCl pn heterojunction for improved visible-light-driven photodegradation of organic pollutants. *Journal of Hazardous Materials*, *432*, 128759.

Radziemska, M., Bęś, A., Gusiatin, Z. M., Cerdà, A., Jeznach, J., Mazur, Z., & Brtnický, M. (2020). Assisted phytostabilization of soil from a former military area with mineral amendments. *Ecotoxicology and Environmental Safety*, *188*, 109934.

Radziemska, M., Gusiatin, Z. M., Cydzik-Kwiatkowska, A., Cerdà, A., Pecina, V., Bęś, A., & Brtnický, M. (2021). Insight into metal immobilization and microbial community structure in soil from a steel disposal dump phytostabilized with composted, pyrolyzed or gasified wastes. *Chemosphere*, *272*, 129576.

Rafique, N., & Tariq, S. R. (2016). Distribution and source apportionment studies of heavy metals in soil of cotton/wheat fields. *Environmental Monitoring and Assessment*, *188*, 1–10.

Ravaoarinorotsihoarana, L. A., Ratefinjanahary, I., Aina, C., Rakotomahazo, C., Glass, L., Ranivoarivelo, L., & Lavitra, T. (2023). Combining traditional ecological knowledge and scientific observations to support mangrove restoration in Madagascar. *Forests*, *14*(7), 1368.

Redmile-Gordon, M., Gregory, A. S., White, R. P., & Watts, C. W. (2020). Soil organic carbon, extracellular polymeric substances (EPS), and soil structural stability as affected by previous and current land-use. *Geoderma*, *363*, 114143.

Rehman, M. Z., Rizwan, M., Ali, S., Ok, Y. S., Ishaque, W., Nawaz, M. F., . . . Waqar, M. (2017). Remediation of heavy metal contaminated soils by using Solanum nigrum: A review. *Ecotoxicology and Environmental Safety*, *143*, 236–248.

Rezania, S., Taib, S. M., Din, M. F. M., Dahalan, F. A., & Kamyab, H. (2016). Comprehensive review on phytotechnology: Heavy metals removal by diverse aquatic plants species from wastewater. *Journal of Hazardous Materials*, *318*, 587–599.

Shen, X., Dai, M., Yang, J., Sun, L., Tan, X., Peng, C., & Naz, I. (2022). A critical review on the phytoremediation of heavy metals from environment: Performance and challenges. *Chemosphere*, *291*, 132979.

Shen, X., Ge, M., Handel, S. N., Wang, W., Jin, Z., & Kirkwood, N. G. (2023). Advancing environmental design with phytoremediation of brownfield soils using spontaneous invasive plants. *Science of the Total Environment*, *883*, 163635.

Shivaraju, H. P., Yashas, S. R., & Harini, R. (2020). Application of Mg-doped TiO2 coated buoyant clay hollow-spheres for photodegradation of organic pollutants in wastewater. *Materials Today: Proceedings*, *27*, 1369–1374.

Shrivastava, M., Khandelwal, A., & Srivastava, S. (2019). Heavy metal hyperaccumulator plants: The resource to understand the extreme adaptations of plants towards heavy metals. *Plant-metal Interactions*, 79–97.

Skuza, L., Szućko-Kociuba, I., Filip, E., & Bożek, I. (2022). Natural molecular mechanisms of plant hyperaccumulation and hypertolerance towards heavy metals. *International Journal of Molecular Sciences*, *23*(16), 9335.

Suman, J., Uhlik, O., Viktorova, J., & Macek, T. (2018). Phytoextraction of heavy metals: A promising tool for clean-up of polluted environment? *Frontiers in Plant Science*, *9*, 1476.

Sun, Y., Xiong, X., He, M., Xu, Z., Hou, D., Zhang, W. & Tsang, D. C. (2021). Roles of biochar-derived dissolved organic matter in soil amendment and environmental remediation: A critical review. *Chemical Engineering Journal*, *424*, 130387.

Sytar, O., Ghosh, S., Malinska, H., Zivcak, M., & Brestic, M. (2021). Physiological and molecular mechanisms of metal accumulation in hyperaccumulator plants. *Physiologia Plantarum*, *173*(1), 148–166.

Thakur, S., Singh, L., Wahid, Z. A., Siddiqui, M. F., Atnaw, S. M., & Din, M. F. M. (2016). Plant-driven removal of heavy metals from soil: Uptake, translocation, tolerance mechanism, challenges, and future perspectives. *Environmental Monitoring and Assessment*, *188*, 1–11.

Valliere, J. M., Wong, W. S., Nevill, P. G., Zhong, H., & Dixon, K. W. (2020). Preparing for the worst: Utilizing stress-tolerant soil microbial communities to aid ecological restoration in the Anthropocene. *Ecological Solutions and Evidence*, *1*(2), e12027.

Visconti, D., Álvarez-Robles, M. J., Fiorentino, N., Fagnano, M., & Clemente, R. (2020). Use of Brassica juncea and Dactylis glomerata for the phytostabilization of mine soils amended with compost or biochar. *Chemosphere*, *260*, 127661.

Xiang, L., Harindintwali, J. D., Wang, F., Redmile-Gordon, M., Chang, S. X., Fu, Y., . . . Xing, B. (2022). Integrating biochar, bacteria, and plants for sustainable remediation of soils contaminated with organic pollutants. *Environmental Science & Technology*, *56*(23), 16546–16566.

Zakari, S., Jiang, X., Zhu, X., Liu, W., Allakonon, M. G. B., Singh, A. K., . . . Yang, B. (2021). Influence of sulfur amendments on heavy metals phytoextraction from agricultural contaminated soils: A meta-analysis. *Environmental Pollution*, *288*, 117820.

Zhao, X., Joo, J. C., Du, D., Li, G., & Kim, J. Y. (2023). Modelling heavy-metal phytoextraction capacities of Helianthus annuus L. and Brassica napus L. *Chemosphere*, *139341*.

Zhu, Q., Ji, J., Tang, X., Wang, C., & Sun, H. (2023). Bioavailability assessment of heavy metals and organic pollutants in water and soil using DGT: A review. *Applied Sciences*, *13*(17), 9760.

5 Environmental Pollution

Causes, Types, Fates, Effects, and Control Strategies

Kingsley Erhons Enerijiofi, Chidozie Ekene, Saheed Ibrahim Musa, Comfort Okoji, Nathan Moses, Parisa Ebrahimbabaie, Iboyi Nathaniel Onuche, and Charles Oluwaseun Adetunji

5.1 INTRODUCTION

The pollution of the environment is as old as the existence of man on Earth, although it was minute in the early days. However, due to the quest for civilization arising from growth and development of industries and scientific inventions, among others, the environment has continued to remain abated by pollutants. The environment is made up of the lithosphere (soil), hydrosphere (water), and atmosphere (air). Environmental pollution arises from mining activities, agricultural runoffs and pesticides, solid wastes, sewage, and heavy metals. Environmental pollution is the unwanted changes in the physical, chemical, and biological components of water, soil, and air (Enerijiofi and Ikhajiagbe, 2021; Gasiński et al., 2022). These pollutants affect the health of people, plants, other animals, and the surrounding environment. The environment, particularly the soil, is regarded as the universal sink that plays host to all contaminants. Soil contamination negatively affects its texture and nutrient composition, which can impair the balance of the soil microorganisms affect plant growth and food production, which can then lead to famine (Enerijiofi and Ekhaise, 2019).

Water occupies 75% of the Earth, and it is known as an essential resource for socio-economic development worldwide. However, its contamination reduces the dissolved oxygen, destroys aquatic life, and causes waterborne diseases that may threaten sustainable livelihood and account for a high number of deaths worldwide (Ifi et al., 2019). Despite strategies put in place to ensure access to potable drinking water, it still remains a daunting challenge as about 1.2 billion do not have access to safe water in developing countries (Bhowmik et al., 2023; Rygaard et al., 2011). The irrigation and industrial sectors were reported to consume 69% and 23% of potable water, respectively, with the remaining left for human consumption (Mahfooz et al., 2022).

The contamination of the air environment can lead to airborne diseases like severe acute respiratory syndrome (SARS) and other forms of life-threatening respiratory infections. The use of 4Rs; Reduce, Reuse, Recycle and Restoration of spoilt resources in the environment should be truly encouraged, practiced, and sustained so as to assure the potable water, a healthy soil and air for future of life on Earth (Enerijiofi and Ekhaise, 2022).

5.2 TYPES OF ENVIRONMENTAL POLLUTION

Recently, global pollution has been currently occurring on an unprecedented scale, marked by two concerning trends. First, there is a substantial and escalating release of chemicals, primarily from fossil fuels, leading to significant alterations in natural systems worldwide. Second, the use of various biocides and toxic substances is steadily increasing, as is their release. Environmental pollution can be said to happen when pollutants or contaminants are present in the natural environment to a level that can cause detrimental effects on the environment and the organisms inhabiting it. It is well recognized that

DOI: 10.1201/9781032683751-6

pollutants in the environment can upset the equilibrium of the natural ecosystem and have an impact on daily life. This situation may give rise to human diseases and global warming. Environmental pollution occurs in all aspects of the environment as a result of industrialization and modernization. The rise in industrialization has now led to pollutants being available in different aspects of the environment. Generally, environmental pollution occurs in different forms that are related to the three components of the environment: soil, air, and water (Enerijiofi et al., 2018a; Femi-Ola et al., 2019).

Pollution arises from two sources: point sources and nonpoint sources. A point source is pollution that occurs from a traceable, identifiable, monitorable, and controllable source. An example of this is oil floating on top of a river. You can say obviously that the oil is from an oil vessel leak. Nonpoint sources of pollution are hard to control and they cannot be traced to a particular source. They have multitude of sources. An example is a polluted drainage. A non-profit environmental organization called Pure Earth stated that approximately 200 million people worldwide are impacted by toxic environmental pollution. Concerning outcomes include youngsters losing 30 to 40 IQ points, babies born with birth deformities, and life expectancy decreasing to as low as 45 years owing to cancer and other ailments in some of the most severely polluted locations.

5.2.1 Soil Pollution

Soil pollution is simply the presence of a pollutant in the soil. This pollutant is present to such an extent that it can affect the environment as well as the organisms inhabiting that environment. Soil pollution poses significant environmental and health risks. High concentrations of benzene in soil increase the risk of contracting leukemia. It is crucial to recognize that all soils contain compounds harmful to humans and other organisms (Ajuzie et al., 2015). However, these substances are found in trace amounts in unpolluted soil, meaning the local ecology is not threatened. When the amount of one or more hazardous substances in soil reaches levels that are harmful to living things, the soil is said to be contaminated. Causes of soil pollution can easily be linked to agricultural usage of pesticides and synthetic fertilizers, excessive industrial activity due to industrialization, and poor management and improper disposal of waste. This has brought about the need for remediation (Enerijiofi et al., 2023). Xenobiotics, synthetic substances that do not occur naturally, constitute some of the most hazardous soil pollutants. Heavy metals like lead and mercury, particularly when exceeding permissible concentrations, are significant contributors to soil pollution. The presence of these chemicals in the soil can lead to an increase in pollutant toxicity due to alterations in soil chemistry (Ikhajiagbe et al., 2021).

5.2.2 Air Pollution

Air pollution is considered one of the most dangerous forms of pollution. When air is polluted, there is high risk that all other components of the environment are affected, including the organisms inhabiting that environment. Air pollution happens when toxic gases, smoke, and dust enter the Earth's atmosphere, contaminating the atmosphere thereby making it difficult for life to exist. It can also be described as the existence of materials, chemicals, or energy in the atmosphere that disrupts normal ecosystem functions, impacting the survival of organisms. Examples of activities contributing to air pollution include the burning of fossil fuels, agricultural practices, mining operations, industrial and factory emissions, and the use of household cleaning products. Individuals release a substantial quantity of chemical substances into the air daily, reaching concentrations that qualify as air pollution. Overall, the effects of air pollution are notably concerning in contemporary times. This is because cases of global warming, acid rain, and respiratory and heart problems are increasing. In addition, aquatic regions are experiencing serious eutrophication as a result of the depletion of the ozone layer and increasing light emission. Looking deeper, it is evidence that air pollution has caused a number of wildlife organisms to change their habitat for survival. Air pollution can occur from either domestic or industrial release of pollutants into the atmosphere.

5.2.3 Water Pollution

It is a general belief that every organism depends either directly or indirectly on water, because no living organism can exist without water. In addition to direct dependencies, over 60% of species inhabit some form of water, emphasizing the need to address water pollution. Factors contributing to water pollution include the disposal of industrial wastes into rivers and seas, leading to a significant imbalance in water properties detrimental to aquatic life (Ifi et al., 2019). A major contributing factor to diseases in non-aquatic species is water pollution. Oil spills in the ocean cause permanent damage to aquatic ecosystems, while insecticides and herbicides used on plants pollute groundwater. Eutrophication results from everyday activities like washing clothes and cutlery beside bodies of water. These activities introduce detergents that block sunlight, lowering oxygen levels and making the water uninhabitable (Enerijiofi et al., 2022). Eighty percent of marine pollution originates on land, mostly from runoff, according to the National Oceanic and Atmospheric Administration (NOAA). This pollution significantly impacts marine life, with sewage fostering pathogen growth and organic and inorganic compounds altering water composition. Reduced dissolved oxygen concentration in water is acknowledged by the Environmental Protection Agency (EPA) as a pollutant resulting from the breakdown of organic compounds, such as sewage. Water pollution is a major environmental problem that arises from a variety of human activities, including home, agricultural, and industrial activities. Excessive use of pesticides and fertilizers in agriculture, hazardous industrial effluents, and sewage water contaminated with human and animal waste are the main causes of widespread pollution (Enerijiofi et al., 2018b; Aransiola et al., 2021). Soil erosion, organic matter decomposition, and minerals leaching from rocks are the main causes of natural water pollution. Pollution from both point and nonpoint sources can affect water bodies such as rivers, lakes, seas, oceans, estuaries, and groundwater sources. Pollutants that are directly released from specific areas, like industrial effluents that are discharged through drainpipes, are referred to as point sources of pollution. On the other hand, nonpoint sources refer to the dispersed release of contaminants across a larger region, such as runoff from roads, streets, building sites, grazing grounds, and abandoned mines.

5.3 CAUSES OF POLLUTION IN THE FOOD CHAIN

Pollution in a food chain can arise from various sources. These include agricultural runoff, indiscriminate use of pesticides and fertilizers, industrial discharges releasing chemicals into water bodies, and improper waste disposal leading to bioaccumulation of contaminants in organisms. These pollutants can adversely impact the health of species within the food chain, posing risks to both wildlife and human consumers (Smith et al., 2016; Wang et al., 2018).

5.3.1 Agricultural Runoff

Runoff from fertilizers and pesticides used in agriculture enters rivers and lakes through irrigation or rainfall, contributing to water pollution. This runoff introduces substances like organophosphates and nitrogen compounds into aquatic ecosystems (Hladik and Kuivila, 2012). Aquatic organisms, especially sensitive species, can suffer from direct toxicity or experience reproductive disruptions. The altered nutrient balance in water bodies can lead to eutrophication, exacerbating the impact on the entire food chain (Carpenter et al., 1998).

5.3.2 Industrial Discharges

Industrial activities release a spectrum of pollutants into water, including heavy metals like mercury, lead, and cadmium. These substances often persist in the environment, posing chronic threats to aquatic life (Grosell et al., 2007). Elements taken up by organisms accumulate as they move up the food chain, leading to the bioaccumulation of heavy metals. This process, known as biomagnification, possesses an increased risk of toxicity for apex predators due to the higher concentrations they experience (Suedel et al., 2002).

5.3.3 Improper Waste Disposal

The persistence of microplastics in water bodies is a result of improper plastic waste disposal. When aquatic organisms consume microplastics with their adsorbed pollutants, it starts the process of transferring toxins into the food chain (Ziajahromi et al., 2017). When species at lower trophic levels consume microplastics, it can have a negative impact on their behavior and overall health. Furthermore, the potential transfer of microplastics and associated pollutants to higher trophic levels raises concerns for consequences across the ecosystem (Wright and Kelly, 2017; Enerijiofi and Ekhaise, 2019).

5.3.4 Airborne Deposition

Pollutants emitted into the air, such as heavy metals and particulate matter, can settle on soil and water surfaces through atmospheric deposition. These pollutants are then absorbed by plants or directly ingested by aquatic organisms, initiating their entry into the food chain (Alloway, 2013).

5.3.5 Soil Contamination

Soil pollution can result from mining operations, improper disposal of industrial waste, and irrigation with water that has been contaminated. Plants growing in polluted soil absorb toxic elements, and herbivores consuming these plants introduce the contaminants into the terrestrial food chain (Bai et al., 2014; Enerijiofi and Ekhaise, 2019).

5.4 ENTRY AND TRANSPORT OF ENVIRONMENTAL POLLUTANTS INTO THE FOOD CHAIN

There are different routes through which environmental pollutants enter and are transported in the food chain.

5.4.1 Water Contamination

Pollutants, including heavy metals, pesticides, and microplastics, enter water bodies through industrial discharges, agricultural runoff, and improper waste disposal (Hartmann et al., 2019; Sarkar et al., 2020). Water serves as a medium for pollutant transport, leading to their dispersion in aquatic ecosystems. Soluble contaminants can be directly absorbed by aquatic organisms, initiating their entry into the food chain (Li et al., 2018).

5.4.2 Bioaccumulation in Aquatic Organisms

Pollutants absorbed by water organisms accumulate in their tissues, a process known as bioaccumulation. This encompasses the uptake of persistent organic pollutants (POPs) and heavy metals (Erickson et al., 2019; Wang et al., 2021). As smaller organisms are consumed by larger ones, the concentration of pollutants increases, initiating the upward transfer of contaminants through trophic levels (Nfon et al., 2018).

5.4.3 Soil Pollution and Terrestrial Pathways

Agricultural practices and atmospheric deposition can introduce pollutants into soil, impacting plants and smaller soil-dwelling organisms. This includes the accumulation of pesticides and metal contaminants (Khan et al., 2021; Rodríguez-Martín et al., 2019). Plants absorb soil contaminants, and herbivores consuming these plants introduce pollutants into the terrestrial food chain. This establishes a route for the transfer of pollutants from soil to higher trophic levels (Wu et al., 2018).

5.4.4 Airborne Pollutants and Atmospheric Deposition

Through atmospheric processes, airborne contaminants, including heavy metals and persistent organic pollutants, can settle on land surfaces and water bodies. This includes the fallout of pollutants from industrial emissions (Masiol and Harrison, 2014; Sun et al., 2017). Deposition of airborne pollutants onto water and land facilitates their incorporation into ecosystems. This atmospheric pathway contributes to the contamination of aquatic and terrestrial organisms, initiating their integration into the respective food chains (Wania et al., 2020).

5.5 FATE OF ENVIRONMENTAL POLLUTANTS IN THE FOOD CHAIN

5.5.1 Absorption by Primary Producers

Environmental pollutants, including heavy metals, pesticides, and organic compounds, are absorbed by plants or algae during their growth. The uptake occurs through roots from contaminated soil, through leaves from polluted air, or from water in the case of aquatic plants. This entry point establishes the foundation for the introduction of pollutants into the food chain (Vidal-Liñán and Raldúa, 2020).

5.5.2 Bioaccumulation in Herbivores

Herbivores consume plants containing pollutants, leading to the accumulation of these contaminants in their tissues. The process of bioaccumulation is influenced by factors such as an organism's feeding habits, metabolic rate, and the nature of the pollutants. Herbivores may experience higher concentrations of pollutants than their surroundings due to continuous exposure through their diet (Javed and Usmani, 2019).

5.5.3 Transfer to Carnivores/Omnivores

The trophic transfer of pollutants occurs as carnivores or omnivores consume contaminated herbivores. The biomagnification phenomenon amplifies pollutant concentrations with each trophic level, posing a higher risk to organisms higher up in the food chain. Predators may experience increased pollutant levels compared to their prey (Kidd et al., 2018).

5.5.4 Accumulation in Top Predators

Apex predators at the top of the food chain accumulate the highest concentrations of pollutants due to biomagnification. POPs and certain heavy metals have slow elimination rates in these organisms. This accumulation raises concerns about the potential health impacts on apex predators and, in some cases, on human populations consuming predator species (Jenssen et al., 2019).

5.6 METABOLISM AND ELIMINATION

Organisms within the food chain metabolize pollutants as a defense mechanism, attempting to eliminate them from their bodies. However, the efficiency of metabolism varies, and certain pollutants, especially those that are persistent, may resist complete elimination. Pollutants persist in organisms due to slow elimination rates, which extend their time in the food chain (Lankadurai et al., 2016).

5.7 HUMAN EXPOSURE

Food items, including fish, meat, and crops, can become contaminated and expose humans to environmental toxins. This exposure pathway raises concerns about potential adverse health effects, as humans may accumulate pollutants from various sources in their bodies. Assessing and controlling the hazards to human health require knowledge of the kinds and quantities of contaminants in the food chain as well as close monitoring of them (Krauss et al., 2017).

5.8 EFFECTS OF ENVIRONMENTAL POLLUTANTS ON THE FOOD CHAIN

5.8.1 Effects on Human Health

The contamination of the environment comprising soil, water, and air has long been reported to cause serious public health challenges to humans. Enerijiofi and Ajuzie (2012) reported that heavy metals reach humans when plants grown on soils polluted by those elements become contaminated upon harvesting and thereafter consumed as food or drugs. These heavy metals are known to affect the reproductive system by reducing libido in males and impairing egg formation in females. Also, these elements are known to affect the ossification of bones in growing children. Contaminated water finds it effects on almost all other facets of the environment inclusive of air and soil. The World Health Organization in Enerijiofi et al. (2018a), stated that more than 3.4 million people die yearly from waterborne diseases, with the majority being young children and 80% of those in ill health, especially in developing nations, stemming from the lack of a potable water supply. This makes infections arising from polluted water a foremost source of illness and death globally (Enerijiofi et al., 2018a). The release of sewage and runoff from informal households has been identified to be the main issue leading to the poor microbiological quality of water (Ifi et al., 2019). Contaminated water causes waterborne diseases like cholera and diarrhea, and other ailments are the outcome of high fecal coliform bacteria in water supplies.

5.8.2 Effects on Biodiversity

Environmental pollution, which is primarily attributed to human activities, has emerged as a global concern resulting in severe repercussions on organisms, their habitats, and biodiversity. Notably, mercury pollution has been linked to a decline in earthworm populations attributed to their incapacity to digest heavy meals (Musa and Ikhajiagbe, 2021). A number of animal species are at risk of extinction due to environmental contamination, which is caused directly by climate change. The disappearance of the golden toad and Monteverde harlequin frog, the threat to polar bear populations from melting Arctic ice, and the possible extinction of North Atlantic whales owing to dwindling plankton, their main food supply, are a few examples. Preliminary research indicates that endemic species like the Nilgiri tahr are at higher risk of going extinct, even though a thorough investigation into the precise effects of climate change on India's natural resources is still needed. Additionally, reports suggest that species such as the black-and-rufous flycatcher (*Ficedula nigrorufa*) are moving uphill from their lower range boundaries. Rising ambient surface temperatures have also been linked to occasional deaths in Shola forest sections.

Temperature fluctuations have contributed to the decline of oceanic coral reef ecosystems. Because of rapid sea level rise – estimated to be 0.1–0.2 m over the past century – coastal regions face an imminent risk of submersion, which might have disastrous effects on a variety of species and ecotone communities. Historical climatic shifts have resulted in ecosystems exhibiting varied species compositions, reflecting differences in species' adaptive capacities to climate changes. For all living things, including humans, to survive collectively, the present biodiversity and ecological balance must be preserved. Moreover, environmental pollution has precipitated significant habitat degradation. Habitat degradation, whether natural or anthropogenic, renders natural habitats incapable of supporting existing species. This phenomenon is influenced by a number of variables, including climate change, geological processes, habitat fragmentation, and human activities, including invasive species introduction and nutrient depletion in ecosystems. Habitat destruction, often driven by human endeavors for industrial production and urbanization, accelerates biodiversity loss as organisms reliant on the site are displaced or eradicated. Agricultural expansion, a primary catalyst for habitat destruction, exemplifies the predominant cause of this environmental degradation.

5.8.3 Effects on Plants

Environmental pollution is at its global perilous point with catastrophic effects in humans, animals, and plants. Summarized below are the impacts of air pollution, land pollution, and various other environmental pollutants on plants. Air pollution interferes with the buildup of resources, which is detrimental to plant growth. Among other pollutants, nitric oxide and ozone disrupt metabolic processes and the plant canopy's ability to fix net carbon (Weber et al., 2021). Reduced levels of foliar sugar and polysaccharides, increased membrane permeability, disruption of membrane-bound photosynthetic processes, and oxidation of fatty acid and sulfhydryl double bonds are among the biological consequences of ozone. Through physiological reductions in net photosynthesis, stomata conductance, water use efficiency, and carbon allocation, ozone causes an imbalance between shoots and roots, accelerates the senescence of leaves, increases visible injuries and leaching on foliar surfaces, reduces floral yield, and delays fruit development (Ikhajiagbe et al., 2021).

Furthermore, decreased wood density and tracheid length are two more ozone-induced effects on plant morphology that lead to decreased wood strength, pulp yield, and quality. Acidification emerges as a critical threat to plant ecosystems in developing countries, while Europe and North America face challenges from increasing ozone and nitrogen deposition. Ozone, a pervasive phytotoxic pollutant in industrialized regions of the Northern Hemisphere, exhibits well-documented detrimental effects on various plant species. Exposure to air pollutants can lead to reduced stem growth, alterations in wood density, and the deposition of pollutants in stems, potentially modifying the defensive capacities of stems through changes in chemical constituents.

Airborne pollutants, particularly heavy metals, initially deposited in the soil impact root function and disrupt the plant's ability to capture soil resources. Plant resource allocation is altered as a result of these disturbances to photosynthesis, the synthesis of carbohydrates, and the uptake of water and mineral nutrients from the soil. These modifications eventually impact the growth of the plant as a whole. Air pollution can change the competitive balance of species at the ecological level, changing the species composition of plant communities. Such modifications may show up in agricultural systems as a decline in economic yield. Twelve herbaceous species were the subject of an investigation into the impacts of diesel exhaust emissions (Honour et al., 2009). The study revealed species-specific changes in growth, phenology, hastened senescence, and delayed flowering. The study highlighted the potential impact of realistic levels of vehicle exhaust on plant responses.

Smog and haze diminish sunlight available for photosynthesis, and the resultant tropospheric ozone harms plants. Soil fertility can decline, rendering it unsuitable for plant growth. Sulfur dioxide and nitrogen oxides induce acid rain, lowering soil pH. Organic pollution in watercourses depletes oxygen levels, reducing species diversity. Nitrogen oxides, removed by rain and fertile land, alter ecosystem species composition. Greenhouse gas emissions contribute to global warming, favoring invasive species over natives, thereby diminishing biodiversity. Invasive plants introduce debris and biomolecules, altering soil and chemical compositions and reducing native species competitiveness.

Improper solid waste management, a major cause of environmental pollution (Kimani, 2007), exacerbates land pollution, a critical global issue (Khan, 2004). The adverse impacts of inappropriate waste disposal on plant health were highlighted by Silveira et al. (2017), who showed a substantial reduction in root growth in onion plants exposed to used potliner and lettuce exposed to cadmium.

5.9 PREVENTION OF ENVIRONMENTAL POLLUTANTS FROM ENTERING THE FOOD CHAIN

5.9.1 Avoid Excessive Application of Pesticides

The overuse of pesticides in agricultural techniques can have detrimental effects on the population of microorganisms that live in the soil, particularly if the chemicals are applied incorrectly or excessively, causing an accumulation of chemical compounds in the soil. While some research has shown that pesticides negatively impact soil microorganisms and biochemical processes, other studies have discovered that some pesticide residues can be broken down and metabolized by microbes (Devi et al., 2022). It is still unclear how exactly pesticides affect soil microorganisms. According to Abdel-Mallek et al. (1994), several environmental factors as well as pesticides' toxicity, persistence, and concentration all have an impact on the microorganisms in soil. Considering a number of factors, the relationship between pesticides and soil ecology is complex, making it difficult to draw firm conclusions. Long-term pesticide usage generally has the potential to interfere with the metabolic pathways involved in the cycling of nutrients (Devi et al., 2022).

Pesticide ingredients frequently include persistent soil contaminants that have the potential to gradually deteriorate soil quality. Using pesticides generally lowers the biodiversity of the soil. Improving soil organic matter helps the soil retain more water, and using non-chemical techniques can improve the condition of the soil. Because organic farms usually yield 20%–40% more than conventional farms, this is advantageous for farmers during drought years. Furthermore, because organic matter facilitates the breakdown of pesticides, a decrease in organic matter in the soil is linked to an increase in pesticide runoff following treatment (Musa and Ikhajiagbe, 2021).

The persistence of pesticides in soil is influenced by both sorption and degradation processes. These mechanisms, directly influenced by the chemical composition of the pesticide, govern the transfer from soil to water, subsequently affecting its movement into the air and potentially our food. Soil-bacterial interactions are triggered by the breakdown of organic compounds. Adsorption affects the bioavailability of pesticides, whose bioaccumulation depends on biological material in the soil. Because of their low pH and mostly acidic structure, weak organic acids are only absorbed by the soil poorly. Compounds that have been sorbed are known to be less understandable to microorganisms. As pesticide residues age and lose biological activity, they become more difficult to extract and degrade due to unknown aging mechanisms.

5.9.2 Waste Management through Composting

Poor waste management is harmful to people's health. Garbage creates an unpleasant environment and contributes to climate change through air pollution, damage to aquatic environments, disruption of the food chain, and ozone depletion. Traditional waste management methods are often inadequate (Aruna et al., 2018; Alam and Ahmade, 2013). Waste is burned, dropped by the side of the road, and dumped in streams and the ocean (Ogwueleka, 2009). Anaerobic or aerobic processes are used to convert organic wastes (degradable wastes). When organic waste is broken down in an aerobic environment, compost is produced (Lasaridi et al., 2018). Biogas and effluents from anaerobic treatment can be used as biological fertilizers (Khan et al., 2018). Composting is an aerobic process in which bacteria convert complicated degradable materials into both organic and inorganic waste (Toledo et al., 2018).

These by-products are distinguished from those found in the soil, fossil fuels, and peats by the presence of "humic-like" components. Composting is a technique that transforms diverse biodegradable waste materials into products suitable for safe and effective incorporation into the soil as nutrients and biofertilizers (Cai et al., 2007; Bai et al., 2010; Yu et al., 2019). The process of composting is used to boost agricultural production and the quality of organic matter in the soil because it consists of necessary nutrients and has organisms that support plant growth (Pane et al., 2014). (Luo et al., 2017).

This makes a big difference in preserving the security of food. Compost can be used for more than just fertilizer; it can also be used for weed control, pollution mitigation, erosion control, environmental cleanup, plant pathogen management, garden design, and wetland restoration (Ventorino et al., 2019). According to Pose-Juan et al. (2017), composting enhances soil ecological diversity and mitigates the environmental risks linked to chemical-based fertilizers.

When it comes to cleaning up soil contaminated with heavy metals, compost works well. It aids in the decomposition of several materials in the soil, including pesticides, petroleum derivatives, explosives, chemical solvents, heavy metals, wood preservatives, and hydrocarbons, both chlorinated and non-chlorinated. Compost has the capacity to diminish the toxicity of certain chemical contaminants by absorbing or decomposing these elements (Huang et al., 2017). Precipitation (Katoh et al., 2014), adsorption (Soares et al., 2016), complexation (Tsang et al., 2014), and redox processes (Zhang et al., 2013) can all render heavy metals inaccessible. The safest way to handle organic waste that is decomposing is through composting. Composting is a better alternative to incineration or disposing of waste by dumping it in bodies of water. These by-products of digested wastes are put to a variety of good uses (Gonawala and Jardosh, 2018; Khater, 2015).

5.9.3 Genetically Modified Crops

Many pesticides put human health at risk or threaten ecosystems, posing dangers to establishments that buy or grow food and other products. The majority of genetically modified organisms grown in the United States have undergone development. As a case study, Roundup Ready (RR) crops are genetically modified to tolerate glyphosate, the active ingredient in Roundup herbicide. In 2015, the world's foremost cancer authority declared that glyphosate, the herbicide most commonly used worldwide, was probably carcinogenic to humans (Whitman, 2000). Farmers use genetically engineered crops that are intended to withstand particular herbicides. This enables them to control weeds using pesticides without endangering the crops. In the past, using herbicides could have killed both the weeds and crops, leading to reduced yields. Herbicide-tolerant crops make weed control more efficient and effective for farmers. Canola, corn, soybeans, and cotton are among the most widely grown plants that are resistant to glyphosate herbicides, also known as Roundup Ready (RR) cultivators. Glyphosate effectively kills sedges, grasses, and broadleaf weeds. Other genetically modified crops that are resistant to herbicides include BXN cotton, which is resistant to bromoxynil, and Liberty Link (LL) maize, which is glufosinate-ammonium resistant. Additionally, some crops have been traditionally developed to be herbicide tolerant, such as soybeans and maize that are resistant to sulfonylurea (STS), imidazolinone (IMI), and sethoxydim (SR; James, 2011).

5.9.4 Integrated Pest Management (IPM)

Integrated pest management (IPM) employs a holistic approach to pest control, utilizing biological, cultural, physical, and chemical tools to minimize risks. This approach helps minimize problems with the environment, economy, and health. IPM contributes to human and environmental health by minimizing the necessity for excessive chemical inputs, which can otherwise lead to soil pollution and entry into the food chain. It is important to note that IPM is a technique that can be applied as appropriate for specific circumstances. For instance, some crops with tolerant or pest-resistant varieties can effectively use host plant resistance, but not all crops can. Pheromones are often used to monitor certain hemipteran pests, but they are not effective for all lepidopteran and coleopteran pests. They may not disrupt mating or trap large numbers of pests. In greenhouses, biological management is a common practice for pest control, but it is less common in the field. Although high-value crops like strawberries use mechanical tools such as insect vacuums, these tools are not practical for non-specialty crops and they do not reduce greenhouse gas emissions as they require fossil fuels. Chemical pesticides should only be applied as a last resort. However, there are situations where they are the only way to protect seeds and transplants from persistent pest issues or to stop the spread of some invasive or endemic pests and diseases. In recent years, chemical pesticides increasingly have been used for seed treatment in many crops as a preventive measure (Dara, 2019). During different phases of crop growth, several pest management methods can be utilized to prevent, minimize, or treat insect infestations. These methods can also help in avoiding soil contamination in agriculture. Although each method can provide some level of control, their collective effect could significantly reduce yield losses.

5.9.4.1 Resistance of the Host Plant

A technique that uses cultivars created through genetic engineering or conventional breeding to produce pest-resistant and tolerant crops is an effective way to minimize output losses. The first line of defense in IPM includes cultivars that possess physical, morphological, or biochemical characteristics that make them repulsive to pests or are inappropriate for their growth, development, or efficient reproduction. Choosing cultivars that are resistant to pest damage is the main tactic in IPM.

5.9.4.2 Cultural Control

Cultural control is the use of trustworthy agronomic methods to prevent or minimize harm caused by insect infestations. Use pest-free seed or plant material to lessen the likelihood of introducing pests at the beginning of crop development. Adjusting planting dates can prevent bug infestations. Early cowpea planting has proved effective in preventing the following pests and diseases: aphids, thrips, pod bugs, legume pod borer (*Maruca vitrata*), legume flower thrips (*Megalurothrips sjostedti*), and pod-sucking insects (*Clavigralla tomentosicollis*; Karungi et al., 2000; Asante et al., 2001). Pest infestations may be impacted by plant density or row spacing. For example, it has been shown that increasing plant density can prevent aphid infestations in Ugandan cowpeas (Karungi et al., 2000). Additionally, in Canada, canola fields with larger row spacing may have fewer instances of root maggot (*Delia* sp.) infestations (Dosdall et al., 1996). Modifications to irrigation schedules, fertilizer applications, and other agronomic practices can reduce the preferred habitat of spider mites. For organic strawberry beds in California, micro-sprinklers are used as a control measure. Strawberry plots equipped with micro-sprinklers showed fewer severe cases of powdery mildew (*Podosphaera aphanis*) and botrytis fruit rot infections (*Botrytis cinerea*) than strawberry plots with overhead aluminum sprinklers (Dara et al., 2016). Jasmonic acid is produced when plants have low potassium levels, which makes them more resistant to various illnesses and insect pests (Davis et al., 2018). Hodson and Lampinen (2018) contend that higher plant nitrogen levels may exacerbate arthropod infestations. Numerous disease problems can also be made worse by elevated (Dara et al., 2016) or minimal (Snoeijers et al., 2000) nitrogen levels in plants. Eradication of insect breeding sites and management of the insects' soil-dwelling phases will necessitate crop devastation and attentive cultivation. Following sanitation protocols to remove diseased or infested plant detritus, cleaning field instruments frequently, and avoiding accidentally contaminating healthy regions with human activity are all essential for stopping the spread of pests. In one instance, winter plowing of orchard floors in Iran led to fewer pistachio psyllids (*Agonoscena pistaciae*) hibernating in the weeds and leaf litter. Plows are an essential pest management tool because they clear away crop leftovers and expose the pests' soil-dwelling stages (Kunjwal and Srivastava, 2018). Reducing infestations of spotted-wing drosophila (*Drosophila suzukii*) can be achieved by changing the harvest cycle from every 3 days to every 2 days or by employing sanitation techniques, including bagging unsalable berries (Leach et al., 2017). An additional method for managing pests is crop rotation. Pests can have their life cycles disrupted and annual pest accumulation can be reduced by planting non-host or tolerant crops. Crop rotation has been effectively applied to manage diseases, weeds, and insects in a variety of cropping systems (Curl, 1963; Wright, 1984; Liebman and Dyck, 1993; Mohler and Johnson, 2009). Furthermore, IPM incorporates cultural management strategies, such as deploying trap crops to keep pests away from the main crop and intercropping non-host or pest-repelling plants with the main crop (Pretty and Bharucha, 2015; Nielsen et al., 2016).

5.9.4.3 Microbial Control

In order to battle plant pathogens, arthropod pests, and plant parasitic nematodes, microbial control involves the application of entomopathogenic bacteria, fungi, microsporidia, nematodes, and viruses (Mankau, 1981; Paulitz and Bélanger, 2001; Dong and Zhang, 2006; Lacey, 2017).

5.9.4.4 Mechanical or Physical Control

Numerous studies on pest management strategies have been carried out (Webb and Linda, 1992; Gamliel and Katan, 2012; Gogo et al., 2014; Dara et al., 2018). These methods entail excluding, trapping, removing, or destroying pests via mechanical or physical means. These techniques include using row covers or netting to exclude pests, hand-picking or vacuuming to eliminate pests, using mechanical weed-control instruments, setting rodent pest traps, and regulating the temperature and humidity in greenhouses. Other methods include solarization or steam sterilization, as well as the application of visible and physical bird deterrents like reflective surfaces or sound-producing equipment.

5.9.4.5 Biological Control

Natural predators, such as predatory arthropods and parasitic wasps, can occasionally be quite successful in drastically lowering pest populations (Hajek and Eilenberg, 2018). There are several widely used methods for managing endemic pests, such as refraining from actions that can endanger the population of naturally occurring predators, periodically releasing naturally occurring predators that are sold commercially, and offering a secure and appropriate habitat for naturally occurring predator populations. Biological control is very useful for specialty crops such as outdoor- and greenhouse-grown strawberries (van Lenteren, 1988; Zalom et al., 2018). A traditional biological control method for invasive pests involves introducing, reproducing, and releasing natural predators in the impacted areas (Kenis et al., 2017; Heimpel and Cock, 2018). One effective method for controlling pests is the use of sterile and irradiated insects. Pest populations have been successfully decreased with this strategy (Klassen and Curtis, 2005). Furthermore, instruments for monitoring and managing pest behavior include behavior modification strategies including baits, traps, and mating disruption (Heinz et al., 1992; Shorey and Gerber, 1996; Foster and Harris, 1997; Vladés et al., 2005; El-Sayed et al., 2009; Morrison et al., 2016). Pests can be drawn in, lured in, and killed by placing deadly baits in traps or on the ground. Certain colors, luminosities, and pheromones can be used to attract, capture, or kill pests that are drawn to these characteristics.

5.9.4.6 Chemical Control

To control pests, chemical control entails applying synthetic chemical pesticides. However, compounds derived from plants or animals should also be included in the definition of "chemical control." Even though microbe-derived toxic metabolites like spinosad and azadirachtin and botanical extracts like pyrethrins and azadirachtin are frequently referred to as "biopesticides," they are still chemical substances that are comparable to synthetic chemicals and present comparable risks to the safety of humans and the environment (Lasota and Dybas, 1991; Sarfraz et al., 2005; Dodia et al., 2010). Government restrictions restrict the frequency and amount of certain chemical pesticides used in order to reduce these dangers.

5.10 CONCLUSION

The entry of toxic elements into the food chain through various environmental pollution pathways poses significant challenges to ecosystems and human health. Airborne deposition, agricultural runoff, industrial discharges, soil contamination, plastic waste, and mining activities contribute to the introduction of contaminants into soil, water, and air. These pollutants are then absorbed by plants and aquatic organisms, initiating their journey through the food chain. The consequences of environmental pollution on the food chain are far-reaching. Bioaccumulation and biomagnification of toxic elements occur as contaminants traverse different trophic levels, leading to elevated concentrations in apex predators. This not only disrupts the delicate balance of ecosystems but also poses risks to biodiversity, with potential species reduction and ecosystem instability. Human health is directly impacted as contaminants enter the food chain. Consuming contaminated organisms can lead to the ingestion of toxic elements, resulting in various health risks, including developmental issues and increased susceptibility to chronic diseases. Addressing the challenges posed by environmental pollutants in the food chain requires a multidimensional approach. Strategies such as sustainable agricultural practices, proper waste management, efficient wastewater treatment, and international collaboration are essential. Furthermore, education and awareness programs can empower communities to adopt responsible practices and contribute to the protection of ecosystems. In the future, rigorous enforcement of environmental regulations, ongoing research into emerging pollutants, and the development of eco-friendly technologies are paramount. By understanding the mechanisms through which toxic elements enter the food chain and implementing proactive measures, we can strive to mitigate the impact of environmental pollution, safeguard ecosystems, and ensure the well-being of both wildlife and human populations.

REFERENCES

Abdel-Mallek, A. Y., Moharram, A. M., Abdel-Kader, M. I. and Omar, S. A. (1994). Effect of soil treatment with the organophosphorus insecticide Profenfos on the fungal flora and some microbial activities. *Microbiol. Res.* 149(2): 167–171.

Ajuzie, C. U., Atuanya, E. I. and Enerijiofi, K. E. (2015). Biodegradation potentials of microorganisms isolated from Eleme petrochemical industrial effluent. *NISEB Journal*, 15 (4): 128–136.

Alam, P. and Ahmade, K. (2013). Impact of solid waste on health and the environment. *Int. J. Sustain. Dev. Green Econ.* 2: 165–168.

Alloway, B. J. (2013). *Heavy metals in soils: Trace metals and metalloids in soils and their bioavailability.* Springer Science & Business Media.

Aransiola, S. A., Ijah, U. J. J., Abioye, O. P. and Victor-Ekwebelem, M. O. (2021). ANAMMOX in wastewater treatment. In N. R. Maddela, L. C. García Cruzatty, and S. Chakraborty (eds.), *Advances in the domain of environmental biotechnology. Environmental and microbial biotechnology*. Springer. https://doi.org/10.1007/978-981-15-8999-7_15

Aruna, G., Kavitha, B., Subashini, N. and Indira, S. (2018). An observational study on practices of disposal of waste Garbage in Kamakshi Nagar at Nellore. *Int. J. Appl. Res.* 4: 392–394.

Asante, S. K., Tamo, M. and Jackai, L. E. N. (2001). Integrated management of cowpea insect pests using elite cultivars, date of planting and minimum insecticide application. *Afr. Crop Sci. J.* 9: 655–665.

Bai, J., et al. (2014). Multivariate geostatistical analysis and source identification of heavy metals in the urban soils of Hangzhou, China. *Environ. Pollut.* 184: 492–499.

Bai, J., Shen, H. and Dong, S. (2010). Study on eco-utilization and treatments of highway greening waste. *Proc. Environ. Sci.* 2: 25–31.

Bhowmik, D., Chetri, S., Enerijiofi, K. E., Naha, A., Kanungo, T. D., Shah, M. P. and Nath, S. (2023). Multitudinous approaches, challenges and opportunities of bioelectrochemical systems in conversion of wastes to energy from wastewater treatment plants. *J CLCB-Elsevier*. 4: 100040.

Cai, Q.-Y., Mo, C.-H., Wu, Q.-T., Zeng, Q.-Y. and Katsoyiannis, A. (2007). Concentration and speciation of heavy metals in six different sewage sludge-composts. *J. Hazard. Mater.* 147: 1063–1072.

Carpenter, S. R., et al. (1998). Nonpoint pollution of surface waters with phosphorus and nitrogen. *Ecol. Appl.* 8(3): 559–568.

Curl, E. A. (1963). Control of plant diseases by crop rotation. *The Bot. Rev.* 29: 413–479.

Dara, S. K., Peck, D. and Murray, D. (2018). Chemical and non-chemical options for managing twospotted spider mite, western tarnished plant bug and other arthropod pests in strawberries. *Insects*. 9: 156.

Dara, S. K., Sandoval-Solis, S. and Peck, D. (2016). Improving strawberry irrigation with micro- sprinklers and their impact on pest management. *Agr. Sci.* 7: 859–868.

Dara, S. K. (2019). The new integrated pest management paradigm for the modern age. *J. Integr. Pest Manag.* 10(1): 12, 1–9.

Davis, J. L., Armengaud, P., Larson, T. R., Graham, I. A., White, P. J., Newton, A. C. and Amtmann, A. (2018). Contrasting nutrient-disease relationships: Potassium gradients in barley leaves have opposite effects on two fungal pathogens with different sensitivities to jasmonic acid. *Plant. Cell Environ.* 41: 2357–2372.

Devi, P. I., Manjula, M. and Bhavani, R. V. (2022). Agrochemicals, environment, and human health. *Annu. Rev. Environ. Resour.* 47:399–421.

Dodia, D. A., Patel, I. S. and Patel, G. M. (2010). *Botanical pesticides for pest management.* Scientific Publishers (India).

Dong, L. Q. and Zhang, K. Q. (2006). Microbial control of plant-parasitic nematodes: A five-party interaction. *Plant Soil.* 288: 31–45.

Dosdall, L. M., Herbut, M. J., Cowle, N. T. and Micklich, T. M. (1996). The effect of seeding date and plant density on infestations of root maggots, Delia spp. (Diptera: Anthomyiidae), in canola. *Can. J. Plant Sci.* 76: 169–177.

El-Sayed, A. M., Suckling, D. M., Byers, J. A., Jang, E. B. and Wearing, C. H. (2009). Potential of "lure and kill" in long-term pest management and eradication of invasive species. *J. Econ. Entomol.* 102: 815–835.

Enerijiofi, K. E. and Ajuzie, C. U. (2012). Health impact of some heavy metals resistance in microbes isolated from Ikhueniro refuse dumpsite, Benin City. *NJM.* 26: 2593–2599.

Enerijiofi, K. E. and Ekhaise, F. O. (2019). Physicochemical and microbiological characterisation of Edaiken and Santana markets open solid waste dumpsites soils in Benin City, Nigeria. *FULafia J. Sci. Technol.* 5(2): 136–149.

Enerijiofi, K. E. and Ekhaise, F. O. (2022). Achieving ecofriendly environment through sustainable management of solid wastes in the ecosystem. In T. Aftab (ed.), *Sustainable Management of Environmental Contaminants: Eco-friendly Remediation Approaches.* Springer, Cham, pp. 451–470. https://doi.org/10.1007/978-3-031-08446-1_17

Enerijiofi, K. E. and Ikhajiagbe, B. (2021). Plant-microbe interaction in attenuation of toxic waste in ecosystem. In V. Kumar, R. Prasad, and M. Kumar (eds.), *Rhizobiont in Bioremediation of Hazardous Waste.* Springer, Singapore, pp. 291–315. https://doi.org/10.1007/978-981-16-0602-1_13

Enerijiofi, K. E., Fakeye, O. D. and Oziegbe, O. S. (2018b). Physicochemical and microbial assessment of sewage from Ogwa and Ebelle communities of Esan land, Edo state. *Bayero J Pure Appl Sci.* 11(1): 202–207. http://dx.doi.org/10.4314/bajopas.v11i1.35

Enerijiofi, K. E., Musa, S. I., Okolafor, F. I., Igiebor, F. I., Odozi, E. B. and Ikhajiagbe, B. (2023). Sustainable approaches for the remediation of agrochemicals in the environment. In M. C. Ogwu and S. C. Izah (eds.), *One Health Implications of Agrochemicals and their Sustainable Alternative.* Springer Nature, Switzerland, pp. 511–543. https://doi.org/10.1007/978-981-99-3439-3_19

Enerijiofi, K. E., Okuguni, N. N. and Ajayi, A. V. (2022). Physicochemical composition and heavy metals tolerance of bacterial isolates in leachate from Solid waste dumpsites in Delta North Senatorial District, Delta State. *J. Environ. Earth Sci – Elsevier.* 13(1): 48–52.

Enerijiofi, K. E., Olatunji, E. O. and Irerua, P. A. (2018a). Effect of human activities on the physicochemical and bacteriological qualities of Ujiogba river, Ujiogba, Edo State, Nigeria. *Journal of Applied Life Sciences International,* 17(1): 1–8. doi: 10.9734/JALSI/2018/40435.

Erickson, M. L., et al. (2019). Bioaccumulation of polychlorinated biphenyls (PCBs) in fish: A bioenergetics modeling approach. *Ecological Modelling.* 408: 108721.

Femi-Ola, T. O., Orjiakor, P. I. Enerijiofi, K. E., Oke, I. O. and Fatoyinbo, A. A. (2019). Kinetic properties of Lipase obtained from *Pseudomonas aeruginosa* isolated from crude oil contaminated soil. *Sci. J. Marit. Univ. Szc.* 59(131): 154–161.

Foster, S. P., and Harris, M. O. (1997). Behavioral manipulation methods for insect pest-management. *Annu. Rev. Entomol.* 42: 123–146. http://66.219.30.210/weekly/science/archive/040214/science13.htm

Gamliel, A., and Katan, J. (2012). *Solarization: Theory and Practice.* American Phytopathological Society, St. Paul, MN.

Gasiński, A., Kawa-Rygielska, J., Mikulski, D., Klosowski, G. and Glowacki, A. (2022). Application of white grape pomace in the brewing technology and its impact on the concentration of esters and alcohols, physicochemical parameteres and antioxidative properties of the beer. *Food Chem.* 367: 130646.

Gogo, E. O., Saidi, M., Ochieng, J. M., Martin, T., Baird, V. and Ngouajio, M. (2014). Microclimate modification and insect pest exclusion using agronet improve pod yield and quality of French bean. *HortScience.* 49: 1298–1304.

Gonawala, S. S. and Jardosh, H. (2018). Organic waste in composting: A brief review. *Int. J. Curr. Eng. Technol.* 8: 36–38.

Grosell, M., et al. (2007). Cellular and subcellular distribution of accumulated cadmium in the hepatopancreas of the Caribbean spiny lobster, Panulirus argus. *Aquatic Toxicol.* 82(2): 99–110.

Hajek, A. E. and Eilenberg, J. (2018). *Natural Enemies: An Introduction to Biological Control.* Cambridge University Press, Cambridge.

Hartmann, N. B., et al. (2019). Are we speaking the same language? Recommendations for a definition and categorization framework for plastic debris. *Environ. Sci. Technol.* 53(3), 1039–1047.

Heimpel, G. E. and Cock, M. J. W. (2018). Shifting paradigms in the history of classical biological control. *BioControl*. 63: 27–37.

Heinz, K. M., Parrella, M. P. and Newman, J. P. (1992). Time-efficient use of yellow sticky traps in monitoring insect populations. *J. Econ. Entomol*. 85: 2263–2269.

Hladik, M. L. and Kuivila, K. M. (2012). Assessing the occurrence and distribution of pyrethroids in water and suspended sediments. *Environ. Sci. Technol*. 46(2): 1118–1125.

Hodson, A. K. and Lampinen, B. D. (2018). Effects of cultivar and leaf traits on the abundance of Pacific spider mites in almond orchards. *Arthropod Plant Interact*: 1–11.

Honour, S. L., Bell, J., Nigel, B., Ashenden, T. A., Cape, J. N. and Power, S. A. (2009). Responses of herbaceous plants to urban air pollution: Effects on growth, phenology and leaf surface characteristics. *Environ. Pollut*. 157(4): 1279–1286.

Huang, J., Yu, Z., Gao, H., Yan, X., Chang, J., Wang, C., Hu, J. and Zhang, L. (2017). Chemical structures and characteristics of animal manures and composts during composting and assessment of maturity indices. *PLoS One*. 12: e0178110.

Ifi, F., Enerijiofi, K. E. and Ekhaise, F. O. (2019). Investigation of the physicochemical and bacteriological qualities of Okokpon River, Edo State, Nigeria for its portability status. *FUDMA J. Sci*. 3(2): 185–194.

Ikhajiagbe, B., Enerijiofi, K. E. and Umendu, P. O. (2021). Mycorestoration of an oil polluted soil. *Studia UBB Biologia*. 66(1): 73–84.

James C. (2010). *Global Status of Commercialized Biotech/GM crops, 2011*. Vol. 44. ISAAA, Ithaca, NY.

Javed, M. and Usmani, N. (2019). Accumulation of heavy metals and human health risk assessment via the consumption of freshwater fish, Tilapia mossambicus, from Al-Hassa, Saudi Arabia. *Chemosphere*. 214: 692–702.

Jenssen, B. M., et al. (2019). Polychlorinated biphenyls and their hydroxylated metabolites in top predators from the Barents Sea. *Environ Sci Pollut Res*. 26(10): 9674–9684.

Karungi, J., Adipala, E., Ogenga-Latigo, M. W., Kyamanywa, S. and Oyobo, N. (2000). Pest management in cowpea. Part 1. Influence of planting time and plant density on cowpea field pests infestation in eastern Uganda. *Crop Prot*. 19: 231–236.

Katoh, M., Kitahara, W. and Sato, T. (2014). Sorption of lead in animal manure compost: Contributions of inorganic and organic fractions. *Water Air Soil Pollut*. 225: 1828.

Kenis, M., Hurley, B. P., Hajek, A. E. and Cock, M. J. W. (2017). Classical biological control of insect pests of trees: Facts and figures. *Biol. Invasions*. 19: 3401–3417.

Khan, M., Chniti, S. and Owaid, M. (2018). An overview on properties and internal characteristics of anaerobic bioreactors of food waste. *J. Nutr. Health Food Eng*. 8: 319–322.

Khan, S. I. (2004). Dumping of solid waste: A threat to environment. *The Dawn*.

Khan, S., et al. (2021). Pollution status and health risk assessment of heavy metals in soils irrigated with wastewater in different agricultural zones of Pakistan. *Environ. Pollut*. 278: 116865.

Khater, E. (2015). Some physical and chemical properties of compost. *Int. J. Waste Resour*. 5: 1–5.

Kidd, K. A., et al. (2018). Direct and indirect responses of a freshwater food web to a potent synthetic oestrogen. *Philos. Trans. R. Soc. Lond. B Biol. Sci*. 374(1768): 20160081.

Kimani, N. G. (2007). *Environmental Pollution and Impacts on Public Health: Implications the Dandora Dumping Site Municipal in Nairobi*. United Nations Environment Programme, pp. 1–31. http://www.korogocho.org/english/index.php?option=com_docman&task=doc_download&gid=5

Klassen, W. and Curtis, C. F. (2005). History of the sterile insect technique. In V. A. Dyck, J. Hendrichs, and A. Robinson (eds.), *Sterile Insect Technique*. Springer, Dordrecht, The Netherlands, pp. 3–36.

Krauss, M., et al. (2017). Human health risks of polycyclic aromatic hydrocarbons (PAHs) via consumption of smoked mussel (Mytilus edulis) and whelk (Buccinum undatum) from the Western Baltic Sea. *Environ. Sci. Pollut. Res*. 24(20): 17244–17257.

Kunjwal, N. and Srivastava, R. M. (2018). Insect pests of vegetables. In Omkar (ed.), *Pests and Their Management*. Springer, Singapore, pp. 163–221.

Lacey, L. A. (2017). *Microbial Control of Insect and Mite Pests: From Theory to Practice*. Academic Press, London.

Lankadurai, B. P., et al. (2016). Accumulation and elimination of polycyclic aromatic hydrocarbons in aquatic insects: Metamorphosis alters trophic magnification patterns. *Environ. Sci. Technol*. 50(11): 5675–5682.

Lasaridi, K.-E., Manios, T., Stamatiadis, S., Chroni, C. and Kyriacou, A. (2018). The evaluation of hazards to man and the environment during the composting of sewage sludge. *Sustainability*. 10: 2618.

Lasota, J. A. and Dybas, R. A. (1991). Avermectins, a novel class of compounds: Implications for use in arthropod pest control. *Annu. Rev. Entomol*. 36: 91–117.

Leach, H., Moses, J., Hanson, E., Fanning, P. and Isaacs, R. (2017). Rapid harvest schedules and fruit removal as non-chemical approaches for managing spotted wing Drosophila. *J. Pest Sci.* 91: 219–226.

Li, J., et al. (2018). Microplastics in freshwater systems: A review on occurrence, environmental effects, and methods for microplastics detection. *Water Res.* 137: 362–374.

Liebman, M. and Dyck, E. (1993). Crop rotation and intercropping strategies for weed management. *Ecol. Appl.* 3: 92–122.

Luo, X., Liu, G., Xia, Y., Chen, L., Jiang, Z., Zheng, H. and Wang, Z. J. (2017). Use of biochar-compost to improve properties and productivity of the degraded coastal soil in the Yellow River Delta, China. *J. Soils Sediments.* 17: 780–789.

Mahfooz, Y., Yasar, A., Tanveer, R. and Tabinda, A. (2022). Challenges and Solutions for Sustainable Urban Water Management. In T. Aftab (ed.), *Sustainable Management of Environmental Contaminants: Eco-friendly Remediation Approaches.* Springer, Cham, pp. 391–423. https://doi.org/10.1007/978-3-031-08446-1_21

Mankau, R. (1981). Microbial control of nematodes. In B. Zuckerman and R. A. Rohde (eds.), *Plant Parasitic Nematodes*, Vol. III. Academic Press, New York, NY, pp. 475–494.

Masiol, M., & Harrison, R. M. (2014). Aircraft engine exhaust emissions and other airport-related contributions to ambient air pollution: A review. *Atmos. Environ.* 95: 409–455.

Mohler, C. L. and Johnson, S. E. (2009). *Crop rotation on Organic Farms a Planning Manual.* Natural Resource, Agriculture, and Engineering Service, Ithaca, NY.

Morrison, W. R., Lee, D.-H., Short, B. D., Khrimian, A. and Leskey, T. C. (2016). Establishing the behavioral basis for an attract-and-kill strategy to manage the invasive Halyomorpha halys in apple orchards. *J. Pest Sci.* 89: 81–96.

Musa, S. I. and Ikhajiagbe, B. (2021). Screening of bacterial isolates for phosphate solubilizing capability in a ferruginous ultisol in Benin city, Edo State Nigeria. *Bayero J. Pure Appl Sci.* 13(2): 94–106.

Nfon, E. R., et al. (2018). Cadmium and lead in fish from the Great Lakes and evaluation of the health risks of fish consumption. *Environ. Monit. Assess.* 190(8): 478.

Nielsen, A. L., Dively, G., Pote, J. M., Zinati, G. and Mathews, C. (2016). Identifying a potential trap crop for a novel insect pest, *Halyomorpha halys* (Hemiptera: Pentatomidae), in organic farms. *Environ. Entomol.* 45: 472–478.

Ogwueleka, T. C. (2009). Municipal solid waste characteristics and management in Nigeria. *Iran. J. Environ. Health Sci. Eng.* 6: 173–180.

Pane, C., Palese, A. M., Celano, G. and Zaccardelli, M. (2014). Effects of compost tea treatments on the productivity of lettuce and kohlrabi systems under organic cropping management. *Ital. J. Agron.* 9: 153–156.

Paulitz, T. C. and Bélanger, R. R. (2001). Biological control in greenhouse systems. *Annu. Rev. Phytopathol.* 39: 103–133.

Pose-Juan, E., Igual, J. M., Sánchez-Martín, M. J. and Rodríguez-Cruz, M. S. (2017). Influence of herbicide triasulfuron on soil microbial community in an unamended soil and a soil amended with organic residues. *Front. Microbiol.* 8: 378.

Pretty, J. and Bharucha, Z. P. (2015). Integrated pest management for sustainable intensification of agriculture in Asia and Africa. *Insects.* 6: 152–182.

Rodríguez-Martín, J. A., et al. (2019). Assessment of heavy metal pollution in agricultural soils: Insights from traditional and multi-model geostatistics. *Land Degrad. Dev.* 30(2): 168–180.

Rygaard, M., Binning, P. J. and Albrechtsen H. J. (2011). Increasing urban water self-sufficiency: New era, new challenges. *J. Environ Manage.* 92(1): 185–194.

Sarfraz, M., Dosdall, L. M. and Keddie, B. A. (2005). Spinosad: A promising tool for integrated pest management. *Outlooks on Pest Manag.* 16: 78–84.

Sarkar, A., et al. (2020). The impact of microplastics on the tropic transfer of associated contaminants in marine organisms. *Environ. Res.* 182: 109114.

Shorey, H. H., and Gerber, R. G. (1996). Use of puffers for disruption of sex pheromone communication among navel orangeworm moths (Lepidoptera: Pyralidae) in almonds, pistachios, and walnuts. *Environ. Entomol.* 25: 1154–1157.

Silveira, G. L., Lima, M. G., Barreto, G. R., Palmieri, M. J. and Andrade-Vieria, L. F. (2017). Toxic effects of environmental pollutants: Comparative investigation using *Allium cepa L.* and *Lactuca sativa L. Chemosphere.* 178: 359–357.

Smith, V. H., et al. (2016). Eutrophication: Impacts of excess nutrient inputs on freshwater, marine, and terrestrial ecosystems. *Environ. Pollut.* 100(1–3): 179–196.

Snoeijers, S. S., Pérez-García, A., Joosten, M. H. A. J. and De Wit, P. J. G. M. (2000). The effect of nitrogen on disease development and gene expression in bacterial and fungal plant pathogens. *Eur. J. Plant Pathol.* 106: 493–506.

Soares, M. A., Marto, S., Quina, M. J., Gando-Ferreira, L. and Quinta-Ferreira, R. (2016). Evaluation of egg-shell-rich compost as biosorbent for removal of Pb (II) from aqueous solutions. *Water Air Soil Pollut.* 227: 150.

Suedel, B. C., et al. (2002). Bioaccumulation of PCBs from sediments to aquatic insects and riparian spiders in the Upper Hudson River, USA. *Environ Toxicol Chem.* 21(11): 2187–2197.

Sun, H., et al. (2017). The transport of airborne persistent toxic substances (PTSs) to the Arctic and potential effects on human health: A review. *Environ. Pollut.* 230: 931–941.

Toledo, M., Siles, J., Gutiérrez, M. and Martín, M. (2018). Monitoring of the composting process of different agroindustrial waste: Influence of the operational variables on the odorous impact. *Waste Manag.* 76: 266–274.

Tsang, D. C., Yip, A. C., Olds, W. E. and Weber, P. A. (2014). Arsenic and copper stabilisation in contaminated soil by coal fly ash and green waste compost. *Environ. Sci. Pollut. Res.*21: 10194–10204.

Van Lenteren, J. C. (1988). Biological and integrated pest control in greenhouses. *Annu. Rev. Entomol.* 33: 239–269.

Ventorino, V., Pascale, A., Fagnano, M., Adamo, P., Faraco, V., Rocco, C., Fiorentino, N. and Pepe, O. (2019). Soil tillage and compost amendment promote bioremediation and biofertility of polluted area. *J. Clean. Prod.* 239: 118087.

Vidal-Liñán, L. and Raldúa, D. (2020). Environmental chemicals and metabolic disruption in fish: Insights from an integrative perspective. *Gen. Comp. Endocrinol.* 291: 113375.

Vladés, E. M. A. E., Aldana, L. L. L., Figueroa, B. R., Gutiérrez, O. M., Hernández, R. M. C. and Chavelas, M. T. (2005). Trapping of *Scyphophorus acupunctatus* (Coleoptera: Curculionidae) with two natural baits in a field of *Polianthes tuberosa* (Liliales: Agavaceae) in the state of Morelos, México. *Fla. Entomol.* 88: 338–340.

Wang, C., et al. (2021). Bioaccumulation of heavy metals and human health risk assessment via consumption of freshwater fish, crustacean, and mollusk in the Yangtze River. *Ecotoxicol Environ Saf.* 207: 111354.

Wang, X., et al. (2018). Bioaccumulation of heavy metals in the soil-maize system and associated potential health risks in Guilin, China. *Environ. Monit. Assess.* 190(7): 413.

Wania, F., et al. (2020). Global application of the trophic magnification factor to microplastics in organisms and implications for human risks. *Environ. Sci. Technol. Lett.* 7(4): 258–263.

Webb, S. E. and Linda, S. B. (1992). Evaluation of spunbonded polyethylene row covers as a method of excluding insects and viruses affecting fallgrown squash in Florida. *J. Econ. Entomol.* 85: 2344–235

Weber, J., Tingey, D. and Anderson, C. (2021). *Plant Response to Air Pollution.* U.S. Environmental Protection Agency, Washington, DC, EEPA/600/A-93/050(NTIS PB93167260).

Whitman, D. B. (2000). Genetically modified foods: Harmful or helpful? *CSA Disc Guide*: 1–13.

Wright, R. J. (1984). Evaluation of crop rotation for control of Colorado potato beetles (Coleoptera: Chrysomelidae) in commercial potato fields on Long Island. *J. Econ. Entomol.* 77: 1254–1259.

Wright, S. L. and Kelly, F. J. (2017). Plastic and human health: A micro issue? *Environ. Sci. Technol.* 51(12): 6634–6647.

Wu, J., et al. (2018). Current status and contamination assessment of soil heavy metals in urban areas of Beijing, China. *J. Geochem. Explor.* 194: 193–201.

Yu, H., Xie, B., Khan, R. and Shen, G. (2019). The changes in carbon, nitrogen components and humic substances during organic-inorganic aerobic co-composting. *Bioresour. Technol.* 271: 228– 235.

Zalom, F. G., Bolda, M. P., Dara, S. K. and Joseph, S. (2018). *UC IPM Pest Management Guidelines: Strawberry (Insects and Mites).* University of California Statewide IPM Program, Oakland, CA, Publication Number 3468.

Zhang, X., Wang, H., He, L., Lu, K., Sarmah, A., Li, J., Bolan, N. S., Pei, J. and Huang, H. J. E. S. (2013). Research, P. Using biochar for remediation of soils contaminated with heavy metals and organic pollutants. *Environ. Sci. Pollut. Res.* 2013, 20, 8472–8483.

Ziajahromi, S., et al. (2017). Synthetic fibers as an indicator of land-based microplastic pollution: A review of methods. *Water Research.* 115: 162–173.

6 Mechanism of Phytoremediation and Plant Uptake of Heavy Metals

Anjaneyulu Musini, P. Gnana Deepu, Jogipeta Harihara, Kathuroju Harikrishna, Chittepu Obula Reddy, Bachha Laxmi Dora, and Bhagyashree Mohakhud

6.1 INTRODUCTION TO HEAVY METALS

6.1.1 Heavy Metals

Heavy metal contamination refers to the presence of heavy metals in the environment. These metals are commonly characterized as having a specific weight exceeding 5 g cm^{-3} (Jeng and Singh, 1995; Aransiola et al., 2013). They occur naturally and possess high atomic weight and density. There are approximately 40 elements that fall into this category, with a significant portion being found in aquatic systems and soil. A smaller portion can be found in the atmosphere as vapors or particulates, as well as dispersed forms within rock formations. The increase in concentrations of heavy metals in the biosphere can be attributed to industrialization and urbanization. The toxicity of these metals varies depending on plant species, specific types of metals, chemical forms, soil composition, pH, and other factors (Wintz et al., 2002). For instance, metals like zinc (Zn) and copper (Cu) can either act as cofactors or exhibit catalytic properties (Mildvan, 1970). On the other hand, metals like cadmium (Cd), arsenic (As), and mercury (Hg) are highly toxic to certain plant species. Some heavy metals are essential nutrients or relatively harmless in small quantities but become toxic in larger amounts.

6.1.2 Sources of Heavy Metals

Heavy metals can enter the environment through both natural and human-made (anthropogenic) sources. An overview of the sources of metals is given in Figure 6.1.

6.1.2.1 Natural Sources

1. *Weathering of rocks and minerals*: Heavy metals are released into the soil by the natural breakdown of rocks and minerals in the Earth's crust.
2. *Volcanic activity*: Volcanic eruptions can release heavy metals into the environment, affecting soil composition.
3. *Weathering of parent material*: The composition of the soil is influenced by the weathering of the underlying geological material.
4. *Erosion and sedimentation*: Natural processes like erosion can transport heavy metals from one location to another, affecting soil and water quality.

DOI: 10.1201/9781032683751-7

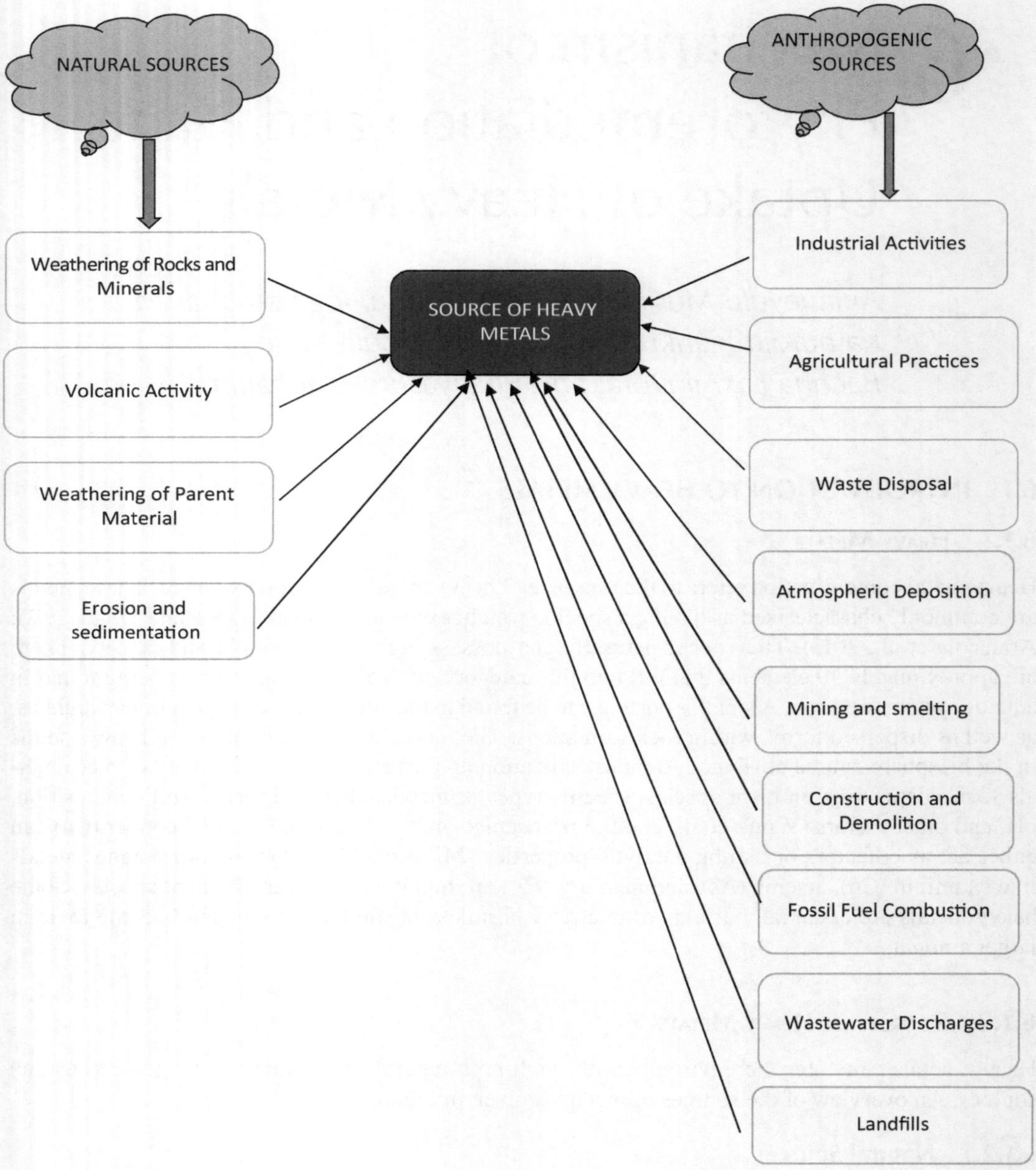

FIGURE 6.1 Overview of sources of heavy metals.

6.1.2.2 Anthropogenic Sources

1. *Industrial activities*: Various industries release heavy metals into the environment through emissions, discharges, and waste disposal. Metal smelting, mining, manufacturing, and processing operations are common contributors.
2. *Agricultural practices*: The use of pesticides, fertilizers, and herbicides can introduce heavy metals into the soil. Additionally, irrigation with contaminated water or the application of sewage sludge as fertilizer can contribute to metal contamination.

3. *Waste disposal*: Heavy metals can be released into land and water when industrial trash, electronic garbage (e-waste), and residential waste are not properly disposed of.
4. *Atmospheric deposition*: Airborne particles from industrial emissions, vehicle exhaust, and other sources can settle on the ground, contributing to soil and water contamination.
5. *Mining and smelting*: Activities related to mining and metal smelting can release large quantities of heavy metals into the environment. Runoff from mining sites can carry metals into nearby water bodies.
6. *Construction and demolition*: Construction activities and the demolition of structures can release heavy metals, particularly if there are materials containing lead (Pb), Cd, or other metals.
7. *Combustion of fossil fuel*: Burning fossil fuels, such as coal and oil, releases heavy metals like mercury into the air. These metals can then be deposited onto land and water surfaces.
8. *Wastewater discharges*: Effluents from industrial facilities and municipal wastewater treatment plants may contain heavy metals. If not properly treated, these discharges can contaminate receiving water bodies.
9. *Landfills*: Landfills can be a source of heavy metal contamination if waste containing metals is not adequately managed.

Heavy metals are classified as trace elements due to their presence in the environment (Kabata-Pendias, 2000). The availability of these metals is influenced by physical factors like sequestration, temperature, and adsorption. Chemical factors like lipid solubility, thermodynamic equilibrium, and complexation kinetics (Hamelink et al., 1994), and biological factors like physiological adaptation, trophic interactions, and species characteristics, further impact their bioavailability (Verkleji, 1993).

6.2 PHYTOREMEDIATION AND ITS SIGNIFICANCE

6.2.1 Phytoremediation

Contamination of soils by potentially toxic elements (PTEs) is a worldwide issue caused by various human activities including industry, agriculture, and mining. These activities release harmful elements into the environment, which can persist for extended periods and have negative impacts on both human health and the environment. Over time, diverse strategies with biological, physical, and chemical procedures have been developed to eradicate these heavy metals from the environment. However, physical and chemical remediation techniques are expensive and also harm the soil's microbial community, thereby affecting its texture. As a result of these concerns, there is an increasing demand for remediation approaches like phytoremediation.

Phytoremediation refers to the utilization of plants and their related microorganisms for environmental treatment (Salt et al., 1998). It is an effective technology for removing various types of inorganic and organic pollutants. The term *phytoremediation* is a combination of the terms *phyton* (plants) and *remediare* (to remediate), describing a process in which specific plants and soil microbes collaborate to convert toxins into safe and often profitable forms (Baker, 1987; Raskin et al., 1997). Soil heavy metal pollution can be remediated through different approaches, including physical (Siyar et al., 2020), chemical, and biological remediation (Liu et al., 2020; Kang et al., 2021). It is widely favored due to its environmentally friendly nature and cost-effectiveness as a bioremediation method. It is commonly used to clean up hazardous organic compounds and heavy metals in contaminated sites, as well as to remove radioactive contaminants from agricultural lands and groundwater. Phytoremediation plays a crucial role in purifying air, water, soil, and sediments (Aransiola et al., 2019; Shikha and Singh, 2021).

6.2.2 Significance

In the past 10 years, phytoremediation has become popular with industries and government agencies due to the relatively low cost of cleaning the environment (Glass, 1999; Tsao, 2003). Since the biological processes rely on solar energy, they are generally ten times more cost-effective than engineering-based methods like soil washing, burning, soil excavation, or pump-and-treat systems (Glass, 1999). The cost-effectiveness of phytoremediation is enhanced by its typical in situ implementation, which also helps to minimize the exposure of contaminated substrate to humans, wildlife, and the environment.

Phytoremediation has applications in both land-based and water-based settings. It can serve as either a preliminary or final measure in conjunction with other methods of environmental cleanup. The success of phytoremediation, which is a cost-effective approach, relies on the presence of pollutants within the root zones of the plants.

6.2.3 Types and Process of Phytoremediation

In areas where there is contamination from metals, plants can be utilized to either stabilize or eliminate the metals from the soil and groundwater. This can be achieved through various mechanisms such as phytoextraction, rhizofiltration, phytodegradation, and phytovolatilization. Types of phytoremediation are listed in Figure 6.2, and a summary appears in Table 6.1.

6.2.3.1 Phytoextraction

Phytoextraction, also known as phytoaccumulation, is a process where the pollutants are absorbed by the root system of plants and transported to the above-ground parts of the plant. This approach uses plants to gather toxic metals from polluted soils, concentrate them, and then store them in the biomass above ground. Phytoextraction offers a relatively affordable approach compared to conventional methods. The success of this remediation technology relies on several factors, including the availability of metals in the soil, the level of contamination, and the ability of the plants to absorb, intercept, and accumulate metals (Ernst, 1996). Metals like Zn, Cu, and nickel (Ni) are particularly suitable for removal through phytoextraction since many plants have demonstrated a preference for absorbing and accumulating these metals in large quantities.

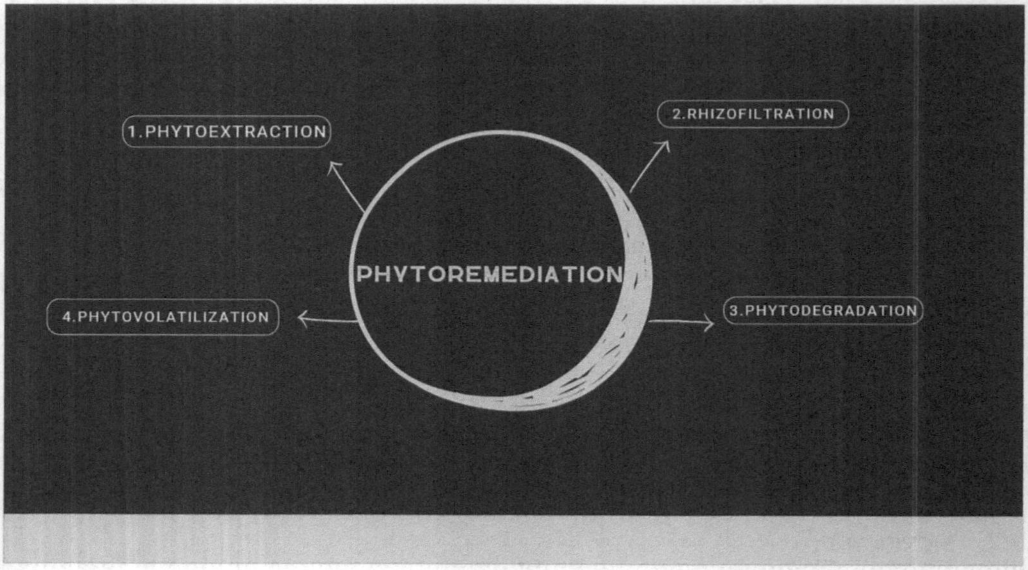

FIGURE 6.2 Types of phytoremediation.

TABLE 6.1
Types of Phytoremediation

Phytoremediation Technique	Mechanism	Media
Phytoextraction	Involves the removal of contaminants through plant tissue	Groundwater, wastewater
Rhizofiltration	Involves the removal of contaminants through plant roots in wastewater and surface water	Soil, sediment
Phytodegradation	Degradation of organic contaminants by enzymes	Soil, sediment
Phytovolatilization	Conversion into the vapor state and release into the atmosphere through transpiration	Soil, sediment

6.2.3.2 Rhizofiltration

Rhizofiltration is a commonly employed technique for eliminating metals or other inorganic substances from groundwater, wastewater, or surface water. Unlike phytoextraction, which focuses on soil remediation, rhizofiltration primarily targets contaminated groundwater.

To ensure successful implementation, the plants must first establish a robust root system. They are then acclimatized by replacing their natural water source with contaminated water. Once ready, the plants are planted in the contaminated region, where their roots absorb both chemicals as well as water. When the roots get saturated with contaminants, they are collected. The effectiveness of this method relies on the ability of plant roots to absorb and isolate metal pollutants from the water. Through this phytoremediation mechanism, it becomes possible to remove metals such as Ni, Cd, Cu, vanadium (V), Pb, and chromium (Cr), as well as radionuclides like strontium (Sr), uranium (U), and cesium (Cs; Jabeen et al., 2009; Singh and Santal, 2015).

6.2.3.3 Phytodegradation

Phytodegradation refers to the enzymatic breakdown of organic contaminants in plants, sediments, and soils. Enzymes such as dehalogenase, laccase, nitrilase, nitroreductase, and peroxidase have been discovered in these environments. Phytodegradation occurs in soil where biodegradation is not feasible. However, one disadvantage of this process is the production of harmful intermediates or products of degradation (Dec et al., 1994; McCutcheon et al., 1995; Susarla et al., 2002; Prakash and Priyanka, 2023).

6.2.3.4 Phytovolatilization

Phytovolatilization is a technique used in phytoremediation, where plants are utilized to mitigate the environmental impact of pollutants. This process involves plants taking contaminants from soil, transforming them into volatile forms, and then releasing them into the environment through transpiration. Phytovolatilization is commonly employed to eliminate pollutants found in sediment, water, or soil. Heavy metals such as As and Hg are often the focus of phytovolatilization studies, as they exist in volatile forms and can be transformed into gaseous species by plants. The advantage of this process lies in its ability to convert pollutants into less harmful chemicals.

6.3 HEAVY METAL UPTAKE IN THE AGRICULTURAL SYSTEM

In recent times, soil contamination through heavy metals has emerged as a major global environmental issue (Gratão et al., 2015). Various factors contribute to the contamination, which can be attributed to a range of causes. These include the expansion of industries, mining operations, the combustion of fossil fuels, unintentional discharge of petrochemicals, incorrect disposal of waste containing high levels of metals (like batteries), as well as agricultural methods and deposition from the atmosphere (Khan et al., 2008). The utilization of anthropogenic materials like fertilizers, wastewater, pesticides, and biosolids has led to an overload of pollutants in the system, disrupting the natural balance. As a result, agricultural practices have seen a significant accumulation of potentially harmful heavy metals in soil, posing a threat to plant health and soil biology, and even entering the food chain through biomagnification.

6.3.1 Sources of Heavy Metals in Agricultural Soil

Anthropogenic activities in recent times have disrupted and accelerated the natural geochemical cycle of metals, resulting in an imbalance. The elevated levels of heavy metals in agricultural soils have been associated with an imbalance resulting from various factors (Aransiola et al., 2022). These include the overuse of inorganic fertilizers, the application of biosolids from sludge manures, the discharge of wastewater, and the usage of pesticides in agricultural practices. Collectively, these practices contribute to the accumulation of heavy metals in the soil, impacting its overall health and productivity. As a result, plant metabolism can be adversely affected. The following are some of the frequently utilized sources contributing to this issue.

6.3.1.1 Inorganic Fertilizers

Macronutrients and micronutrients play a vital role in the complete life cycle of plants. In commercial farming, agricultural soils often lack essential nutrients, necessitating the addition of large quantities of fertilizers to promote optimal plant growth. However, improper or excessive fertilizer use contributes to environmental harm and pollution (Mirzaei and Sakizadeh, 2015). For example, nitrogen fertilizers can elevate Cd levels in both soil and plants. Additionally, phosphate fertilizers may introduce toxic elements like Pb, Cd, As, fluorine (F), and Hg (Radis Steinmetz et al., 2009; Guo et al., 2019) into the soil.

6.3.1.2 Biosolids

Biosolids, such as livestock manure, composts, and sewage sludge, are organic fertilizers widely used across the world to improve soil quality and restore damaged regions (Larney and Angers, 2012; Lu et al., 2012). These biosolids serve as a valuable source of organic carbon (Torri, 2014) and contribute to improving soil fertility (Scharenbroch et al., 2013; Tytła et al., 2016). However, improper application of biosolids in agricultural soils can result in the accumulation of heavy metals like Ni, Cu, As, Cr, antimony (Sb), Hg, thallium (Tl), Cd, and Pb (Basta et al., 2005).

6.3.1.3 Pesticides

The use of pesticides has significantly boosted agricultural and horticultural output by safeguarding plants from diseases and pests transmitted by vectors. However, an increased frequency of pesticide usage also poses a disadvantage to humanity. Compounds like Hg, Cu, Zn, Pb, and manganese (Mn) were commonly authorized for their insecticidal and fungicidal properties. Nevertheless, the unregulated application of pesticides can lead to environmental issues, particularly the contamination of agricultural soils with heavy metals.

6.3.2 Uptake of Heavy Metals in Plants

The interaction between plant species and the specific metal contaminants in the soil has a significant influence on the uptake of heavy metals by plants. Different plant species have varying abilities to accumulate and tolerate heavy metals. Some plants, known as hyperaccumulators, have the unique ability to concentrate metals at levels much higher than typical plants. This makes them highly suitable for phytoremediation purposes.

Plant roots play a crucial role in the initial absorption of heavy metals. Through root hairs and cell membranes, plants absorb water and dissolved nutrients, including heavy metals. This absorption can occur through active transport or passive diffusion. Active transport involves the use of specific transporters that facilitate the movement of heavy metals across the cell membranes. These transporters tend to be specific to certain types of heavy metals, allowing plants to selectively take up or exclude particular metals. The regulation of these transporters can be influenced by various factors, such as the concentration of heavy metals in the soil or the presence of other essential elements.

There are several commonly employed methods for measuring the quantity of heavy metals that accumulate in plant tissues. One such method is ICP-MS (inductively coupled plasma mass spectrometry), which is widely utilized for simultaneously detecting and quantifying multiple heavy metals. AAS (atomic absorption spectroscopy) and XRF (x-ray fluorescence spectroscopy) are other techniques that provide valuable information about the concentration of specific heavy metals in plant tissues. These methods rely on measuring the absorption or emission of light by heavy metal atoms, thereby offering precise quantitative data.

6.4 PLANT AND HEAVY METAL INTERACTIONS

Throughout the evolutionary process of angiosperms, a select group of 19 elements have played a crucial role in their basic metabolism. These elements include carbon (C), calcium (Ca), hydrogen (H), magnesium (Mg), nitrogen (N), oxygen (O), potassium (K), phosphorus (P), and sulfur (S), which are classified as macronutrients. Additionally, there are micronutrients such as Cu, boron (B), cobalt (Co), chlorine (Cl), Mn, Ni, iron (Fe), molybdenum (Mo), and Zn that have also been identified as essential for angiosperms' metabolic processes. One element that warrants attention is silicon (Si), which is considered beneficial for plant development. Studies have shown its involvement in maintaining plant structures in certain species. The presence of both macronutrients and micronutrients is vital for the physiological and biochemical functions of plants. These elements participate in various processes, including nitrogen fixation, photosynthesis, chlorophyll biosynthesis, redox reactions in chloroplasts and mitochondria, sugar metabolism, DNA synthesis, and protein modifications.

In recent years, there has been an increase in human actions, swift industrialization, and modern agricultural practices. Unfortunately, this has led to an increase in the contamination of the environment with heavy metals, causing harm to living organisms (Eapen and D'souza, 2005; Kavamura et al., 2010). Extensive areas of land have become polluted with heavy metals as a result of the use of pesticides, fertilizers, compost, and municipal wastes. Additionally, heavy metal release from metalliferous mines and smelting industries has contributed to this problem (Yang et al., 2005).

While certain heavy metals are naturally present in the crust of the Earth, their excessive release into the environment as a result of natural or human activity is harmful. High quantities of heavy metals can induce serious toxicity in plants. As a result, plants have developed tight mechanisms to regulate the uptake and utilization of these metals within their cells (Janicka-Russak et al., 2008; Singh et al., 2009; Saito et al., 2010; Farias et al., 2013). Despite being non-essential and lacking physiological function, some heavy metals such as Hg, aluminum (Al), Cd, Pb, and Cr are highly toxic even at very low concentrations (Ernst, 2006; Janicka-Russak et al., 2008). Types of heavy metals and their impact plants are provided in Table 6.2.

6.4.1 Uptake of Heavy Metals in Roots

The process governing the tolerance of plant cells to heavy metals consists of multiple stages. Initially, the plant roots absorb heavy metals from the soil. Subsequently, these heavy metals become bound to the cell walls. Following this, they are actively transported into the vacuole. To aid in the binding of metals, peptides are activated and metal complexes are synthesized

TABLE 6.2
Types of Heavy Metals and Their Impact on Plants

Heavy Metal	Occurrence	Impact on Plants	References
Cadmium (Cd)	Found in nature at low concentrations, mainly with lead, copper, and ores of zinc	• Plants exposed to high levels of Cd experience a reduction in photosynthesis, water uptake, and nutrient uptake • Causes visible symptoms of injury reflected in terms of chlorosis, growth inhibition, browning root tips, and finally death of the plant	(Salt et al., 1995; Wojcik and Tukiendorf, 2004; Mohanpuria et al., 2007)
Zinc (Zn)	• The 24th most common element in Earth's crust • There are many types of Zn ore, but the most common is sphalerite (Zn blend), made of Zn and sulfur.	• Results in stunted development and senescence • Zn poisoning leads to chlorosis in younger leaves, which can spread to older leaves following extended exposure to high soil Zn levels	(de Vries et al., 2002, Warne et al., 2008; Ebbs and Kochian, 1997; Fontes and Cox, 1998; Ebbs and Kochian, 1997)
Copper (Cu)	• Mixed with a variety of minerals, including chalcocite, chalcopyrite, bornite, cuprite, malachite, and azurite • Found in seaweed ashes, many marine corals, the human liver, and several mollusks and arthropods	• Excess Cu in soil may adversely affect plants by causing cytotoxicity, stress, and oxidative stress (reactive oxygen species [ROS]). • ROS disrupts metabolic processes and damages macromolecules	(Demirevska-kepova et al., 2004)
Mercury (Hg)	Occurs abundantly in the environment, chiefly as the minerals cinnabar and metacinnabar, and as impurities in other minerals in their inorganic state	• High levels of Hg^{2+} are very harmful to plant cells • High doses of Hg^{2+} can cause apparent damage and physiological abnormalities in plants	(Israr et al., 2006; Cargnelutti et al., 2006)
Chromium (Cr)	The 21st most commonly detected element in the earth's crust, with a mean concentration in US soils of around 40 mg/kg	• Excess Cr inhibits plant development, produces chlorosis in early leaves, and leads to nutritional imbalance, top wilting, and root damage • Impacts the overall dry matter production and yield of plants	(Sharma and Dubey, 2005; Chatterjee and Chatterjee, 2000; Scoccianti et al., 2008)
Nickel (Ni)	A transition metal that occurs naturally in the environment and is present in rocks, soils, and terrestrial and aquatic species	• In different plant species, it causes a variety of physiological changes and toxicity symptoms, including chlorosis and necrosis • It also shows disruption of nutritional balance, leading to disturbance of cell membrane function	(Pandey and Sharma, 2002)
Pb	• A natural bluish-grey metal found in the earth's crust • May be found in *a*ll parts of the environment • Much of it originates from human activities like burning fossil fuels, mining, and manufacturing	• Affects plant morphology, growth, and photosynthetic processes • Inhibits enzyme activity and changes membrane permeability • Causes oxidative stress in plants by raising ROS	(Sharma and Dubey, 2005)

(Holmberg et al., 1998; Memon and Schröder, 2009). The absorption of heavy metal ions is facilitated by channel proteins (Greipsson, 2011; Nguyen et al., 2021). When plants are exposed to heavy metals, they produce oligopeptide ligands such as phytochelatin (PC) and metallothionein (MT) in their tissues. These ligands are derived from glutathione and are subsequently converted into peptides. The ability to tolerate heavy metals is a crucial requirement for successful phytoremediation. Plants employ various methods to eliminate and tolerate heavy metals, each possessing unique characteristics that contribute to detoxification. These mechanisms assist in maintaining heavy metal concentrations within plant cells at levels that are not hazardous (Hall, 2002; Sharma and Dietz, 2006). Plants have various mechanisms for mitigating the effects of heavy metals in the rhizosphere, one of which involves the production of organic acids to detoxify them and decrease their availability. Additionally, heavy metals can be immobilized in the root system by binding them to negatively charged pectins present in cell walls and the interfaces between the cytoplasm and the membrane, thanks to their electrochemical potential. Some plants also possess the ability to reduce the valence of heavy metals through the production of redox enzymes, converting them into less harmful forms.

When it comes to selecting the most suitable plant species for phytoremediation, several factors must be taken into consideration. These include the plant's capacity to handle specific metals, its restorative properties, and its adaptability to site-specific conditions (Cunningham and Ow, 1996). Furthermore, desirable traits for phytoremediation plants include high biomass production, a well-developed root system, robust growth and productivity, tolerance to pollutants, susceptibility to genetic modification, and easy harvestability.

6.4.2 Heavy Metal Stress and Plant Responses

The occurrence of elevated levels of heavy metals in plants results in various alterations to their physical and metabolic functions (Villiers et al., 2011). However, because different heavy metals target distinct areas within the plant, the overall observable response to toxicity differs among them. The most prevalent visual indication of heavy metal toxicity is a reduction in plant growth (Sharma and Dubey, 2007), which encompasses leaf discoloration, wilting, loss of internal pressure, diminished seed germination, and impairment to the photosynthetic system, frequently leading to aging or plant demise (Foy et al., 1978; DalCorso et al., 2008; Tan et al., 2010; DalCorso et al., 2010). These effects are all associated with structural, biochemical, and molecular modifications in plant cells and tissues caused by the presence of heavy metals (Sandalio et al., 2001). Furthermore, heavy metals disrupt homeostatic processes such as water transport, evaporation, uptake, and nutrient metabolism (Carrier et al., 2003; Gamalero et al., 2009) and interfere with the absorption of essential elements like Ca, Mg, K, and P (Fodor, 2002).

The presence of high levels of heavy metals typically leads to a reduction in photosynthesis by directly impacting the photosynthetic machinery, particularly the thylakoids. This can result in the inhibition of enzymes that are involved in the fixation of carbon dioxide, leading to a decrease in C assimilation (Böddi et al., 1995). Moreover, heavy metals can disrupt carbohydrate metabolism and its distribution within growing plants. When heavy metals bind to the cell nucleus, they can cause damage to DNA, which includes modifications to DNA bases, crosslinking of DNA and proteins both within and between molecules, DNA strand breaks, rearrangements, and depurination (Skórzyńska-Polit et al., 2006). Additionally, heavy metals can impact the assembly and disassembly of microtubules, thereby altering the cell cycle and division (Zhen et al., 2007). When *Arabidopsis* plants are exposed to Cd, they exhibit a high rate of mutation and develop malformed embryos (Fusconi et al., 2006). Under stressful conditions, plants experience an increase in ethylene levels, which impacts various plant responses such as senescence and stress (Ernst et al., 2008). In higher plants, ethylene production induced by Cu can accelerate senescence (Deikman and Jill, 1997), increase cell wall rigidity, and inhibit cell growth through lignification (Maksymiec, 1998). The presence of heavy metals in plants can result in the accumulation of reactive oxygen species (ROS), which are harmful to plant cells. The production of ROS is dependent on the specific type of heavy metal present. The activity of ROS has the potential to disrupt cellular processes and contribute to senescence, along with ethylene and jasmonic acid. However, plants have the ability to activate antioxidant mechanisms to counteract the stress caused by heavy metals (Skórzyńska-Polit et al., 2006). The molecular response of plants to heavy metal stress involves the generation of stress-related amino acids, proteins, genes, and signaling molecules (Zhen et al., 2007; Zhao et al., 2011).

6.4.3 Molecular Mechanisms of Heavy Metal Tolerance in Plants

The ability of plants to tolerate specific heavy metals is regulated by a complex network of physiological, and molecular mechanisms. It is crucial to understand these mechanisms and their genetic foundation in order to develop plants for phytoremediation purposes (Clemens et al., 2002; DalCorso et al., 2008; Hossain et al., 2010; DalCorso et al., 2010). Different plant species may employ different mechanisms to tolerate excessive heavy metals, and even within the same species, multiple mechanisms may be at play. Plants possess both inherent and adaptive mechanisms to withstand high heavy metal levels (Meharg, 1994). Various physiological, biochemical, and molecular approaches are being used to determine the underlying mechanisms of heavy metal accumulation, tolerance, and adaptation to heavy metal stress. Some examples of adaptive mechanisms in tolerant plants include immobilization, exclusion of heavy metals from the cell membrane, restriction of heavy metal uptake and transport, production of specific heavy metal transporters, chelation and sequestration of heavy metals using ligands like PCs and MTs, and induction of mechanisms that counteract the effects of ROS and methylglyoxal (e.g., increased production of antioxidants and glyoxalase system). Other responses include the production of stress proteins, biosynthesis of proline and polyamines, and the release of signaling molecules such as salicylic acid and nitric oxide.

6.5 FACTORS AFFECTING HEAVY METALS UPTAKE BY PLANTS

As the primary class of soil contaminants, heavy metals, also known as PTEs, are extremely significant since they are non-degradable and their accumulation might harm the soil ecosystem (Muthusaravanan et al., 2018). As a result, because phytoremediation procedures are both ecologically benign and economically viable, great emphasis has been given to the remediation of PTE-polluted soils. The concentration of metals in soils, pH of soil, capability of cation exchange, organic matter content, plant species and varieties, microbial activity, water availability, and plant age are some of the variables that might influence plant absorption of metals (Jung, 2008).

6.5.1 Metals in Soils

Plants have differing affinities for different metals. Certain metals can be absorbed easily, while others might need special transport systems. For instance, hyperaccumulating plants, such as some *Brassica* species, can store significant amounts of heavy metals (Zn and Cd; Prasad and De Oliveira Freitas, 2003).

6.5.1.1 Soil pH

The pH of soil can also affect the metal uptake by plants by influencing its availability. Translocation factors values show that the pH of the soil has relatively little effect on the transport of metals inside plant tissues. The roots of dandelions and basil were able to absorb the metals due to the acidic soil's high mobility. On the other hand, the simplest environment was where the dandelion roots accumulated Mn the best. Within the neutral and basic treatments, lemon balm stimulated the absorption of Cu in above-ground portions (Adamczyk-Szabela et al., 2022).

6.5.1.2 Plant Species

The affinities of different plant species for various metals differ. Phytoremediation can only be effective if the right plant species are chosen based on the specific metal being targeted. It has been observed that certain *Thlaspi* species, such as *T. goesingense* for Ni and Zn; *T. rotundifolium* for Ni, Pb, and Zn; *T. caerulescens* for Ni, Cd, Pb, and Zn; and *T. ochroleucum* for Ni and Zn, have the ability to hyperaccumulate multiple metals (Prasad and Freitas, 2005; Lone et al., 2008). To ensure the success of the phytoextraction process, it is crucial to use established crop production and management techniques to identify suitable plant species that can hyperaccumulate heavy metals and produce significant biomass (Rodriguez et al., 2005).

6.5.1.3 Soil Organic Matter

The mobility and bioavailability of metals can be impacted by organic matter. It can create complexes with metals that can either help or impede plant roots in absorbing the metal (Kabata-Pendias, 2000; Angelova et al., 2010).

6.5.1.4 Root Characteristics

Surface area and the release of substances by roots have an impact on the uptake of metals. Research, such as the studies conducted on rhizofiltration, highlights the importance of root characteristics in metal absorption (Salt et al., 1995). Phytoremediation relies heavily on the root zone, which has the ability to absorb and either store or decompose pollutants within plant tissues. Another approach to phytoremediation involves the use of enzymes released by plant roots to break down pollutants in the soil. When plants experience drought stress, they adapt by increasing the diameter of their roots and decreasing their extension, as dry soil becomes less permeable (Merkel et al., 2005).

6.5.1.5 Concentration of Metals

Plant absorption of metals is directly impacted by the concentration of those metals in the soil. While low quantities may result in inadequate absorption for efficient phytoremediation, high amounts of metals may cause toxicity (Ma et al., 2001). Plant cells produce ROS and free radicals when accessible forms of heavy metals are present in high concentrations. Next, unchecked oxidation occurs, which sets off a series of events involving cellular macromolecules like nucleic acids, proteins, and lipids that ultimately result in oxidative stress and cellular damage (Goyal et al. (2020).

6.5.1.6 Chelating Agents

By adding biodegradable physicochemical components like micronutrients and chelating agents, one can improve the bioavailability of heavy metals and hence increase the absorption of those metals by energy crops. Research indicates that extending the duration of EDTA (ethylenediamine-tetraacetic acid) exposure to plants to 2 weeks can enhance both the overall phytoextraction efficacy and metal translocation inside plant tissue (Bieby Voijant Tangahu et al., 2011).

6.5.1.7 Competition with Other Ions

For absorption, plants may face competition from various ions. Elevated levels of vital nutrients or analogous ions may obstruct the absorption of the intended metal, which can lead to lower uptake of metals by plants (Leitenmaier and Küpper, 2013).

6.5.1.8 Temperature and Climate

Temperature and environment affect plant growth, which in turn affects plant absorption of metals. Temperature has an impact on soil microbial activity, which in turn influences metal availability and speciation (Giller et al., 1998). Temperature has an impact on growth factors, which in turn impacts root length. Under outdoor circumstances, the root structure is different from that in a greenhouse (Merkel et al., 2005).

6.5.1.9 Water Availability

Water availability is a crucial factor that affects the uptake of heavy metals by plants in phytoremediation. Sufficient water availability is essential for plant development and, by extension, for the absorption of metals. Metal translocation inside the plant is impacted by the availability of water.

6.5.1.10 Microbial Activity

By promoting plant development and assisting with metal absorption, transport, and dispersion, microbes can be extremely important to phytoremediation. Exudates from plant roots give soil microorganisms energy and nutrients, and they also form complex communication networks with them. As plant growth–promoting microorganisms (PGPMs), certain beneficial bacteria and fungi can reduce metal phytotoxicity and promote plant growth in two ways: either directly by solubilizing mineral nutrients (e.g., N, P, K, iron [Fe]), or indirectly by inducing defense mechanisms against phytopathogens and secreting particular enzymes. Through several processes, including acidification, precipitation, chelation, complexation, and redox reactions, PGPMs can also alter the bioavailability of metals in soil (Ma et al., 2016).

The effectiveness of phytoremediation activities depends on an understanding of and optimization of these variables since they impact how well plants remove metal from polluted environments.

6.6 PHYTOREMEDIATION APPROACHES FOR DIFFERENT METALS

The issue of heavy metal pollution caused by human activities is a significant global concern. Not only do heavy metals have negative effects on the environment and human health, but these effects have been extensively documented. The entry of heavy metals into the food chain poses several risks to human health (Shah and Daverey, 2020). Different techniques, including physical, chemical, and biological methods, are used to eliminate heavy metals and metalloids from soils (Sarwar et al., 2010). Among these techniques, phytoremediation has been proven to be an effective and cost-efficient strategy for removing heavy metals from soils. Phytoremediation involves using plants that can absorb and accumulate contaminants to remove, relocate, or reduce them. Traditional phytoremediation methods can be categorized into sub-classes based on their mode of action, such as phytofiltration, phytovolatilization, phytoextraction, and phytostabilization. Modern approaches to phytoremediation include chemical-assisted phytoremediation using non-hyperaccumulator plants, biochar-assisted phytoremediation, microbial-assisted phytoremediation, and the use of transgenic plants (Ali et al., 2013).

The traditional method of phytoremediation is more feasible than any other engineering remediation methods like soil incineration, flushing, excavation, soil washing, and solidification.

The traditional approach has been widely accepted by the public and can be implemented on a large-scale field site, unlike other physical remediation methods that are ineffective and expensive. The economic benefits of establishing green plants on contaminated soil are evident in several ways, including (1) phytostabilization; (2) phytoextraction of valuable metals such as Hg, silver (Ag), and Ni; and (3) sustainable land management.

6.6.1 Phytostabilization

Phytoremediation involves technologies that use higher plants to clean up and revegetate contaminated sites. Phytostabilization provides a plant cover on the surface of contaminated sites. That process is carried out to minimize the flow of contaminants within the zone affected by vadose by either accumulation by roots or immobilization inside the rhizosphere to decrease off-site contamination (Bolan et al., 2011). The process involves immobilizing contaminants via transpiration and root growth, lowering leaching, restricting erosion, providing an aerobic environment, and associating contaminants with organic matter in the substrate. Microbial activity in plant roots can reduce organic pollutants like pesticides and hydrocarbons into harmless forms. To improve phytostabilization, soil additives that immobilize metal(loid)s can be paired with plant species that can withstand high levels of contamination and low fertility, such as tailings and the uptake and absorption of heavy metals in the root system. Changes in soil parameters including pH, organic matter, and redox levels may impact pollutant speciation and immobilization. Root exudates govern precipitation and immobilize contaminants (Sarwar et al., 2017). Vegetation as a barrier reduces physical interactions with soil by animals and humans. Mechanical stabilizing minimizes erosion by wind and water, increasing evapotranspiration and reducing contaminant leaching. Table 6.3 provides details of some plants that can be used for phytostabilization for various heavy metals. Phytostabilization is not a permanent solution as heavy metals remain in the soil, it involves only restricted movement, and it needs to be monitored regularly.

TABLE 6.3
Phytoremediation of Heavy Metals with Different Plant Species

Heavy Metal	Plant Species	Substrate	Containment Process	References
Arsenic (As)	• Pome fruit trees • Mustard (*Brassica juncea*)	• Orchard soil • Tailings	• Phosphate-induced desorption followed by plant uptake • Phytoextraction	(Salt et al., 1995; Wojcik and Tukiendorf, 2004; Mohanpuria et al., 2007)
Cadmium (Cd)	• *B. juncea* • *Solanum nigrum* L. • Canola (*Brassica napus*) • Radish (*Raphanus sativus*) • *Brassica rapa* • *Cannabis sativa* • *Sedum alfredii* • *Echinochloa*	• Variable charge soil • Farmland soil containing Cd in the unpolluted surface layer • Multimetal contaminated • Calcareous dredged sediment • Mining soil	• Phytoimmobilization • Improved plant growth and Cd uptake by fungi and citric acid • Phytoextraction • Chelation followed by uptake	
Chromium (Cr)	• *B. napus* • *R. sativus* • *B. juncea*	• Multimetal contaminated • Radium (Ra) production site contaminated with Ra, uranium (U) and heavy metals	Chelation followed by uptake and phytoextraction	(Bolan et al., 2003)
Cr(VI)	*B. juncea*	Mineral soil	Reduction	
Copper (Cu)	• *B. napus* • *R. sativus* • *B. juncea* • *B. rapa* • *C. sativa* • *Agrostis tenuis* • Rice (*Oryza sativa* L.) • *Echinochloacrus-galli*	• Multimetal contaminated • Calcareous dredged sediment • Mining and pasture soil • Paddy soils contaminated by several toxic metals under aerobic soil conditions • Ra production site contaminated with Ra, U, and heavy metals	• Chelation followed by uptake • Root exudates–enhanced phytoextraction	(Demirevska-kepova et al., 2004)
Mercury (Hg)	• *B. juncea* • Transgenic tobacco • Ryegrass (*Lolium perenne*)	• Base metal tailings • Metal mine soil • Tailings dam of the abandoned	• Chelation followed by uptake • Phytoextraction followed by phytovolatilization	(Moreno et al., 2005)
Nickel (Ni)	• *B. napus* • *R. sativus* • *B. rapa* • *C. sativa*	• Multimetal contaminated • Calcareous dredged sediment	• Phytoextraction • Chelation followed by uptake	(Pandey and Sharma, 2002)

(Continued)

TABLE 6.3 *(Continued)*
Phytoremediation of Heavy Metals with Different Plant Species

Heavy Metal	Plant Species	Substrate	Containment Process	References
Lead (Pb)	• *B. napus* • *R. sativus* • *B. juncea* • *B. rapa* • *C. sativa* • *S. alfredii* • *O. sativa* L. • Alfalfa (*Medicago sativa*) • Maize (*Zea mays* L.) • *E. crus-galli*	• Multimetal contaminated • Calcareous dredged sediment • Mining soil • Paddy soils contaminated by several toxic metals under aerobic soil conditions • Hydroponics media • Artificially polluted by 800 mg (N03)2.Pb(N03)2. • Ra production site contaminated with Ra. U. and heavy metals	• Improved metal uptake by plant growth regulators (GA3 and IAA) and ethylenediaminetetraacetic acid (EDTA) • Root exudates–enhanced phytoextraction • Chelation followed by uptake and phytoextraction	(Sharma and Dubey, 2005)
Se	• *Lycopersicon esculentum* • *Brassica oleracea* • *Festuca arundinacea* • *M. sativa* • *Astragalus bisulcatus* • Indian mustard (*Brassica* L.)	• Se added soil • Soil amended with dimethylselenoniopropanoate, selenomethionine, sodium selenite, or sodium selenate	Phytovolatilization	(Souza et al., 2000)
Uranium (U)	• *B. juncea*	• Ra production site contaminated with Ra, U, and heavy metals	Chelation followed by uptake and phytoextraction	(Duquène et al., 2009)
Zinc (Zn)	• *B. napus* • *R. sativus* • *B. juncea* • *S. alfredii* • *O. sativa* L. • *Helianthus annuus* • *Z. mays* • *L. perenne*	• Multimetal contaminated • Mining soil • Paddy soils contaminated by several toxic metals under aerobic soil conditions • Derived surface soil • Tailings dam of the abandoned	Chelation followed by uptake and phytoextraction	(Duquène et al., 2009)

6.6.2 Phytovolatilization and Phytoextraction

Phytovolatilization converts metal into a volatile form and releases it into the atmosphere through stomata (Ghosh and Singh, 2005). This process converts mercuric ions into less hazardous elemental forms, making it especially advantageous for Hg. This technology offers an interim solution as the volatile form of Hg released into the atmosphere can be recycled back into the soil through precipitation.

Phytoextraction is the process by which plant roots absorb pollutants from soil or water and transfer them to any harvestable plant section. Phytoextraction has the potential to remove toxins from soil or wastewater, promoting long-term cleaning. The viability of phytoextraction as a potential environmental remediation solution is determined by elements such as metal availability for uptake and plants' ability to absorb and deposit metals in aerial portions.

Traditional methods of phytoremediation come with certain limitations that include the following:

1. It takes time to remediate contaminated soil.
2. Hyperaccumulator plants have limited ability to extract contaminants due to low aboveground biomass.

3. Only a small portion of metals in the soil are bioavailable, and this fluctuates depending on factors such as soil pH, organic matter content, competitive cations, and calcareousness.
4. It can only be applied to locations with minimal or moderate contamination.
5. There is a lack of knowledge about agronomy, potential development, and insect pests.

However, modern solutions have been developed to address these challenges. These include chemical-assisted phytoremediation using non-hyperaccumulator plants, biochar-assisted phytoremediation, microbial-assisted phytoremediation, and the use of transgenic plants.

6.6.3 Chemical-Assisted Phytoremediation Using Non-Hyperaccumulator Plants

Plants that have the ability to accumulate high levels of heavy metals in arid regions are known as hyperaccumulator plants. These plants tend to store more metals in their shoots rather than their roots. However, one major drawback of using these plants is their slow growth and low above-ground biomass production. Due to these limitations, hyperaccumulator plants are not feasible for large-scale phytoremediation projects. Nevertheless, they can serve as a valuable model for developing genetically modified plants that can both accumulate higher metal concentrations and produce larger amounts of biomass. Therefore, it is important to study non-hyperaccumulator plants that have higher above-ground biomass for phytoremediation purposes. To enhance the extraction of heavy metals from these plants, chelate-assisted or induced phytoextraction methods can be employed, which involve the use of synthetic or organic chelating agents. Synthetic chelating compounds such as EDTA, diethylenetriamine pentaacetic acid (DTPA), and ethylene glycol tetraacetic acid (EGTA) have been successfully utilized to increase metal bioavailability and improve plant uptake. However, it is important to note that this technique has its limitations, particularly due to the high mobility of heavy metals.

6.6.4 Biochar-Assisted Phytoremediation

Biochar is a carbonaceous porous substance generated through the combustion of organic sources of biomass such as plant materials, organic manures, and sludges. Biochar possesses some specific physicochemical properties, i.e. high pH, large surface area for sorption of metals, alkaline nature, ash, carbon contents, and ability to immobilize toxic heavy metals, which place it among good heavy metal remediators (Serrano et al., 2024). The large surface area available for sorption aids in the formation of heavy metal-biochar complexes, either through the exchange of heavy metal cations with other metal cations (e.g., sodium [Na], K, Ca^{2+}, Mg^{2+}) at functional groups found in biochar or through physical adsorption (Lebrun et al., 2022; Lu et al., 2012). Biochar aids in the phytoremediation of heavy metal–contaminated wastes. Pyrolysis is used to dispose of polluted biomass (Ghosh and Maiti, 2020).

6.6.5 Use of Transgenic Plants for Phytoremediation

Genetic manipulation has significantly contributed to the field of phytoremediation. Several transgenic plants have been created for this purpose, as detailed in Table 6.4, which provides a list of these plants and their corresponding outcomes. Generally, transgenic plants are engineered to enhance metal immobilization and increase the plant's ability to tolerate heavy metals. This enables better translocation and accumulation of these metals in the plant's above-ground parts. As shown in Table 6.4, certain plants have demonstrated an enhanced capacity to absorb h0eavy metals, while others have displayed resistance to pesticides that contain such metals.

TABLE 6.4
Different Transgenic Plants Involved in Phytoremediation

Recipient Plant	Target Gene	Result in Field Trials	References
Cottonwood	• Mercuric ion reductase (merA)	• Volatilization and increase of mercury (Hg) resistance	(Ruiz and Daniell, 2009)
Indian mustard	• ATP sulfurylase • Gamma-glutamylcysteine synthetase • Glutathione synthetase • Selenocysteine lyase (cpSL) • Selenocysteine methyltransferase (SMT)	• APS (ATP Sulfurylase), ECS (g-glutamyl cysteine synthetase), and GS (Glutathione Synthetases) increases of selenium (Se) accumulation	(Misra and Gedamu, 1989)
Poplar	• Glutathione biosynthesis (gsh1)	• Change in morphology of chloroplast	(Moreno et al., 2005)
Rice	• Cytochrome (CYP2B6) • Nicotianamine synthase (NAS) • Cytochromes (CYP1A1, 2B6, 2C19) • Iron (Fe) transporter (OsIRT1) • Cytochromes (CYP1A1, 2B6, 2C19)	• Herbicide removal from soil • Tolerance to low Fe in soil • Removal of herbicides from soil • Increase in Fe and zinc (Zn) content • Removal of herbicides from soil	(Ruiz and Daniell, 2009)
Tobacco	• Mammalian metallothionein	• Decrease of cadmium (Cd) content in leaf	(De Borne et al., 1998)
Tomato	• Esterase	• Resistance to thiazopyr	(Feng et al. 1997)

6.7 CONCLUSION AND FUTURE PERSPECTIVES

This chapter dives into the changing mechanisms of phytoremediation, as well as the complicated processes that govern heavy metal uptake by plants. In future research, significant discoveries will offer fresh routes that have the potential to change the phytoremediation industry. Integrating nanotechnology, genetic engineering, microbiome-induced phytoremediation, smart sensors, monitoring technologies, and climate change adaptation in phytoremediation highlights a multifaceted approach to environmental remediation. This comprehensive strategy underscores the importance of policy and societal integration to ensure the successful implementation and acceptance of these innovative solutions.

In a nutshell, there are many exciting things that lie ahead for phytoremediation, such as developing genetically modified plants and integrating state-of-the-art technologies. The changing processes and plant uptake of heavy metals in phytoremediation will surely help to shape a better and cleaner environment for future generations as we move closer to a more sustainable future.

REFERENCES

Adamczyk-Szabela, D., and W. M. Wolf. "The impact of soil pH on heavy metals uptake and photosynthesis efficiency in melissa officinalis, taraxacum officinalis, ocimum basilicum." *Molecules (Basel, Switzerland)* 27, no. 15 (2022): 4671. https://doi.org/10.3390/molecules27154671

Ali, Hazrat, Ezzat Khan, and Muhammad Anwar Sajad. "Phytoremediation of heavy metals – concepts and applications." *Chemosphere* 91, no. 7 (May 2013): 869–881. https://doi.org/10.1016/j.chemosphere.2013.01.075.

Angelova, V., R. Ivanova, G. Pevicharova, K. Ivanov, R. J. Gilkes, and N. Prakongkep. "Effect of organic amendments on heavy metals uptake by potato plants," January 1, 2010, 84–87. https://www.cabdirect.org/abstracts/20113337421.html.

Aransiola, S.A., U. J. J. Ijah, and O. P. Abioye. "Phytoremediation of lead polluted soil by glycine max L." *Applied and Environmental Soil Science* (2013), Article ID 631619. doi: 10.1155/2013/631619. https://www.hindawi.com/journals/aess/2013/631619/abs/.

Aransiola, S.A., U. J. J. Ijah, O. P. Abioye, and J. D. Bala. "Microbial-aided phytoremediation of heavy metals contaminated soil: A review." *European Journal of Biological Research* 9, no. 2 (2019): 104–125. http://www.journals.tmkarpinski.com/index.php/ejbr/article/view/157.

Aransiola, S. A., U. J. J. Ijah, O. P. Abioye, and J. D. Bala. "Vermicompost-assisted phytoremediation of toxic trace element-contaminated soil in Madaka, Nigeria using Melissa officinalis L and Sida acuta." *International Journal of Environmental Science and Technology* (2022). http://dx.doi.org/10.1007/s13762-022-04105-y

Baker, Alan J. M. "Metal tolerance." *New Phytologist* 106 (1987): 93–111.

Basta, N. T., J. A. Ryan, and R. L. Chaney. "Trace element chemistry in residual-treated soil: Key concepts and metal bioavailability." *Journal of Environmental Quality* 34, no. 1 (2005): 49–63.

Bieby Voijant Tangahu, Siti Rozaimah Sheikh Abdullah, Hassan Basri, Mushrifah Idris, Nurina Anuar, Muhammad Mukhlisin. "A review on heavy metals (As, Pb, and Hg) Uptake by plants through phytoremediation." *International Journal of Chemical Engineering* (2011), Article ID 939161, 31 pages, 2011. https://doi.org/10.1155/2011/939161

Böddi, B., A. R. Oravecz, and E. Lehoczki. "Effect of cadmium on organization and photoreduction of protochlorophyllide in dark-grown leaves and etioplast inner membrane preparations of wheat." *Photosynthetica (Czech Republic)* (1995): 67–72.

Bolan, N. S., D. C. Adriano, R. Natesan, and B.-J. Koo. "Effects of organic amendments on the reduction and phytoavailability of chromate in mineral soil." *Journal of Environmental Quality* 32, no. 1 (January 2003): 120–128. https://doi.org/10.2134/jeq2003.1200.

Bolan, N. S., Jin Hee Park, Brett Robinson, Ravi Naidu, and Keun Young Huh. "Phytostabilization." *Advances in Agronomy* (2011): 145–204. https://doi.org/10.1016/b978-0-12-385538-1.00004-4.

Cargnelutti, Denise, Luciane Almeri Tabaldi, Rosélia Maria Spanevello, Gladis de Oliveira Jucoski, Vanessa Battisti, Marciel Redin, Carlos Eduardo Blanco Linares et al. "Mercury toxicity induces oxidative stress in growing cucumber seedlings." *Chemosphere* 65, no. 6 (2006): 999–1006.

Carrier, Patrick, Aurore Baryla, and Michel Havaux. "Cadmium distribution and microlocalization in oilseed rape (Brassica napus) after long-term growth on cadmium-contaminated soil." *Planta* 216 (2003): 345–939.

Chatterjee, J., and C. Chatterjee. "Phytotoxicity of cobalt, chromium and copper in cauliflower." *Environmental Pollution* 109, no. 1 (2000): 69–74.

Clemens, Stephan, Michael G. Palmgren, and Ute Krämer. "A long way ahead: Understanding and engineering plant metal accumulation." *Trends in Plant Science* 7, no. 7 (2002): 309–315.

Cunningham, Scott D., and David W. Ow. "Promises and prospects of phytoremediation." *Plant Physiology* 110, no. 3 (1996): 715.

DalCorso, Giovanni, Silvia Farinati, and Antonella Furini. "Regulatory networks of cadmium stress in plants." *Plant Signaling & Behavior* 5, no. 6 (2010): 663–667.

DalCorso, Giovanni, Silvia Farinati, Silvia Maistri, and Antonella Furini. "How plants cope with cadmium: Staking all on metabolism and gene expression." *Journal of Integrative Plant Biology* 50, no. 10 (2008): 1268–1280.

De Borne, François Dorlhac, Taline Elmayan, Christian De Roton, Louis De Hys, and Mark Tepfer. "No Title." Molecular Breeding 4, no. 2 (January 1, 1998): 83–90. https://doi.org/10.1023/a:1009669412489.

De Vries, Jacobus J., and Ian Simmers. "Groundwater recharge: An overview of processes and challenges." *Hydrogeology Journal* 10 (2002): 5–17.

Dec, Jerzy, and Jean-Marc Bollag. "Use of plant material for the decontamination of water polluted with phenols." *Biotechnology and Bioengineering* 44, no. 9 (1994): 1132–1139.

Deikman, and Jill. "Molecular mechanisms of ethylene regulation of gene transcription." *Physiologia Plantarum* 100, no. 3 (1997): 561–566.

Demirevska-Kepova, K., Lyudmila Simova-Stoilova, Z. Stoyanova, Regina Hölzer, and Urs Feller. "Biochemical changes in barley plants after excessive supply of copper and manganese." *Environmental and Experimental Botany* 52, no. 3 (2004): 253–266.

Duquène, L., H. Vandenhove, F. Tack, E. Meers, J. Baeten, and J. Wannijn. "Enhanced phytoextraction of uranium and selected heavy metals by Indian mustard and ryegrass using biodegradable soil amendments." *Science of the Total Environment* 407, no. 5 (February 2009): 1496–1505. https://doi.org/10.1016/j.scitotenv.2008.10.049.

Eapen, Susan, and S. F. D'souza. "Prospects of genetic engineering of plants for phytoremediation of toxic metals." *Biotechnology Advances* 23, no. 2 (2005): 97–114.

Ebbs, Stephen D., and Leon V. Kochian. "Toxicity of zinc and copper to Brassica species: Implications for phytoremediation." *Journal of Environmental Quality* 26, no. 3 (May 1997): 776–781. https://doi.org/10.2134/jeq1997.00472425002600030026x.

Ernst, W. H. O. "Bioavailability of heavy metals and decontamination of soils by plants." *Applied Geochemistry* 11, no. 1–2 (1996): 163–167.

Ernst, Wilfried H. O. "Evolution of metal tolerance in higher plants." *Forest Snow and Landscape Research* 80, no. 3 (2006): 251–274.

Ernst, Wilfried H. O., Gerd-Joachim Krauss, Jos AC Verkleij, and Dirk Wesenberg. "Interaction of heavy metals with the sulphur metabolism in angiosperms from an ecological point of view." *Plant, Cell & Environment* 31, no. 1 (2008): 123–143.

Farias, Júlia G., Fabiane L. G. Antes, Pedro A. A. Nunes, Sibila T. Nunes, Gabriel Schaich, Liana V. Rossato, and Alcione Miotto et al. "Effects of excess copper in vineyard soils on the mineral nutrition of potato genotypes." *Food and Energy Security* 2, no. 1 (2013): 49–69.

Feng, Paul C.C, Thomas G Ruff, Shaukat H Rangwala, and Sudabathula R Rao. "Engineering Plant Resistance to Thiazopyr Herbicide via Expression of a Novel Esterase Deactivation Enzyme." *Pesticide Biochemistry and Physiology* 59, no. 2 (October 1, 1997): 89–103. https://doi.org/10.1006/pest.1997.2312.

Fodor, F. "Physiological responses of vascular plants to heavy metals." *Physiology and Biochemistry of Metal Toxicity and Tolerance in Plants* (2002): 51–67, 149–177.

Fontes, R. L. F., and F. R. Cox. "Zinc toxicity in soybean grown at high iron concentration in nutrient solution." *Journal of Plant Nutrition* 21, no. 8 (1998): 1723–1730.

Foy, C. D., RL T. Chaney, and M. C. White. "The physiology of metal toxicity in plants." *Annual Review of Plant Physiology* 29, no. 1 (1978): 511–566.

Fusconi, Anna, Ombretta Repetto, Elisa Bona, Nadia Massa, Cristina Gallo, Eliane Dumas-Gaudot, and Graziella Berta. "Effects of cadmium on meristem activity and nucleus ploidy in roots of Pisum sativum L. cv. Frisson seedlings." *Environmental and Experimental Botany* 58, no. 1–3 (2006): 253–260.

Gamalero, Elisa, Guido Lingua, Graziella Berta, and Bernard R. Glick. "Beneficial role of plant growth promoting bacteria and arbuscular mycorrhizal fungi on plant responses to heavy metal stress." *Canadian Journal of Microbiology* 55, no. 5 (2009): 501–514.

Ghosh, Dipita, and Subodh Kumar Maiti. "Biochar assisted phytoremediation and biomass disposal in heavy metal contaminated mine soils: A review." *International Journal of Phytoremediation* (November 11, 2020): 1–18. https://doi.org/10.1080/15226514.2020.1840510.

Ghosh, Mridul, and S. P. Singh. "A review on phytoremediation of heavy metals and utilization of it's by products." *Asia Pacific Journal of Energy and Environment* 6, no. 4 (2005): 18.

Giller, K. E., E. Witter, and S. P. Mcgrath. "Toxicity of heavy metals to microorganisms and microbial processes in agricultural soils: A review." *Soil Biology and Biochemistry* 30, no. 10–11 (1998): 1389–1414. https://doi.org/10.1016/S0038–0717(97)00270–8

Glass, David J. "Current market trends in phytoremediation." *International Journal of Phytoremediation* 1, no. 1 (1999): 1–8.

Goyal, D., et al. "Effect of heavy metals on plant growth: An overview." In: M. Naeem, A. Ansari, and S. Gill (eds.), *Contaminants in Agriculture*. Springer, Cham, 2020. https://doi.org/10.1007/978-3-030-41552-5_4

Gratão, Priscila Lupino, Carolina Cristina Monteiro, Tiago Tezotto, Rogério Falleiros Carvalho, Letícia Rodrigues Alves, Leila Priscila Peters, and Ricardo Antunes Azevedo. "Cadmium stress antioxidant responses and root-to-shoot communication in grafted tomato plants." *Biometals* 28, no. 5 (2015): 803–816.

Greipsson, S. "Phytoremediation. Nat. Educ." *Google Scholar* (2011): 64–72.

Guo, Qian, Sabahat Majeed, Rong Xu, Kai Zhang, Apurva Kakade, Aman Khan, Fauzia Yusuf Hafeez, Chunlan Mao, Pu Liu, and Xiangkai Li. "Heavy metals interact with the microbial community and affect biogas production in anaerobic digestion: A review." *Journal of Environmental Management* 240 (2019): 266–272.

Hall, J. áL. "Cellular mechanisms for heavy metal detoxification and tolerance." *Journal of Experimental Botany* 53, no. 366 (2002): 1–11.

Hamelink, Jerry, Harold Bergman, and Peter F. Landrum. *Bioavailability: Physical, Chemical, and Biological Interactions*, CRC Press, 1994. http://ci.nii.ac.jp/ncid/BA23248199.

Holmberg, Niklas, and Leif Bülow. "Improving stress tolerance in plants by gene transfer." *Trends in Plant Science* 3, no. 2 (1998): 61–66.

Hossain, Mohammad Anwar, Mirza Hasanuzzaman, and Masayuki Fujita. "Up-regulation of antioxidant and glyoxalase systems by exogenous glycine betaine and proline in mung bean confer tolerance to cadmium stress." *Physiology and Molecular Biology of Plants* 16 (2010): 259–272.

Israr, M., S. V. Sahi, and J. Jain. "Cadmium accumulation and antioxidative responses in the Sesbania drummondii callus." *Archives of environmental contamination and toxicology* 50 (2006): 121–127.

Jabeen, Ruqaya, Altaf Ahmad, and Muhammad Iqbal. "Phytoremediation of heavy metals: Physiological and molecular mechanisms." *The Botanical Review* 75 (2009): 339–364.

Janicka-Russak, Małgorzata, Katarzyna Kabała, Marek Burzyński, and Grażyna Kłobus. "Response of plasma membrane H+-ATPase to heavy metal stress in Cucumis sativu s roots." *Journal of Experimental Botany* 59, no. 13 (2008): 3721–3728.

Jeng, A. S., and B. R. Singh. "Cadmium status of soils and plants from a long-term fertility experiment in southeast Norway." *Plant and Soil* 175 (1995): 67–74.

Jung, M. C. "Heavy metal concentrations in soils and factors affecting metal uptake by plants in the vicinity of a Korean Cu-W Mine." *Sensors (Basel, Switzerland)* 8, no. 4 (2008): 2413–2423. https://doi.org/10.3390/s8042413

Kabata-Pendias, A. *Trace Elements in Soils and Plants* (3rd ed.). CRC Press, 2000. https://doi.org/10.1201/9781420039900

Kang, Zhiming, Maojian Gong, Yinshi Li, Weizhen Chen, Yanan Yang, Junhao Qin, and Huashou Li. "Low Cd-accumulating rice intercropping with Sesbania cannabina L. reduces grain Cd while promoting phytoremediation of Cd-contaminated soil." *Science of the Total Environment* 800 (2021): 149600.

Kavamura, Vanessa Nessner, and Elisa Esposito. "Biotechnological strategies applied to the decontamination of soils polluted with heavy metals." *Biotechnology Advances* 28, no. 1 (2010): 61–69.

Khan, S., Q. Cao, Y. M. Zheng, Y. Z. Huang, and Y. G. Zhu. "Health risks of heavy metals in contaminated soils and food crops irrigated with wastewater in Beijing, China." *Environmental pollution* 152, no. 3 (2008): 686–692.

Larney, Francis J., and Denis A. Angers. "The role of organic amendments in soil reclamation: A review." *Canadian Journal of Soil Science* 92, no. 1 (2012): 19–38.

Lebrun, Manhattan, Romain Nandillon, Florie Miard, Sylvain Bourgerie, and Domenico Morabito. "Biochar assisted phytoremediation for metal(loid) contaminated soils." *Assisted Phytoremediation* (2022): 101–130. https://doi.org/10.1016/b978-0-12-822893-7.00010-0.

Leitenmaier, B., and H. Küpper. "Compartmentation and complexation of metals in hyperaccumulator plants." *Frontiers in Plant Science* 4 (2013): 374. https://doi.org/10.3389/fpls.2013.00374

Liu, Zhongchuang, Boning Chen, Li-ao Wang, Oksana Urbanovich, Liubov Nagorskaya, Xiang Li, and Li Tang. "A review on phytoremediation of mercury contaminated soils." *Journal of Hazardous Materials* 400 (2020): 123138.

Lone, M. I., Z. L. He, P. J. Stoffella, and X. E. Yang. "Phytoremediation of heavy metal polluted soils and water: Progresses and perspectives." *Journal of Zhejiang University. Science. B* 9, no. 3 (2008): 210–220. https://doi.org/10.1631/jzus.B0710633

Lu, Anxiang, Jihua Wang, Xiangyang Qin, Kaiyi Wang, Ping Han, and Shuzhen Zhang. "Multivariate and geostatistical analyses of the spatial distribution and origin of heavy metals in the agricultural soils in Shunyi, Beijing, China." *Science of the Total Environment* 425 (2012): 66–74.

Ma, L., K. Komar, C. Tu et al. "A fern that hyperaccumulates arsenic." *Nature* 409, no. 579 (2001). https://doi.org/10.1038/35054664

Ma, Y., R. S. Oliveira, H. Freitas, and C. Zhang. "Biochemical and molecular mechanisms of plant-microbe-metal interactions: Relevance for phytoremediation." *Frontiers in Plant Science* 7 (2016): 918. doi: 10.3389/fpls.2016.00918

Maksymiec, W. "Effect of copper on cellular processes in higher plants." *Google Scholar* (1998): 341–342.

McCutcheon, Steven C., N. Lee Wolfe, L. Carreria, and T. Ou. "Phytoremediation of hazardous wastes." In *Innovative Technologies for Site Remediation and Hazardous Waste Management: Proceedings of the National Conference*. American Society of Civil Engineers, New York, 1995, 597–604.

Meharg, A. A. "Integrated tolerance mechanisms: Constitutive and adaptive plant responses to elevated metal concentrations in the environment." *Plant, Cell & Environment* 17, no. 9 (1994): 989–993.

Memon, Abdul R., and Peter Schröder. "Implications of metal accumulation mechanisms to phytoremediation." *Environmental Science and Pollution Research* 16 (2009): 162–175.

Merkel, N., R. Schultze-Kraft, and C. Infante. "Phytoremediation in the tropics–influence of heavy crude oil on root morphological characteristics of graminoids." *Environmental Pollution* 138, no. 1 (2005): 86–91https://doi.org/10.1016/j.envpol.2005.02.023

Mildvan, Albert S. *9 Metals in Enzyme Catalsis*. Academic Press, Chicago, 1970.

Mirzaei, Rouhollah, and Mohamad Sakizadeh. "Comparison of interpolation methods for the estimation of groundwater contamination in Andimeshk-Shush Plain, Southwest of Iran." *Environmental Science and Pollution Research* 23, no. 3 (October 7, 2015): 2758–69. https://doi.org/10.1007/s11356-015-5507-2.

Misra, S., and L. Gedamu. "Heavy metal tolerant transgenic Brassica napus L. and Nicotiana tabacum L. plants." *Theoretical and Applied Genetics* 78 (1989): 161–168.

Mohanpuria, Prashant, Nisha K. Rana, and Sudesh Kumar Yadav. "Cadmium induced oxidative stress influence on glutathione metabolic genes of Camellia sinensis (L.) O. Kuntze." *Environmental Toxicology: An International Journal* 22, no. 4 (2007): 368–374.

Moreno, Fabio N., Chris W.N. Anderson, Robert B. Stewart, and Brett H. Robinson. "Mercury volatilisation and phytoextraction from base-metal mine tailings." *Environmental Pollution* 136, no. 2 (July 2005): 341–352. https://doi.org/10.1016/j.envpol.2004.11.020.

Muthusaravanan, S., N. Sivarajasekar, J. S. Vivek et al. "Phytoremediation of heavy metals: Mechanisms, methods, and enhancements." *Environmental Chemistry Letters* 16 (2018): 1339–1359. https://doi.org/10.1007/s10311-018-0762-3

Nguyen, Thien Q., Verena Sesin, Anna Kisiala, and R. J. Neil Emery. "Phytohormonal roles in plant responses to heavy metal stress: Implications for using macrophytes in phytoremediation of aquatic ecosystems." *Environmental Toxicology and Chemistry* 40 (2021): 7–21.

Pandey, Nalini, and Chandra Prakash Sharma. "Effect of heavy metals Co2+, Ni2+ and Cd2+ on growth and metabolism of cabbage." *Plant Science* 163, no. 4 (2002): 753–758.

Prakash, Priyanka. "Nano-phytoremediation of heavy metals from soil: A critical review." *Pollutants* 3, no. 3 (2023): 360–380.

Prasad, M.N.V., Kenneth S. Sajwan, and Ravi Naidu. *Trace Elements in the Environment*. CRC Press, 2005. http://books.google.ie/books?id=TRzNBQAAQBAJ&printsec=frontcover&dq=9780429141560&hl=&cd=1&source=gbs_api.

Prasad, M. N. V., and H. M. De Oliveira Freitas. "Metal hyperaccumulation in plants – biodiversity prospecting for phytoremediation technology." *Electronic Journal of Biotechnology* 6, no. 3 (2003): 110–146. ISSN: 0717–3458.

Radis Steinmetz, Ricardo Luis, Airton Kunz, Valderi Luiz Dressler, Érico Marlon de Moraes Flores, and Ayrton Figueiredo Martins. "Study of metal distribution in raw and screened swine manure." *CLEAN–Soil, Air, Water* 37, no. 3 (2009): 239–244.

Raskin, Ilya, Robert D. Smith, and David E. Salt. "Phytoremediation of metals: Using plants to remove pollutants from the environment." *Current Opinion in Biotechnology* 8, no. 2 (1997): 221–226.

Rodriguez, L., F. J. Lopez-Bellido, A. Carnicer, F. Recreo, A. Tallos, and J. M. Monteagudo. "Mercury recovery from soils by phytoremediation." In: E. Lichtfouse, J. Schwarzbauer, and D. Robert (eds.), *Environmental Chemistry*. Springer, Berlin, Heidelberg, 2005. https://doi.org/10.1007/3-540-26531-7_18

Ruiz, Oscar N, and Henry Daniell. "Genetic engineering to enhance mercury phytoremediation." *Current Opinion in Biotechnology* 20, no. 2 (April 2009): 213–219. https://doi.org/10.1016/j.copbio.2009.02.010.

Saito, Akihiro, Misa Saito, Yusuke Ichikawa, Masaaki Yoshiba, Toshiaki Tadano, Eitaro Miwa, and Kyoko Higuchi. "Difference in the distribution and speciation of cellular nickel between nickel-tolerant and non-tolerant Nicotiana tabacum L. cv. BY-2 cells." *Plant, Cell & Environment* 33, no. 2 (2010): 174–187.

Salt, David E., Michael Blaylock, Nanda P. B. A. Kumar, Viatcheslav Dushenkov, Burt D. Ensley, Ilan Chet, and Ilya Raskin. "Phytoremediation: A novel strategy for the removal of toxic metals from the environment using plants." *Bio/technology* 13, no. 5 (1995): 468–474.

Salt, David E., R. D. Smith, and Ilya Raskin. "Phytoremediation." *Annual Review of Plant Biology* 49, no. 1 (1998): 643–668.

Sandalio, L. M., H. C. Dalurzo, M. Gomez, M. C. Romero-Puertas, and L. A. Del Rio. "Cadmium-induced changes in the growth and oxidative metabolism of pea plants." *Journal of Experimental Botany* 52, no. 364 (2001): 2115–2126.

Sarwar, Nadeem, Muhammad Imran, Muhammad Rashid Shaheen, Wajid Ishaque, Muhammad Asif Kamran, Amar Matloob, Abdur Rehim, and Saddam Hussain. "Phytoremediation strategies for soils contaminated with heavy metals: Modifications and future perspectives." Chemosphere 171 (March 2017): 710–721. https://doi.org/10.1016/j.chemosphere.2016.12.116.

Sarwar, Nadeem, Saifullah, Sukhdev S Malhi, Munir Hussain Zia, Asif Naeem, Sadia Bibi, and Ghulam Farid. "Role of mineral nutrition in minimizing cadmium accumulation by plants." *Journal of the Science of Food and Agriculture* 90, no. 6 (March 3, 2010): 925–937. https://doi.org/10.1002/jsfa.3916.

Scharenbroch, Bryant C., Elsa N. Meza, Michelle Catania, and Kelby Fite. "Biochar and biosolids increase tree growth and improve soil quality for urban landscapes." *Journal of Environmental Quality* 42, no. 5 (2013): 1372–1385.

Scoccianti, Valeria, Marta Iacobucci, Maria Filomena Paoletti, Alessandra Fraternale, and Anna Speranza. "Species-dependent chromium accumulation, lipid peroxidation, and glutathione levels in germinating kiwifruit pollen under Cr(III) and Cr(VI) stress." *Chemosphere* 73, no. 7 (October, 2008): 1042–1048. https://doi.org/10.1016/j.chemosphere.2008.07.083.

Serrano, María F., Julián E. López, Nancy Henao, and Juan F. Saldarriaga. "Phosphorus-loaded biochar-assisted phytoremediation to immobilize cadmium, chromium, and lead in soils." *ACS Omega* (January 9, 2024). https://doi.org/10.1021/acsomega.3c07433.

Shah, Vijendra, and Achlesh Daverey. "Phytoremediation: A multidisciplinary approach to clean up heavy metal contaminated soil." *Environmental Technology & Innovation* 18 (May 2020): 100774. https://doi.org/10.1016/j.eti.2020.100774.

Sharma, Pallavi, and R. S. Dubey. "Involvement of oxidative stress and role of the antioxidative defense system in growing rice seedlings exposed to toxic concentrations of aluminum." *Plant Cell Reports* 26 (2007): 2027–2028.

Sharma, Pallavi, and Rama Shanker Dubey. "Lead toxicity in plants." *Brazilian Journal of Plant Physiology* 17 (2005): 35–52.

Sharma, Shanti S., and Karl-Josef Dietz. "The significance of amino acids and amino acid-derived molecules in plant responses and adaptation to heavy metal stress." *Journal of Experimental Botany* 57, no. 4 (2006): 711–726.

Shikha, Deep, and Prasoon Kumar Singh. "In situ phytoremediation of heavy metal–contaminated soil and groundwater: A green inventive approach." *Environmental Science and Pollution Research* 28 (2021): 4104–4124.

Singh, N. P., and Anita Rani Santal. "Phytoremediation of heavy metals: The use of green approaches to clean the environment." *Phytoremediation: Management of Environmental Contaminants, Volume 2* (2015): 115–129.

Singh, Nandita, Lena Q. Ma, Joseph C. Vu, and Anshita Raj. "Effects of arsenic on nitrate metabolism in arsenic hyperaccumulating and non-hyperaccumulating ferns." *Environmental Pollution* 157, no. 8–9 (2009): 2300–2305.

Siyar, Raheleh, Faramarz Doulati Ardejani, Mohsen Farahbakhsh, Parviz Norouzi, Mohammad Yavarzadeh, and Soroush Maghsoudy. "Potential of Vetiver grass for the phytoremediation of a real multi-contaminated soil, assisted by electrokinetic." *Chemosphere* 246 (2020): 125802.

Skórzyńska-Polit, Ewa, Bożena Pawlikowska-Pawlęga, Ewa Szczuka, Maria Drążkiewicz, and Zbigniew Krupa. "The activity and localization of lipoxygenases in Arabidopsis thaliana under cadmium and copper stresses." *Plant Growth Regulation* 48 (2006): 29–39.

Souza, Mark P. de, C. Mel Lytle, Maria M. Mulholland, Marinus L. Otte, and Norman Terry. "Selenium assimilation and volatilization from dimethylselenoniopropionate by Indian Mustard." *Plant Physiology* 122, no. 4 (April 1, 2000): 1281–1288. https://doi.org/10.1104/pp.122.4.1281.

Susarla, Sridhar, Victor F. Medina, and Steven C. McCutcheon. "Phytoremediation: An ecological solution to organic chemical contamination." *Ecological Engineering* 18, no. 5 (2002): 647–658.

Tan, Yew-Foon Nicholas O'Toole, Nicolas L. Taylor, and A. Harvey Millar. "Divalent metal ions in plant mitochondria and their role in interactions with proteins and oxidative stress-induced damage to respiratory function." *Plant Physiology* 152, no. 2 (2010): 747–761.

Torri, Silvana I., Rodrigo Studart Corrêa, and Giancarlo Renella. "Soil carbon sequestration resulting from biosolids application." *Applied and Environmental Soil Science* 2014 (2014): 1–9.

Tsao, David T. "Overview of phytotechnologies." *Phytoremediation* (2003): 1–50.

Tytła, Malwina, Kamila Widziewicz, and Ewa Zielewicz. "Heavy metals and its chemical speciation in sewage sludge at different stages of processing." *Environmental Technology* 37, no. 7 (2016): 899–908.

Verkleji, J. A. S. *The Effects of Heavy Metals Stress on Higher Plants and Their Use as Biomonitors*. VCH, New York, 1993, 415–424.

Villiers, Florent, Céline Ducruix, Véronique Hugouvieux, Nolwenn Jarno, Eric Ezan, Jérôme Garin, Christophe Junot, and Jacques Bourguignon. "Investigating the plant response to cadmium exposure by proteomic and metabolomic approaches." *Proteomics* 11, no. 9 (2011): 1650–1663.

Warne, Michael St John, Diane Heemsbergen, Darryl Stevens, Mike McLaughlin, Gillian Cozens, Mark Whatmuff, Kris Broos et al. "Modeling the toxicity of copper and zinc salts to wheat in 14 soils." *Environmental Toxicology and Chemistry: An International Journal* 27, no. 4 (2008): 786–792.

Wintz, H., T. Fox, and C. Vulpe. "Responses of plants to iron, zinc and copper deficiencies." *Biochemical Society Transactions* 30, no. 4 (2002): 766–768.

Wojcik, Małgorzata, and Anna Tukiendorf. "Phytochelatin synthesis and cadmium localization in wild type of Arabidopsis thaliana." *Plant Growth Regulation* 44, no. 1 (2004): 71–80.

Yang, Xiaoe, Ying Feng, Zhenli He, and Peter J. Stoffella. "Molecular mechanisms of heavy metal hyperaccumulation and phytoremediation." *Journal of Trace Elements in Medicine and Biology* 18, no. 4 (2005): 339–353.

Zhao, Le, Yong-Le Sun, Su-Xia Cui, Mei Chen, Hao-Meng Yang, Hui-Min Liu, Tuan-Yao Chai, and Fang Huang. "Cd-induced changes in leaf proteome of the hyperaccumulator plant Phytolacca americana." *Chemosphere* 85, no. 1 (2011): 56–66.

Zhen, Yan, Jin-Liang Qi, Si-Si Wang, Jing Su, Guo-Hua Xu, Ming-Sheng Zhang, L. V. Miao, Xin-Xiang Peng, Dacheng Tian, and Yong-Hua Yang. "Comparative proteome analysis of differentially expressed proteins induced by Al toxicity in soybean." *Physiologia Plantarum* 131, no. 4 (2007): 542–554.

Section 2

The Risks of Phytoremediation in Food Safety

7 Danger of Contaminants in Foods through Phytoremediation

Kingsley Chijioke Eze, Nnenna Patrick Obasi, Otobong Anwanabasi Inyang, and Shine Chikaodi Ewa

7.1 INTRODUCTION

Phytoremediation, an environmental-friendly technique, uses plants to remove contaminants from soil, air, and water. Plants naturally help in cleaning up of several types of contaminants such as metals, explosives, pesticides, and oil. The presence of these contaminants, such as heavy metals in agricultural soils (in high concentrations), have great impact on human health (Bakshi & Sharma, 2014). The focus here will be on heavy metals, as they are persistent environmental pollutants with well documented adverse health effects (Aransiola et al., 2013).

Since it has proven potential in mitigating environmental pollution, there are also concerns regarding the possible danger of contaminants accumulating in crops used for phytoremediation as well as entering the food chain. Contaminants like radioactive elements, organic pollutants, and heavy metals can be absorbed by plants during phytoremediation (Liu et al., 2016). When these plants are consumed by animals or humans, there is a risk of those contaminants being transferred to the food chain, thereby posing health hazards (Li et al., 2009).

Health dangers of contaminants in foods used in phytoremediation can differ based on the concentration and type of contaminants that are present. When foods that are exposed to high levels of contaminants through phytoremediation are consumed, this can cause different adverse health effects. For illustration, heavy metals like lead (Pb), mercury (Hg), and cadmium (Cd) are common contaminants that can accumulate in plants during phytoremediation (Shukla et al., 2015; Babaniyi et al., 2023). Once the plants are taken into the body, the contaminants can cause significant risk to health, such as damage to the nervous system, kidney, and other organs, as well as developmental problems, mostly in children (Jarup, 2003).

Likewise, organic pollutants that include pesticides and polycyclic aromatic hydrocarbons (PAHs) can also be taken up by plants during phytoremediation. These pollutants have been connected to different health problems, including carcinogenic and mutagenic effects, as well as endocrine system disruption (Eljarrat & Barcelo, 2003).

Future research should focus on the advancement of plant selection via genetic engineering and exploring the potential of native hyperaccumulators for diverse contaminants and environmental conditions. Also, studies on the long-term impacts of phytoremediation, including its effects on biodiversity, soil health, and ecosystem resilience, are essential for phytoremediation sustainability.

Phytoremediation holds great promise as a sustainable solution for remediating contaminated soils and improving environmental quality. Integrating advanced technologies, such as machine learning approaches, remote sensing, and biosensors, offers good opportunities for real-time monitoring as well as adaptive management practice (Garbisu & Alkorta, 2001; Abioye et al., 2017). The cumulative effects of chronic exposure to these contaminants through phytoremediation derived

DOI: 10.1201/9781032683751-9

foods are discussed, raising concerns about their impact to human health. We cannot ignore the dangers that are associated with the inadvertent transfer of contaminants to the food chain.

The intricate interplay of plant physiology and environmental factors is examined in order to have an understanding of the mechanisms that governs the transfer of pollutants from soil to tissues of plant. Furthermore, this review will investigate the gaps that exist in the regulations and guidelines that governs phytoremediation, emphasizing the need for robust monitoring and assessment protocols in order to ensure the safety of food produced by this method (Isaac & Johnson, 2017). As the field of phytoremediation evolves, interdisciplinary collaboration among scientists, regulators, engineers, and communities is crucial in addressing the complex challenges associated with this approach of remediation that is environmentally safe.

7.2 CONTAMINANT UPTAKE AND TRANSLOCATION IN PLANTS: MECHANISMS, PATHWAYS, AND INFLUENCING FACTORS

Here we will be discussing the mechanisms, pathways through which these contaminants enter plants and the factors that influence it.

7.2.1 Mechanisms of Contaminant Uptake

One will be wondering how does plants take in these contaminants. Plants employ various mechanisms to absorb contaminants from the soil, water, or air.

Key uptake mechanisms include:

Root uptake: These contaminants can enter the plants via their roots two ways: through active transport or by passive diffusion processes (Hinsinger, 2001).

Foliar uptake: Some contaminants are absorbed directly via the leaves of the plant (Isaac & Johnson, 2017).

Endocytosis: Certain contaminants enter plant cells by endocytosis, which is a cellular process by which particles are engulfed (Liu et al., 2017).

7.2.2 Translocation Pathways

The pathways through which these contaminants enter the plants are very important to know. Once they are absorbed, these contaminants are translocated within the plant; most times they move from roots to shoots.

Key translocation pathways include:

Xylem transport: Contaminants, metals in particular, are commonly transported upward through the xylem and they reach aerial parts of the plant (Lux et al., 2011).

Phloem transport: Certain contaminants move via the phloem, facilitating its distribution to the different plant tissues (Kramer, 2010).

Symplastic and apoplastic movement: Here, contaminants can either move within the cell interiors known as symplastic or via the cell walls of plant known as apoplastic (Marschner, 2012).

7.2.3 Factors Influencing Contaminant Accumulation

Several factors play a crucial role in determining the extent of contaminant accumulation in plants:

Plant species and varieties: Different plant species exhibit varying tolerance levels and accumulation capacities for contaminants (Clemens, 2006).

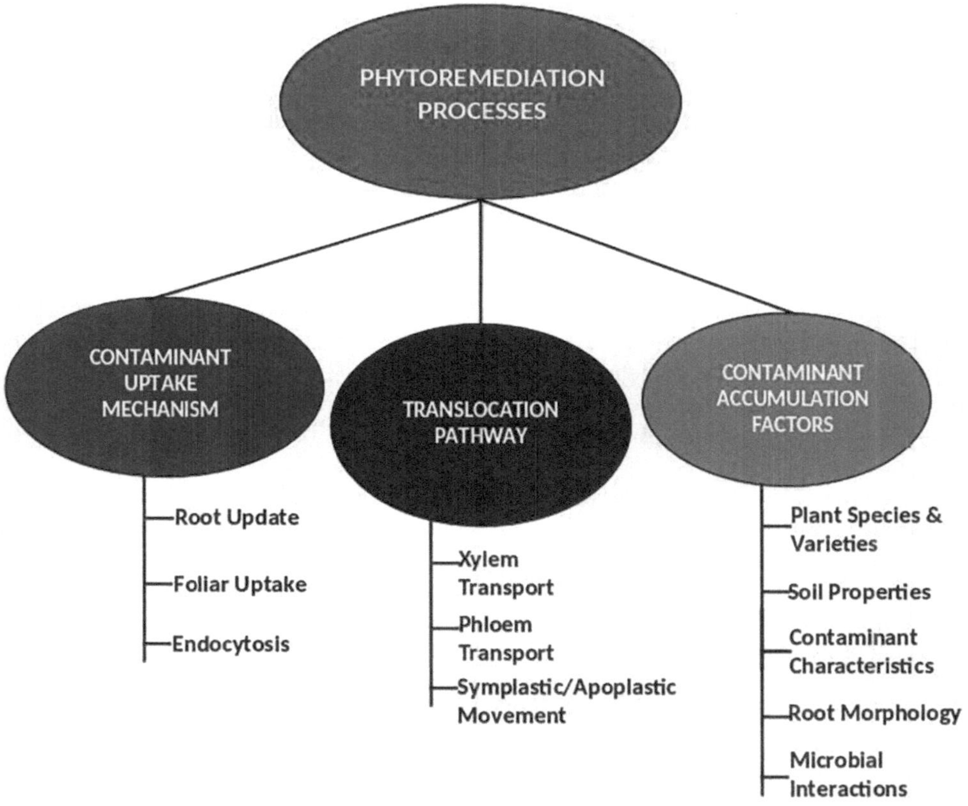

FIGURE 7.1 Mechanisms, pathways, and influencing factors of plant contaminants in phytoremediation processes.

Soil properties: Soil properties such as pH, mineral composition, and organic matter content can influence contaminant availability and plant uptake (Kumpiene et al., 2008).

Contaminant characteristics: The contaminant characteristics such as chemical form, solubility, and speciation of contaminants affect the mobility and uptake (Salt et al., 1995).

Root morphology: Root structure and growth of plant roots affect the efficiency of contaminant uptake (Ma et al., 2009).

Microbial interactions: Microorganisms in the soil can affect contaminant availability and transformation, thereby affecting plant uptake (Weyens et al., 2009; Aransiola et al., 2022).

Understanding these mechanisms, pathways, and influencing factors is essential for assessing the risks associated with contaminant uptake and translocation in plants during phytoremediation. Future research should focus on unraveling the interplay of these factors to enhance the effectiveness and safety of phytoremediation strategies. Figure 7.1 highlights the processes involved in phytoremediation, including the mechanisms through which phytoremediation plants absorb contaminants: root uptake, foliar uptake, and endocytosis. The translocation/transportation pathways through which these contaminants are absorbed include xylem transport, phloem transport, and symplastic/apoplastic movement. Factors that affect the accumulation of contaminants plants include plant species/varieties, soil properties, contaminant characteristics, root morphology, and microbial interaction.

7.3 DANGERS OF HEAVY METALS AND OTHER CONTAMINANTS ON HUMAN HEALTH

The main threats to human health from heavy metals are associated with exposure to Pb, Cd, Hg, and arsenic (As, which is a metalloid but is usually classified as a heavy metal). Other contaminants such as copper (Cu), chromium (Cr), nickel (Ni), bismuth (Bi), and cobalt (Co) will also be discussed.

7.3.1 Cadmium

Here we will learn what Cd is and how it endangers human health.

Cadmium is a heavy metal that endangers health when it is exposed to humans, mostly through food, air, and water that is contaminated. Continued exposure to cadmium have been linked to numerous contrary health effects including the following:

Cancer risk: Prolonged exposure to cadmium is associated with increased risk of developing diverse types of cancer, such as prostate cancer, lung cancer, and kidney cancer (Nordberg et al., 2018).

Kidney damage: Cadmium accumulates in the kidneys for a long time, causing damage to the kidney and also increasing the risk of renal dysfunction as well as diseases such as nephropathy (Nordberg et al., 2018).

Bone health impairment: Exposure to Cd can lead to the displacement of essential minerals (calcium [Ca] and zinc [Zn]) from bones, thereby weakening the skeletal structure and potentially causing osteoporosis (Satarug et al., 2003).

Reproductive and developmental effects: Exposure to Cd can adversely affect the reproductive system and health, leading to issues like infertility problems, reduction in birth weight, and developmental delays (Sallmen & Lindbohm, 2000).

Cardiovascular effects: Exposure to Cd leads to increased blood pressure, which contributes to cardiovascular disease and some other related health problems (Tellez-Plaza et al., 2013).

7.3.2 Lead

Lead is a toxic heavy metal that when exposed to humans can cause severe danger to health, especially by damaging many organs and developmental processes. Exposure to Pb can also have a detrimental effect across different age grades. The impact is commonly pronounced among children because of their developing physiology. Some of these impacts include the following:

Neurological effects: Exposure to Pb causes cognitive impairments, neurodevelopmental disorders, and also behavioral problems, mostly among children (Lanphear et al., 2005).

Cardiovascular effects: Chronic Pb exposure causes adverse effects on cardiovascular health, such as hypertension and high risk of heart disease (Navas-Acien et al., 2007).

Renal effects: Exposure to Pb can cause kidney failure, kidney damage, and impaired renal function (Trachtenberg et al., 2010).

Reproductive and developmental effects: Lead exposure affects reproductive health and fetal development, leading to negative results like preterm birth and developmental delays (Jusko et al., 2008).

Gastrointestinal effects: Exposure to Pb can lead to gastrointestinal issues like nausea, abdominal pains, and constipation (Patrick, 2006).

7.3.3 Mercury

Mercury is a toxic heavy metal that poses significant health problems in humans, thereby affecting different systems and causing serious health challenges. Therefore, it is necessary to note that health dangers associated with Hg exposure can differ depending on the form (inorganic, organic, elemental) and the route of exposure. Some of these health issues include the following:

Neurological effects: Exposure to Hg, especially methylmercury, which is found in some seafoods, has neurotoxic effects that lead to sensory disturbances, cognitive impairments, and motor deficits (Clarkson & Magos, 2006).

Developmental effects: In pregnancy, fetal exposure to Hg can lead to developmental disorder, thereby causing problems in the nervous system that lead to cognitive and behavioral problems, particularly among children (Grandjean et al., 1997).

Cardiovascular effects: Chronic exposure to Hg has been linked to cardiovascular problems, including hypertension, heart disease, and disturbances in heart rhythm (Houston, 2011).

Renal effects: Mercury exposure causes damage to the kidney and impaired renal functions (Satoh & Suzuki, 2006).

Gastrointestinal effects: Exposure to some forms of Hg causes gastrointestinal symptoms like diarrhea, abdominal pains, nausea, and vomiting (Carocci et al., 2014).

7.3.4 Arsenic

Exposure to As, which is a highly toxic metalloid, can cause severe dangers to human health. This is usually associated with contaminated food, air, and water. Arsenic can also have a detrimental effect on different systems and organs of the body. Some of these health issues include the following:

Cancer risk: Long-term exposure to As, especially by contaminated drinking water, is linked to high risk of cancer of the skin, lung, bladder, and kidney (International Agency for Research on Cancer (IARC), 2004).

Skin effects: Arsenic exposure leads to skin lesion, hyperpigmentation, and other dermatological issues (Rahman et al., 1999).

Neurological effects: Chronic exposure to As is associated with neuropathy, neurodevelopmental disorders, and cognitive impairment (Tyler & Allan, 2014).

Cardiovascular effects: Exposure to As is linked to cardiovascular diseases including heart disease, hypertension, and atherosclerosis (Navas-Acien et al., 2008).

Respiratory effects: Inhaled As causes respiratory issues such as coughing, chronic bronchitis, and impairment of lung function (Milton et al., 2001).

7.3.5 Copper

Excessive exposure to Cu, which is a trace essential element, might cause dangers to human health. Since Cu is very important for different physiological processes, inhaling or consuming excess amounts of it can lead to adverse health effects. Some of these health issues includes the following:

Gastrointestinal effects: Ingestion of Cu in high amounts can cause gastrointestinal symptoms like vomiting, abdominal pain, and nausea (Agency for Toxic Substances and Disease Registry [ATSDR], 2004).

Liver damage: Prolonged exposure to copper at high levels causes liver damage and impaired liver function (Klassen et al., 2001)

Neurological effects: Long-term exposure to excess Cu leads to neurotoxic effects that results in symptoms such as cognitive impairments, tremors, and convulsions (Brewer, 2019).

Hepatic encephalopathy: Chronic Cu exposure leads to a health condition called hepatic encephalopathy, which is characterized by neurological symptoms that result in liver dysfunction (Scheiber & Mercer, 2010).

Hemolytic anemia: Occasionally, excess exposure to Cu leads to hemolytic anemia, a condition whereby red blood cells are destroyed prematurely (Elsaid et al., 2018).

7.3.6 Chromium

Chromium is an essential trace element needed for normal physiological processes in humans. However, exposure to some kinds of Cr, mostly hexavalent Cr, poses danger to health. Some of these health issues include the following:

Carcinogenic effects: Hexavalent chromium (Cr(VI)), which is a known carcinogen, is associated with an increased risk in cancer of the lung, especially through inhalation of fine particles in the air (International Agency for Research on Cancer, 2012).

Respiratory effects: When inhaled, Cr(VI) compounds can cause respiratory issues that include increased risk of lung cancer, irritation of the respiratory tract, and shortness of breath (Occupational Safety and Health Administration (OSHA), 2006).

Skin effects: Direct skin contact with some kinds of Cr(VI) causes skin ulcer, skin irritation, and dermatitis (Lodi & Caouto, 2001).

Gastrointestinal effects: Ingestion of high levels of Cr(VI) leads to gastrointestinal issues like abdominal pain, diarrhea, and vomiting (Banfalvi & Schlegel, 2007).

Liver and kidney damage: Chronic exposure to Cr(VI) leads to liver and kidney damage by affecting their normal functions (Costa, 2003).

7.3.7 Nickel

Exposure to Ni, a naturally occurring metal, leads to different dangers in human health. This can take place by ingestion, inhalation, and skin contact, mostly in occupational and environmental settings. Some of these health issues include the following:

Respiratory effects: Inhalation of Ni particles leads to respiratory symptoms like coughing, wheezing, and shortness of breath. Prolonged exposure can lead to a condition called occupational asthma (Anderson et al., 2017).

Skin effects: Direct skin contact with Ni causes allergic contact dermatitis, which results in redness of skin, itching, and inflammation in individuals that are susceptible (Thyssen et al., 2010).

Carcinogenic potential: Prolonged inhalation of some forms of Ni, including Ni compounds and nickel subsulfide, is associated with an increased risk of cancer of the lung (IARC, 2004).

Kidney and respiratory effects: Occupational exposure to Ni is associated with adverse effects on the kidney and respiratory system, especially among individuals working in industries in which Ni is used (Donaldson et al., 2001).

Genotoxicity and DNA damage: Nickel exposure is linked to genotoxic effects, such as DNA damage and chromosomal abnormality (Costa & Klein, 2006).

7.3.8 Bismuth

Bismuth is a heavy metal that has limited exposure as compared to other heavy metals. However, exposure to excess Bi, especially from contaminated sources or certain medical treatments, poses dangers to human health. Some of these health issues include the following:

Gastrointestinal effects: Bismuth compounds, particularly when taken in high doses as medications, lead to gastrointestinal symptoms like vomiting, nausea, and diarrhea (Ghosh & Bist, 2012).

Nephrotoxicity: Prolonged exposure to Bi at high levels from some medications leads to kidney damage and impaired renal function (Kuang et al., 2017).

Neurological effects: Exposure to Bi at high levels results in neurological symptoms like headache, confusion, and cognitive impairment (Breuer et al., 2015).

Cardiovascular effects: Bismuth exposure in some medications leads to cardiovascular issues like disturbances in heart rhythm (Wolf & Bouchard, 2016).

Metabolic effects: High levels of Bi exposure may affect metabolic processes, thereby leading to disruptions in the normal function of the body (De-Bellis & Coomba, 2014).

7.3.9 Cobalt

Exposure to Co, a transition metal, may cause a variety of dangers to human health. Cobalt exposure can occur by inhalation, skin contact, and ingestion, mostly in occupational and environmental settings. Some of these health issues include the following:

Respiratory effects: Inhalation of particles that contain Co especially in the form of hard metal dust, may cause respiratory symptoms such as coughing, wheezing, and shortness of breath (OSHA, 2006).

Cardiovascular effects: Occupational exposure to Co in high amounts, especially in the context of some industrial processes, can be associated with cardiovascular issues like hypertension and other heart problems (Johnson-Arbor & Friedrich, 2013).

Pulmonary effects: Prolonged inhalation of Co dust and fumes may lead to a condition called "hard metal lung disease," which is characterized by lung fibrosis and impaired lung functions (Nijjhawan & Monlnar, 2016).

Hypersensitivity reactions: Direct skin contact with Co may lead to allergic contact dermatitis in susceptible individuals, leading to redness of skin, itching, and inflammation (Uter et al., 2004).

Neurological effects: Chronic exposure to Co at high levels, particularly in occupational settings, has been linked with neurological symptoms like nerve damage, hand tremors, and cognitive impairments (Lucchini et al., 2012).

It is worthy to note that several factors contribute to the danger of contaminants in food through phytoremediation, such as contaminant type, plant species, and contamination level in the environment. Proper selection of plant species that can effectively accumulate the contaminants without transferring them to edible plant parts is therefore crucial. Additionally, the potential for the bioaccumulation and biomagnification of contaminants can be amplified by their presence in the food chain.

To address these dangers, rigorous testing and monitoring protocols are necessary in order to ensure that phytoremediation activities do not inadvertently cause food contamination. Regular analysis of plant tissues, soil, and water samples is essential in tracking contaminant levels and assessing the safety of harvested crops.

While phytoremediation has great potential as a sustainable method for environmental cleanup, careful consideration of the potential risks is also vital to prevent unintentional contamination of the food supply and also protect human health (Figure 7.2). This figure exposes the various health

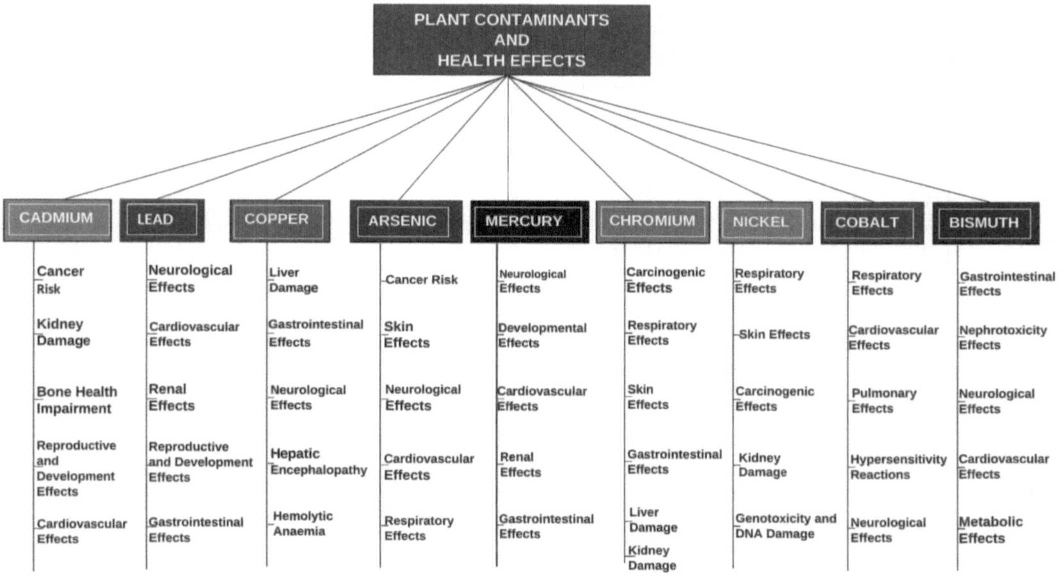

FIGURE 7.2 Health effects of phytoremediation of heavy metal contaminants.

effects of phytoremediation of heavy metals contaminants on individuals. These heavy metal contaminants include Cd, Pb, Cu, As, Hg, Cr, Ni, Co, and Bi. Harmful effects on human health include the risk of cancer development (carcinogenic effects), kidney damage, bone health impairment, reproductive and development effects, cardiovascular effects, neurological effects, renal damage, gastrointestinal effects, skin effects, liver damage, hepatic encephalopathy, genotoxicity/DNA damage, nephrotoxicity, metabolic effects, and hypersensitivity reactions.

7.4 FACTORS INFLUENCING CONTAMINANT TRANSFER TO FOOD CROPS

The factors that influence contaminant transfer to food crops are plant species selection, soil conditions, and influence of contaminant types, as discussed herein.

7.4.1 Plant Species Selection

The choice of plant species in phytoremediation plays a vital part in determining the effectiveness of contaminant removal and also the potential risks of transfer to food crops. The different species of plant exhibits different abilities in accumulating, tolerating, and translocating contaminants. Some plants are known for high heavy metals uptake; these are called hyperaccumulator plants. Examples of hyperaccumulator plants include *Brassica juncea* and *Thlaspi caerulescens* (Baker & Brooks, 1989). Careful consideration is needed in the selection of non-edible or low-value crops to be used for phytoremediation so as to reduce the risk of contaminants that will enter the food chain (Cundy et al., 2008).

7.4.2 Soil Conditions

The physicochemical properties of the soil significantly influence contaminant transfer to food crops during phytoremediation. Key factors are as follows:

Soil pH: The acidity or alkalinity (pH) of the soil affects contaminant mobility as well as plant uptake (Alloway, 2013).

Organic matter content: Soil organic matter can bind contaminants, thereby influencing their availability to plants (Bolan et al., 2003).

Cation exchange capacity (CEC): This influences the retention and also the exchange of contaminants in the soil (Mench et al., 2003).

Soil texture: The texture of the soil determines water retention and drainage, thereby influencing contaminant transport (Kabata-Pendias, 2010).

7.4.3 Influence of Contaminant Types

Various contaminants pose varying risks and show distinct behaviors in the soil-plant system. Factors to consider include the following:

Heavy metals: The chemical form and speciation of heavy metals influence translocation and plant uptake (Lombi & Susini, 2009).

Organic contaminants: The persistence and mobility of organic contaminants such as hydrocarbons and pesticides depend on soil characteristics and microbial activities (Barriuso & Houot, 1996).

Radionuclides: Soil to plant transfer of radionuclides is influenced by soil composition and plant specific characteristics (Strandberg, 1988).

Understanding the interplay of these factors is essential for designing phytoremediation strategies that prioritize both environmental remediation and food safety.

7.5 NOTABLE EXAMPLES OF PHYTOREMEDIATION PROJECTS; LESSONS LEARNED FROM SUCCESSES AND FAILURES

There are some case studies of phytoremediation projects in some parts of the world; here we will learn from their mistakes and success record and improve on cases where necessary.

7.5.1 Notable Phytoremediation Projects

7.5.1.1 The Rhine River Phytoremediation Project (Germany)

In the 1990s, a complete phytoremediation initiative was implemented along the Rhine River to address heavy metal contamination. Willow trees (*Salix viminalis*) were planted strategically to accumulate and absorb heavy metals significantly, thereby reducing pollution levels (Zacchini et al., 2009).

7.5.1.2 Cheonggyecheon Stream Restoration (South Korea)

Phytoremediation played an excellent role in restoring Cheonggyecheon Stream in Seoul. Various plant species were utilized in order to improve the quality of water by absorbing pollutants and enhancing the overall health of the ecosystem (Kim et al., 2007).

7.5.1.3 Chernobyl Exclusion Zone (Ukraine)

Sunflowers (*Helianthus annuus*) were employed in remediating radionuclide-contaminated soil in the Chernobyl Exclusion Zone. This plant exhibits the ability to accumulate cesium (Cs) and strontium (Sr), showcasing the potential of phytoremediation in nuclear disaster sites (Dushenkov et al., 1997).

7.5.2 Successes and Failures

Successes in phytoremediation efforts

Effective plant species selection: Successful projects often emphasize the importance of selecting plant species with high metal-accumulating capacities aligning with site-specific contaminants (Maestri et al., 2010).

Holistic site management: Integrating phytoremediation with other remediation techniques and considering soil amendments can enhance overall efficacy (Epelde et al., 2009).

Community engagement: In most successful cases of phytoremediation, community involvement and awareness contribute greatly to project sustainability and acceptance (Moseman-Valtierra et al., 2014).

Failures in phytoremediation efforts

Insufficient monitoring: Insufficient monitoring of soil and plant conditions during and after phytoremediation might lead to underestimation of project success or failure (Kidd et al., 2015).

Challenges with complex contaminants: Projects dealing with mixed or complex contaminants may face difficulties in achieving complete remediation, highlighting the need for tailored approaches (Glick et al., 2010).

Long-term maintenance: Failure to plan for long-term maintenance and monitoring post-phytoremediation can result in recontamination and project setbacks (Kumpiene et al., 2014).

Note that the success or failure of phytoremediation projects is influenced by myriad factors, including site-specific conditions, contaminant types, and project management strategies. Learning from both successes and failures contributes to the refinement and optimization of phytoremediation practices for future applications.

7.6 REGULATORY GAPS AND MONITORING CHALLENGES IN PHYTOREMEDIATION

7.6.1 Existing Regulations on Phytoremediation

Despite the increasing use of phytoremediation, regulatory frameworks specifically addressing its application are limited. Existing environmental regulations often lack detailed guidelines on the implementation, monitoring, and assessment of phytoremediation projects (EPA, 2013).

The lack of standardized protocols for phytoremediation within regulatory frameworks poses challenges in ensuring consistent and effective practices across different regions and contaminant types (Salt et al., 1998).

7.6.2 Limitations in Monitoring and Assessment

Inadequate sampling techniques: Traditional soil and plant sampling methods may not capture the spatial variability of contaminants, leading to inaccurate assessments of remediation success (Kidd et al., 2015).

Insufficient biomonitoring: Biomonitoring tools, such as remote sensing and molecular techniques, face challenges in providing real-time, site-specific data on plant health and contaminant uptake (Soleimani et al., 2019).

Long-term monitoring challenges: Many regulatory frameworks lack provisions for long-term monitoring, which is crucial for assessing the sustainability and potential risks associated with phytoremediation over time (Deng et al., 2017).

7.6.3 Recommendations for Improved Guidelines

Standardized monitoring protocols: Establishing standardized protocols for monitoring contaminant levels in soil, water, and plant tissues is essential for consistent and reliable data (EPRI, 2001).

Incorporation of biomarker analysis: Guidelines should encourage the use of biomarkers and molecular techniques to assess plant stress, contaminant uptake, and overall remediation efficacy (Liu et al., 2017).

Long-term monitoring requirements: Regulatory frameworks should incorporate provisions for long-term monitoring to evaluate the persistence of contaminants and potential ecological shifts (EPA, 2000).

Adaptive management approaches: Encouraging adaptive management strategies allows for continuous adjustments based on monitoring data, enhancing the effectiveness of phytoremediation projects (Cundy et al., 2008).

7.7 POTENTIAL RISKS TO HUMAN HEALTH IN PHYTOREMEDIATION

7.7.1 Chronic Exposure Scenarios

Contaminant accumulation in edible parts: The uptake of contaminants by food crops during phytoremediation raises concerns about chronic exposure through the consumption of contaminated plant parts (Alloway, 2013).

Bioavailability dynamics: The bioavailability of contaminants by food crops during phytoremediation raises concerns about chronic exposure through the consumption of contaminated plant parts (Alloway, 2013). Bioavailability of contaminants in plants, influenced by soil conditions and plant species, contributes to variations in the degree of chronic exposure (Lombi & Susini, 2009).

Long-term soil contamination: Phytoremediation may alter soil properties and maintain a reservoir of contaminants, leading to prolonged chronic exposure scenarios for nearby communities (Wenzel et al., 1999).

7.7.2 Cumulative Effects on Health

Multiple contaminant exposure: Phytoremediation often involves addressing multiple contaminants simultaneously, raising concerns about the cumulative health effects resulting from exposure to a combination of pollutants (Liu et al., 2016).

Synergistic interactions: Some contaminants may exhibit synergistic interactions, amplifying their toxic effects when present together, which can contribute to increased health risks (Ma et al., 2009).

Delayed onset of health effects: Cumulative exposure to contaminants over time may lead to delayed onset of health effects, complicating the identification and management of associated health risks (NRC, 2003).

7.7.3 Vulnerable Populations

Children and pregnant women: Vulnerable populations, such as children and pregnant women, may be at higher risk due to their increased susceptibility to the adverse effects of contaminants (Grandjean & Landrigan, 2006).

Local communities: Proximity to phytoremediation sites exposes local communities to potential health risks, necessitating careful consideration of the socio-economic and demographic characteristics of affected populations (Menzel et al., 2019).

Occupational exposure: Workers involved in phytoremediation activities may face elevated risks, emphasizing the need for occupational health and safety measures (Alkorta et al., 2004).

7.8 MITIGATION STRATEGIES AND FUTURE DIRECTIONS IN PHYTOREMEDIATION

7.8.1 Enhancing Plant Selection for Phytoremediation

Genetic engineering: Advances in genetic engineering offer the potential to enhance the phytoremediation capabilities of plants, enabling them to efficiently accumulate and tolerate higher levels of contaminants (Rugh et al., 1996).

Native hyperaccumulator exploration: Identifying and utilizing native hyperaccumulator plants that naturally accumulate high levels of specific contaminants can enhance the effectiveness of phytoremediation projects (Baker & Brooks, 1989).

Endophyte-assisted phytoremediation: Harnessing the symbiotic relationship between plants and endophytic microorganisms can improve contaminant uptake and plant stress tolerance, contributing to enhanced phytoremediation efficiency (Weyens et al., 2009).

7.8.2 Soil Management Practices

Amendments and soil conditioners: Incorporating soil amendments, such as organic matter and specific minerals, can influence soil properties to enhance contaminant immobilization or uptake by plants (Bolan et al., 2003).

Phytomanagement approaches: Integrating phytoremediation with other management practices, such as agroforestry and cover cropping, can contribute to sustainable soil management while addressing contamination issues (Cundy et al., 2008).

Rhizosphere engineering: Manipulating the rhizosphere through the addition of specific microbial communities can optimize nutrient cycling and enhance the efficiency of contaminant removal (Epelde et al., 2009).

7.8.3 Advances in Monitoring Technologies

Remote sensing: Remote sensing technologies, including satellite imagery and drones, provide real-time data on the health of plants, contaminants distribution, and overall project performance (Soleimani et al., 2019).

Biosensors and molecular techniques: Advancements in biosensors and molecular techniques allows more precise and rapid monitoring of contaminants levels in both soil and plant tissues (Schreiber, 2019).

Machine learning and data integration: Implementing machine learning algorithms and integrating data from various sources enhances the predictive capabilities of monitoring systems and aids in decision-making and adaptive management (Le et al., 2020).

7.9 CONCLUSION

Phytoremediation is a promising and environmentally friendly approach that is used for mitigating soil contaminations by various pollutants, including heavy metals and organic contaminants.

The success of phytoremediation schemes depends on factors such as plant selection, soil management activities, and monitoring technologies, as well as genetic engineering and native hyperaccumulator exploration that emerge as promising avenues.

Even though phytoremediation as a technique offers environmental benefits, there are potential risks to human health, mostly via chronic exposure scenarios and vulnerable populations. There is a need for careful considerations and regulatory oversights.

Notable phytoremediation projects, both successful and unsuccessful, provide valuable insight into the complexities and challenges associated with this approach.

Investigating the interactions between plants and endophytic microorganisms, as well as optimizing rhizosphere engineering, can contribute to enhancing the efficiency of phytoremediation.

Addressing regulatory gaps and developing standardized monitoring protocols are crucial for ensuring the safe and effective implementation of phytoremediation projects.

REFERENCES

Abioye, O.P., Ijah, U. J. J. & Aransiola, S. A. (2017). Phytoremediation of soil contaminants by biodiesel plant *Jatropha curcas*. In Chapter 4 of K. Bauddh et al. (Eds.), *Phytoremediation Potential of Bioenergy Plants, DOI 10.1007/978-981-10-3084-0_4*. Springer Nature Singapore Pte Ltd. https://link.springer.com/book/10.1007%2F978-981-10-3084-0.

Agency for Toxic Substances and Disease Registry (ATSDR). (2004). *Toxicological Profile for Copper*. Department of Health and Human Services, Public Health Service, Atlanta, GA.

Alkorta, I., Hernandez-Allica, J., Becerril, J. M., Amezaga, I., Albizu, I., Onaindia, M. & Garbisu, C. (2004). Recent findings in phytoremediation of soils contaminated with environmentally toxic heavy metals and metalloids such as Zinc, Cadmium, Lead and Arsenic. *Reviews in Environmental Science and Biotechnology*, 3(1), 71–90.

Alloway, B, J. (2013). *Heavy Metals in Soils: Trace Metals and Metalloids in Soils and their Bioavailability*. Springer.

Anderson, U., Hedberg, Y., Liden, C. & Meding, B. (2017). Contact allergy to nickel. An epidemiological and clinical study from the Swedish Contact Dermatitis Research Group. *Acta Dermato-Venereologica*, 97(5), 585–591.

Aransiola, S. A., Ijah, U.J.J. & Abioye, O.P. (2013). Phytoremediation of Lead Polluted Soil by *Glycine max* L. *Applied and Environmental Soil Science*, 2013, Article ID 631619, doi: 10.1155/2013/631619. https://www.hindawi.com/journals/aess/2013/631619/abs/.

Aransiola, S. A, Ijah, U. J. J., Abioye, O. P. & Bala, J. D. (2022). Vermicompost-assisted phytoremediation of toxic trace element-contaminated soil in Madaka, Nigeria using *Melissa officinalis* L and *Sida acuta. International Journal of Environmental Science and Technology*. http://dx.doi.org/10.1007/s13762-022-04105-y

Babaniyi, B. R., Ogundele, O. D., Bisi-omotosho, A., Babaniyi, E. E. & Aransiola, S. A. (2023). Remediation approaches in environmental sustainability. In Naga Raju Maddela, Lizziane Kretli Winkelstroter Eller, & Ram Prasad (Eds.), *Microbiology for Cleaner Production and Environmental Sustainability*. CRC Press. ISBN: 9781032496061. https://www.routledge.com/Microbiology-for-Cleaner-Production-and-Environmental-Sustainability/Maddela-Eller-Prasad/p/book/9781032496061

Baker, A. J. & Brooks, R. R. (1989). Terrestrial higher plants which hyperaccumulate metallic elements – A review of their distribution, ecology and phytochemistry. *Biorecovery*, 1(2), 81–126.

Bakshi, M. & Sharma, S. (2014). Phytoremediation: A novel approach for utilization of iron and steel slag. *International Journal of Phytoremediation*, 16(9), 940–951.

Banfalvi, G. & Schlegel, K. (2007). Estimation of the toxic potential of hexavalent chromium after oral administration. *Archives of Environmental Contamination and Toxicology*, 53(3), 383–388.

Barriuso, E. & Houot, S. (1996). Influence of soil pH on adsorption of pesticides phenoxyacetic acid, 2, 4-Dichlorophenoxyacetic acid, and 4-Chloro-2-Methylphenoxyacetic acid by clays and organo mineral complexes. *Environmental Science & Technology*, 30(5), 1630–1638.

Bolan, N. S., Adriano, D. C., Mahimairaja, S. & Naidu, R, (2003). Role of phosphorus in immobilization and bioavailability of heavy metals in the soil-plant system. *Reviews of Environmental Contamination and Toxicology*, 177, 1–44.

Breuer, K., Gobel, U., Martin, D. & Gerding, H. (2015). Safety aspects of bismuth in medical use. Schweizerische Rundschau for Medizin Praxis = *Rvue Suisse de Medicine Praxis*, 104(5), 247–253.

Brewer, G. J. (2019). Copper excess, Zinc deficiency, and cognition loss in Alzheimer's disease. *Biofactors*, 45(5), 856–871.

Carocci, A., Rovito, N., Sinicropi, M. S. & Genchi, G. (2014). Mercury toxicity and neurodegenerative effects. *Reviews of Environmental Contamination and Toxicology*, 229, 1–18.

Clarkson, T. W. & Magos, L. (2006). The toxicology of mercury and its chemical compounds. *Critical Reviews in Toxicology*, 36(8), 609–662.

Clemens, S. (2006). Toxic metal accumulation, responses to exposure and mechanisms of tolerance in plants. *Biochimie*, 88(11), 1707–1719.

Costa, M. (2003). Toxicity and carcinogenicity of Cr (IV) in animal models and humans. *Critical Reviews in Toxicology*, 33(2), 137–197.

Costa, M, & Klein, C. B. (2006). Toxicity and carcinogenicity of chromium compounds in humans. *Critical Reviews in Toxicology*, 36(2), 155–163.

Cundy, A. B., Bardos, R. P., Puscheneiter, M., Mench, M., Bert, V., Friesl-Hanl, W., . . . Muller, I. (2008). Brownfields to green fields: Realising wider benefits from practical contaminant phytomanagement strategies. *Journal of Environmental Management*, 87(1), 120–143.

De-Bellis, E. & Coomba, P. (2014). Children's health and heavy metals in contaminated areas: A review of impacts, methods of detection and prevention. *Environmental Chemistry Letters*, 12(4), 477–492.

Deng, L., Zhang, W. & Zheng, J. (2017). A review on heavy metal phytoremediation from contaminated soil. In *Soil Heavy Metals* (pp. 327–357). Springer.

Donaldson, K., Stone, V. & Clouter, A. (2001). Renal and systemic effects of lung inflammation in rats. *Environmental Health Perspectives*, 109(Suppl 4), 683–686.

Dushenkov, V., Mikheev, A., Prokhneysky, A., Ruchko, M., Sorochinsky, B. & Sorochinsky, S. (1997). Phytoremediation of radiocesium-contaminated soil in the Chernobyl exclusion zone: From pilot research to the first experience of practical. *Application Environmental Science & Technology*, 31(1), 165–173.

Eljarrat, E. & Barcelo, D. (2003). Priority lists of hazardous substances in the marine environment: Polycyclic aromatic hydrocarbons and polychlorinated biphenyls. *Environment International*, 29(7), 771–783.

Elsaid, M. F., Chelko, S. P., Kilie, A., Hickey, K. T., Montgomery, K., Lee, E. S., . . . Carnes, C. A. (2018). Effects of copper exposure on cardiac electrophysiology and arrhythmia. *Nature Communication*, 9(1), 4686.

EPA (U.S. Environmental Protection Agency) (2000). *Phytoremediation Technical/Regulatory Guidance Document*. EPA 600/R-99/107.

EPA (U.S. Environmental Protection Agency) (2013). *Phytotechnologies: Regulatory Status, Implementation and Research Needs*. EPA 542-R-13–017.

Epelde, L., Becerril, J. M., Barrutia, O. & Garbisu, C. (2009). Interaction between plant and rhizosphere microbial communities in phytoremediation of hydrocarbons: A review. *European Journal of Soil Biology*, 45(5–6), 327–344.

EPRI (Electric Power Research Institute) (2001). *Guidelines for Phytoremediation*. EPRI TR-113623.

Garbisu, C. & Alkorta, I. (2001). Phytoextraction: A cost-effective plant-based technology for the removal of metals from the environment. *Bioresource Technology*, 77(3), 229–236.

Ghosh, S. & Bist, S. S. (2012). Bismuth subsalicylate: A comprehensive review. *Journal of Pharmacy and Bio allied Sciences*, 4(4), 266–270.

Glick, B. R., Todorovic, B., Czarny, J., Cheng, Z., Duan, J. & McConkey, B. (2010). Promotion of plant growth by bacterial ACC deaminase. *Critical Reviews in Plant Science*, 29(6), 360–393.

Grandjean, P. & Landrigan, P. J. (2006). Developmental neurotoxicity of industrial chemicals. *The Lancet*, 368(9553), 2167–2178.

Grandjean, P., Weihe, P., White, R. F., Debes, F., Araki, S., Yokoyama, K., . . . Murata, K. (1997). Cognitive deficit in 7-year-old children with prenatal exposure to methylmercury. *Neurotoxicology and Teratology*, 19(6), 417–428.

Hinsinger, P. (2001). Bioavailability of soil inorganic P in the rhizosphere as affected by root-induced chemical changes: A review. *Plant and Soil*, 237(2), 173–195.

Houston, M. C. (2011). Role of mercury toxicity in hypertension, cardiovascular disease and stroke. *Journal of Clinical Hypertension*, 13(8), 621–627.

International Agency for Research on Cancer (IARC) (2004). *IARC Monographs on the Evaluation of Carcinogenic Risks to Humans*. Some Drinking-Water Disinfectants and Contaminants, Including Arsenic. Vol. 84.

International Agency for Research on Cancer (IARC). (2012). *IARC Monographs on the Evaluation of Carcinogenic Risks to Humans*: Chemical Agents and Related Occupations. World Health Organization, International Agency for Research on Cancer, Lyon, France.

Isaac, R. A. & Johnson, W. C. (2017). Foliar absorption of atmospheric lead by plants: A critical review. *Critical Reviews in Environmental Science and Technology*, 47(10), 859–901.

Jarup, l. (2003). Hazards of heavy metal contamination. *British Medical Bulletin*, 68(1), 167–182.

Johnson-Arbor, K. & Friedrich, J. (2013). Systemic cobalt toxicity from total hip arthroplasties: Review of a rare condition part 2: Measurement, risk factors and step-wise approach to treatment. *Current Reviews in Musculoskeletal Medicine*, 6(4), 342–348.

Jusko, T. A., Henderson, C. R., Lanphear, B.P., Cory-Slechta, D. A., Parsons, P. J., Canfield, R. L., . . . Bellinger, D. C. (2008). Blood Lead concentrations <10 μg/dL and child intelligence at 6 years of age. *Environmental Health Perspective*, 116(2), 243–248.

Kabata-Pendias, A. (2010). *Trace Elements in Soils and Plants* (4th ed.). CRC Press.

Kidd, P., Barcelo, J., Bernal, M. P., Navari-Izzo, F., Poschenrieder, C., Shiley, S., . . . Monterroso, C. (2015). Trace elements behaviour at the root-soil interface: Implications in phytoremediation. *Environmental and Experimental Botany*, 114, 21–38.

Kim, J. H., Kim, K. R., Owens, G. & Kim, M. S. (2007). Application of biochar to soil reduces cancer risk via rice consumption: A case study in maddy fields in Northern Iran. *Environmental Geochemistry and Health*, 29(3), 167–178.

Klassen, C. D., Amdur, M. O. & Doull, J. (Eds.) (2001). *Casarett and Doull's Toxicology: The Basic Science of Poison* (6th ed.). McGraw-Hill.

Kramer, U. (2010). Metal hyperaccumulation in plants. *Animal Review of Plant Biology*, 61, 517–534.

Kuang, L., Li, X., Yuan, T., Tang, Y., Xie, X., . . . Zhang, S. (2017). A high performance liquid chromatography method for the simultaneous determination of four bismuth compounds in urine and blood samples. *Journal of Analytical Toxicology*, 41(2), 156–165.

Kumpiene, J., Lagerkvist, A. & Maurice, C. (2008). Stabilization of As, Cr, Cu, Pb and Zn in Soil using Amendments: A Review. *Waste Management*, 28(1), 215–225.

Kumpiene, J., Lagervist, A. & Maurice, C. (2014). Stabilization of As, Cr, Cu, Pb and Zn in Soil using Amendments: A review. *Waste Management*, 34(1), 1–18.

Lanphear, B. P., Hornung, R., Khoury, J., Yolton, K., Baghurst, P., Bellinger, D. C., . . . Canfield, R. L. (2005). Low-level environmental lead exposure and children's intellectual function: An international pooled analysis. *Environmental Health Perspectives*, 113(7), 894–899.

Le, T. T., Granger, S., Strobl, B. & Strobl, C. (2020). Time series analysis in environmental science: Machine learning methods and applications. *Environmental Modelling & Software*, 133, 104868.

Li T., Yang, X. & Islam, E. (2009). Effects of crop management practices on uptake and accumulation of heavy metals in paddy rice. *Journal of Environmental Sciences*, 21(10), 1450–1454.

Liu, J., Luo, Y., Leung, H. M., Zhang, Y., Jiang, L. & Wu, S. (2016). Phytotoxicity and accumulation of nanoparticles in wheat. *Chemosphere*, 144, 23–29.

Liu, J., Wu, H., Zhang, L., Zhao, L. & Liu, Y. (2017). Endocytosis of nanoparticles in plant cells. *Trends in Plant Science*, 22(11), 1007–1010.

Lodi, A. & Caouto, R. (2001). Hexavalent chromium causes the oxidation of NADH in the plasma membrane of S. cerevisiae. *Yeast*, 18(14), 1234–1241.

Lombi, E. & Susini, J. (2009). Synchrotron-based techniques for plant and soil science: Opportunities, challenges and future perspectives. *Plant and Soil*, 320(1–2), 1–35.

Lucchini, R. G., Zoni, S., Guazzetti, S., Bontempi, E., Micheletti, S., Broberg, K., . . . Ferri, F. (2012). Neurological and neuropsychological functions in metal workers: A controlled study. *International Archives of Occupational and Environmental Health*, 85(8), 939–951.

Lux, A., Martinka, M., Vaculík, M. & White, P. J. (2011). Root responses to cadmium in the rhizosphere: A review. *Journal of Experimental Botany*, 62(1), 21–37.

Ma, Y., Oliveira, R. S., Freitas, H. & Zhang, C. (2009). Biochemical and molecular mechanisms of plant-microbe-metal interactions: Relevance for phytoremediation. In *Reviews of Environmental Contamination and Toxicology* (Vol. 198, pp. 1–37). Springer.

Maestri, E., Marmiroli, M., Visioli, G., Marmiroli, N. & Yoon, M. (2010). Metal tolerance and hyperaccumulation: Costs and trade-offs between traits and environment. *Environmental and Experimental Botany*, 68(1), 1–13.

Marschner, H. (2012). *Marschner's Mineral Nutrition of Higher Plants*. Academic Press.

Mench, M., Manceau, A., Vangronsveld, J., Clijsters, H. & Mercier, G. (2003). Assessment of the bioavailability of soil trace elements by selective chemical extraction procedures. In S. K. Gupta (Ed.), *Trace Elements as Contaminants and Nutrients: Consequences in Ecosystems and Human Health* (pp. 387–424). Wiley.

Menzel, M., Jiang, W. & Wagner, G. (2019). The role of socio-economic and demographic factors for the acceptance of phytoremediation: A case study in China. *Sustainability*, 11(19), 5365.

Milton, A. H., Hasan, Z., Rahman, A., Rahman, M. & Zaman, M. R. (2001). Respiratory effects and arsenic contaminated well water in Bangladesh. *International Journal of Environmental Health Research*, 11(2), 131–141.

Moseman-Valtierra, S., Goncalves, B., Mangiafico, S. & Restrepo, P. (2014). Phytoremediation potential of spartina alterniflora in moderately contaminated salt marshes. *Ecological Engineering*, 67, 245–254.

Navas-Acien, A., Guallar, E., Silbergeld, E. K., Rothenberg, S. J., (2007). Lead exposure and cardiovascular disease – a systematic review. *Environmental Health Perspectives*, 115(3), 472–482.

Navas-Acien, A., Silbergeld, E. K., Pastor-Barriuso, R. & Guallar, E. (2008). Arsenic exposure and prevalence of type 2 diabetes in US adults. *JAMA*, 300(7), 814–822.

Nijjhawan, R. I. & Monlnar, E. T. (2016). Hard metal lung disease: Radiologic-pathologic correlation. *Clinical Imaging*, 40(3), 496–499.

Nordberg, G. F., Jin, T. & Nordberg, M. (2018). Cadmium. In *Handbook on the Toxicology of Metals* (pp. 667–708). Academic Press.

NRC (National Research Council) (2003). *Bioavailability of Contaminants in Soils and Sediments: Processes, Tools, and Applications*. National Academies Press.

Occupational Safety and Health Administration (OSHA) (2006). *Cobalt and Inorganic Cobalt Compounds*. Health Hazards Information Bulletin.

Patrick, L. (2006). Lead toxicity, a review of the literature. Part 1: Exposure, evaluation, and treatment. *Alternative Medicine Review*, 11(1), 2–22.

Rahman, M., Tondel, M., Ahmad, S. A., Chowdhury, I. A., Faruquee, M. H. & Axelson, O. (1999). Hypertension and arsenic exposure in Bangladesh. *Hypertension*, 33(1), 74–78.

Rugh, C. L., Wilde, H. D., Stack, N. M., Thompson, D. M., Summers, A. O., Meagher, R. B., . . . Salt, D. E. (1996). Mercuric ion reduction and resistance in transgenic arabidopsis thaliana plants expressing a modified bacterial MerA Gene. *Proceedings of the National Academy of Sciences*, 93(8), 3182–3187.

Sallmen, M. & Lindbohm, M. L. (2000). Effects of parental exposure to lead and cadmium on pregnancy outcome and infant development. *Scandinavian Journal of Work, Environment & Health*, 26(5), 431–437.

Salt, D. E., Blaylock, M., Kumar, N. P. B. A., Dushenkov, V., Ensley, B. D., Chet, I. & Raskin, I. (1998). Phytoremediation: A novel strategy for the removal of toxic metals from the environment using plants. *Biotechnology*, 13(5), 468–474.

Salt, D. E., Smith, R. D., . . . Raskin, I. (1995). Phytoremediation. *Annual Review of Plant Biology*, 49, 643–668.

Satarug, S., Baker, J. R. & Reilly, P. E. (2003). Long-term exposure to cadmium leads to the suppression of osteoblast functions and the promotion of osteoclastogenesis: Likely roles of the p38 MAPK pathway. *Toxicology and Applied Pharmacology*, 188(3), 203–214.

Satoh, H. & Suzuki, T. (2006). Review of analytical strategies for the determination of toxic metals in environmental and biological samples. *Analytical and Bioanalytical Chemistry*, 386(4), 769–790.

Scheiber, I. F. & Mercer, J. F. (2010). Drugging copper metabolism. In *Neurodegenerative Diseases* (pp. 161–183). Springer.

Schreiber, L. (2019). Field-scale application of phytoremediation. *Frontiers in Environmental Science*, 7, 10.

Shukla, O. P., Rai, U. N. & Dubey, S. (2015). Heavy metal contamination in vegetables grown around peri-urban and urban-industrial clusters: A case study. *Ecotoxicology and Environmental Safety*, 113, 1–6.

Soleimani, M., Hajabbasi, M. A., Afyuni, M. & Nourbakhsh, F. (2019). Assessment of phytoremediation efficiency and phytoextraction of heavy metals from a multi-metal contaminated soil. *International Journal of Phytoremediation*, 21(3), 239–247.

Strandberg, M. (1988). *Radionuclides in Terrestrial Ecosystems: A Guide for Ecologists and Radionuclide Chemists*. Springer.

Tellez-Plaza, M., Navas-Acien, A., Menke, A. & Crainiceanu, C. M. (2013). Cadmium exposure and all-cause and cardiovascular mortality in the US general population. *Environmental Health Perspectives*, 121(10), 1017–1022.

Thyssen, J. P., Menne, T. & Johansen, J. D. (2010). Nickel allergy in patch-tested female hairdressers and assessment of nickel release from hairdressers' scissors and crochet hooks. *Contact Dermatitis*, 63(6), 325–332.

Trachtenberg, F., Barregard, L. & Akerström, M. (2010). The relationship between cadmium, lead, and mercury in blood and the renal function in a non-smoking population. *Kidney International*, 78(6), 589–596.

Tyler, C. R. & Allan, A. M. (2014). The effects of arsenic exposure on neurological and cognitive dysfunction in human and rodent studies: A review. *Current Environmental Health Reports*, 1(2), 132–147.

Uter, W., Geler, J., Lessmann, H., Schnuch, A. & Frosch, P. J. (2004). Contact allergy to metals patch test results of the information networks of department of dermatology 2001–2003. *Contact Dermatitis*, 51(5–6), 111–117.

Wenzel, W. W., Unterbrunner, R., Sommer, P., Sessitsch, A., Bandte, M. & Stingeder, G. (1999). Heavy metal accumulation in trees planted on contaminated soil. *Environmental Pollution*, 104(2), 231–240.

Weyens, N., van der Lelie, D., Taghavi, S., Newman, L. & Vangronsveld, J. (2009). Exploiting plant–microbe partnerships to improve biomass production and remediation. *Trends in Biotechnology*, 27(10), 591–598.

Wolf, K. & Bouchard, M. F. (2016). Cardiovascular consequences of childhood lead exposure and obesity. *Clinical Epidemiology*, 8, 45–46.

Zacchini, M., Pietrini, F., Mugnozza, G. S., Iori, V., Pietrosanti, L., Massacci, A. & Cardarelli, M. (2009). Metal accumulation and damage in rice (Oryza sativa L.) Seedlings treated with Cd, Ni and Cu. *Plant Stress*, 3(1), 47–56.

8 Microplastics and Nanoplastics in Soil
Plant Uptake and Impact on Food Insecurity

Innocent Ojeba Musa, Job Oloruntoba Samuel, Mustahpa Adams, Vivian Nathaniel, Asmau M. Maude, Mohammed Evuti Mahmud, and Abd'Gafar Tunde Tiamiyu

8.1 INTRODUCTION

The extensive use of plastics has improved our daily lives, but it has also led to a huge increase in plastic waste that has been dumped into the environment (Musa et al., 2024). Plastic trash can be broken down into very small pieces thanks to factors such as mechanical abrasion, ultraviolet (UV) deterioration, oxidation, and biotic action (He et al., 2021). Microplastics (MPs) are plastic particles less than 1 mm in size, whereas nanoplastics (NPs) are less than 1100 nm in size. Both marine and terrestrial environments include microplastics and nanoplastics (MNPs; Adhikari et al., 2022). There was also a lot of evidence that MNPs were present in living species and that they had negative effects like mortality, reproductive and developmental mental toxicity, intestinal oxidative damages, immunological toxicity, and neurotoxicity (Wang et al., 2022). One of the most pressing environmental problems of the future is the emergence of new pollutants with long-lasting effects: MNPs (Auta et al., 2022; Perković et al., 2022).

While there has been a lot of research done on MPs in marine systems, it turns out that the land is a significant sink and source of environmental MPs as well (Sorasan et al., 2022). It has been estimated that there are 4 to 23 times as many MPs in marine systems than there are in soil (Alimi et al., 2018). MPs have been found everywhere over the planet, from metropolitan water systems to agricultural soils to remote regions (Adhikari et al., 2022). Various plastic polymers (e.g., polyethylene, polystyrene, polyurethane, polypropylene, polyvinylchloride), forms (e.g., particles, fibers, and films), and sizes are characteristic of MNP contaminations (Sorasan et al., 2022). These are the results of all the MPs created when humans utilize plastics in the ground (Rubio et al., 2020).

Humans are able to take in MNPs in terrestrial environments, with the average yearly MP consumption per person ranging from 39,000 to 52,000 particles (Song et al., 2019), with those numbers increasing to 74,000 and 121,400, respectively, when inhalation is factored in. Products including salt, honey, sugar, beer, drinking water, and even traditional medicinal remedies derived from animals have been found to be polluted with MPs, pointing to pervasive contamination in terrestrial environments. MPs even have been found in human feces. Concern regarding the effects of MNPs on human health is developing as a result (He et al., 2021; Aransiola, Victor-Ekwebelem et al., 2023).

The soil-plant system is a crucial aspect of Earth's ecosystem since it supports all life on Earth and provides sustenance for human beings. It was predicted that the prevalence of MPs in soils,

DOI: 10.1201/9781032683751-10

especially agricultural soils, would increase as a result of the widespread use of plastics on the land. To add insult to injury, MNPs can be absorbed by plants from soils and make their way up the human food chain. There is some disagreement on where MNPs go in the soil-plant system, but this is an essential question because of the correlation between MNPs and food pollution and human health (Adhikari et al., 2022; Aransiola, Ikhumetse et al., 2023). This chapter examines the intricate interplay between MNPs within soil, their absorption by plants, and the potential ramifications for food security.

8.2 MICROPLASTICS AND NANOPLASTICS IN SOIL

8.2.1 Sources and Pathways

MNPs, which are tiny particles of plastic that have become widespread in our environment, are introduced into soil ecosystems through a complex network of sources and pathways (Iqbal et al., 2020; Musa et al., 2024). The high occurrence of these organisms in soil gives rise to significant concerns regarding soil quality, agricultural practices, and potential implications for food security. In order to effectively mitigate these risks, it is imperative to get a comprehensive understanding of the diverse origins and paths through which these plastics infiltrate the soil ecosystem (Allouzi et al., 2021). The following are sources and pathways of MNPs into the soil.

8.2.1.1 Agricultural Runoff

The discharge of agricultural substances into the environment is a significant contributor to the presence of MNPs within soil ecosystems (Hurley and Nizzetto, 2018). According to Hurley and Nizzetto (2018), the implementation of plastic-based mulching techniques or the usage of plastic items in crop management on farms may inadvertently result in the introduction of plastic particles into the soil. As a consequence of environmental stresses, such as exposure to sunlight and mechanical pressures, it is possible for these polymers to undergo degradation, resulting in the formation of smaller fragments over a period of time. The aforementioned fragments are subsequently conveyed by the runoff of precipitation and ultimately deposited into the soil (Kumar et al., 2023).

8.2.1.2 Sewage Sludge Application

The application of sewage sludge on agricultural fields represents an additional avenue through which MNPs might infiltrate soil ecosystems, as highlighted by Bandini et al. (2022). The sludge derived from wastewater treatment facilities encompasses a broad spectrum of contaminants, which encompasses MPs. According to Bandini et al. (2022), the application of sewage sludge as a soil conditioner or fertilizer has the potential to introduce MPs into the soil, hence posing a risk of crop contamination in the treated soil.

8.2.1.3 Atmospheric Deposition

The atmosphere can carry MNPs and deposit them on soil surfaces. Vehicle tire wear, industrial processes, and atmospheric transport from far-flung locations are just a few of the many potential origins of MP particles in the air. Soil particles become embedded in the soil's structure when they fall to the ground, building up over time (van den Berg et al., 2020).

8.2.1.4 Breakdown of Plastic Mulch Films

Plastic mulch films are frequently utilized in contemporary agriculture to enhance crop productivity and effectively control weed growth (Astner et al., 2019). However, these films are susceptible to weathering and degradation. According to Ullah et al. (2023), the degradation of these materials results in the release of MNP particles into the soil. The deterioration process is accelerated by various factors commonly found in agricultural environments, such as exposure to sunshine, temperature fluctuations, and mechanical strains (Astner et al., 2019).

8.2.2 Persistence and Accumulation of Microplastics and Nanoplastics in Soil

According to Cao et al. (2023), the permanency and accumulation of plastics in soil ecosystems is becoming an increasingly significant cause for worry. Plastics are well-known for their extraordinary durability. Plastics, particularly MNPs, can remain in the environment for hundreds of years, which has significant repercussions for the soil's ability to perform its function. Organic matter, on the other hand, disintegrates over the course of time (Qi et al., 2020).

8.2.2.1 Durability of Microplastics and Nanoplastics in Soil

According to Bratovcic (2019), MNPs possess a synthetic polymer composition that renders them impervious to natural breakdown mechanisms. The formation of these minuscule plastic particles often occurs through the degradation of larger plastic items, which undergo fragmentation into smaller fragments due to weathering, mechanical stress, and exposure to UV radiation. Notwithstanding their diminutive dimensions, these particles possess an identical polymer composition to their parent plastics, rendering them highly unaffected by degradation (Peng et al., 2022).

8.2.2.2 Accumulation in Soil

The enduring presence of MNPs in soil can be attributed to their inherent physical characteristics, as well as the continuous influx from various sources. Plastics have a tendency to accumulate in the soil environment following their introduction by many mechanisms, including agricultural runoff, sewage sludge application, and air deposition (as discussed in section 8.2.1). The accumulation process is influenced by various factors, including the following.

8.2.2.3 Adsorption of Soil Particles

Adsorption is a phenomenon that occurs when MNPs adhere to soil particles. When plastic particles interact with soil, they have the ability to adhere to soil particles and become deeply embedded inside the soil matrix. The occurrence of this interaction arises from various factors that encompass the electrostatic attraction between the charged surfaces of plastics and soil particles, together with the surface chemistry of both materials (Agboola and Benson, 2021).

The adsorption process facilitates the profound incorporation of MNPs inside the soil structure. Distinct particles are not present, but rather they exist as a constituent of the soil matrix. The strong adhesion of plastics to soil particles poses significant challenges in the extraction process. Furthermore, it may be inferred that the introduction of extra plastic particles into the soil will exacerbate the issue of accumulation (Guerrera et al., 2021). According to Guerrera et al. (2021), the presence of adsorbed plastic particles in the soil can serve as a persistent repository of plastics, thereby impacting soil health and potentially being taken up by plants.

8.2.2.4 Reduced Mobility

One of the distinctive characteristics of MNPs is their restricted movement inside soil. The plastic particles exhibit a tendency to remain immobile within the soil environment owing to their minuscule dimensions and absence of active locomotion mechanisms. According to Gong et al. (2023), MNPs exhibit a distinct behavior in comparison to other contaminants that have the capacity to permeate through soil layers and potentially contaminate groundwater. These plastics prefer to persist within the uppermost layers of soil, where they were originally deposited (Kiran et al., 2022). The limited range of movement exhibited by MNPs is of utmost significance, since it primarily results in their predominant accumulation within plant root systems. Plastic particles have a tendency to accumulate inside the upper layers of soil, which coincide with the regions where plant roots exhibit heightened activity in nutrient and water uptake. Consequently, there is an increased probability that plants may encounter MNPs, which may be absorbed by them, thereby giving rise to worries regarding their possible integration into the food chain (Kiran et al., 2022).

8.2.2.5 Continuous Input

The persistent influx of plastics originating from many sources intensifies the issue of plastic accumulation in soil. As previously mentioned, the environment is consistently exposed to MNPs as a result of plastic waste disposal, degradation of plastic mulch film, and various other factors. The concentrations of plastic particles in soil have the potential to increase over time, provided that the sources of these particles persist (Bhat et al., 2023). The continuous influx of plastic waste poses a notable challenge as it implies that the enduring presence of plastics in the soil would endure, despite any mitigation measures aimed at reducing future plastic pollution. The importance of implementing comprehensive policies that effectively tackle the prevention of plastic pollution as well as the management of current plastic contamination in soil. This approach is crucial in mitigating the possible long-term consequences on soil health, agriculture, and food security (Tian et al., 2022).

8.2.3 Factors Influencing Uptake

The extent to which plants absorb MNPs from the soil is influenced by a complex interplay of multiple factors. It is vital to comprehend these attributes as they play a crucial role in the variability of plastic absorption among different plant species and environmental situations (Chen et al., 2022). This section examines the primary factors that impact the absorption of plastic by plants.

8.2.3.1 Plastic Particle Size

The absorption of MNP particles by plants is influenced by their size. The infiltration of plant tissues by plastic particles, particularly nanoparticles, is facilitated to a greater extent by their smaller size. Due to their diminutive dimensions, these organisms possess the ability to infiltrate plant cells by exploiting minute apertures, such as stomata found on leaves or the cell walls of roots (Allouzi et al., 2021). The enhanced surface area–to–volume ratio of diminutive plastic particles augments their capacity to absorb plant-accessible nutrients and other chemical substances, hence potentially influencing the well-being of plants (Allouzi et al., 2021).

8.2.3.2 Soil Type

The soil's texture, such as sandy, loamy, or clayey, and its composition, including the concentration of organic matter, significantly influence the process of plastic uptake. Sandy soils possess larger interstitial gaps between particles, which can facilitate the rapid downward movement of plastics, hence potentially reducing the likelihood of root exposure (Maity et al., 2022).

According to Wang et al. (2022), soils characterized by a greater cation exchange capacity (CEC), which refers to their capacity to retain and exchange ions, have enhanced plastic binding capabilities. Consequently, this phenomenon leads to a reduction in the bioavailability of plastics to plants.

The interaction between soil organic matter and plastics has the potential to modify the mobility of plastics and their uptake by plants. Soil possessing a higher organic matter content may have an enhanced potential to retain plastics due to the presence of an increased number of adsorption sites (Maity et al., 2022).

8.2.3.3 Plant Species

Various plant types exhibit varying levels of plastic absorption. Certain plant species demonstrate a higher level of efficacy in the absorption of MNPs as compared to others. The observed diversity can be attributed to alterations in root morphology, physiological characteristics, and metabolic pathways. Plants that possess specialist root structures, such as fibrous roots, may exhibit an increased surface area, hence facilitating plastic contact and uptake (Lynch, 2022).

8.2.3.4 Environmental Conditions

The moisture content of soil is a determining factor in the movement of polymers inside the soil matrix. The transportation of plastics through soil may be facilitated to a greater extent under wetter conditions, thereby leading to an enhanced accessibility for plant uptake (Kumar et al., 2023). The degradation of plastics and their interactions with soil and plant roots can be influenced by fluctuations in temperature. Annual variations in temperature have the potential to impact the pace at which plastic is absorbed, as suggested by Kumar et al. (2023). Soil pH can exert an influence on the chemical speciation of plastics and their interactions with soil minerals. The mobility and bioavailability of plastic can be influenced by fluctuations in pH (Azeem et al., 2021).

8.3 IMPACT OF MICROPLASTICS AND NANOPLASTICS ON FOOD INSECURITY

8.3.1 Crop Contamination by Microplastics and Nanoplastics

MNPs, which accumulate in edible sections of crops, represent a direct threat to food safety and quality. Plastic particles are taken up by plant roots and then translocated within the plant, causing pollution (Wang et al., 2023). Crop contamination by MNPs has a wide range of consequences:

Food safety concerns: The accumulation of plastics in edible crops presents a potential hazard for food chain contamination. According to Focker et al. (2022), MNPs possess the capacity to facilitate the transportation of various chemical additives and contaminants from the surrounding environment. Contamination can manifest in the form of heavy metals, plasticizers, and various other deleterious substances. Consuming foods that are contaminated can potentially present health risks to individuals (Ng et al., 2018).

Impact on nutritional quality: MNPs in crops can also have an impact on their nutritional content. Plastics may disrupt nutrient uptake and movement inside plants, potentially lowering the nutritional value of harvested crops. This could have consequences for dietary adequacy, particularly in countries where staple crops are damaged (Chen et al., 2022).

Consumer perception: The presence of MNP pollutants in agricultural produce has the potential to erode consumer confidence in the safety and integrity of food items. The presence of plastic contaminants in food has the potential to negatively impact perceptions of the safety of agricultural practices, hence altering consumer preferences and trust in the food industry (Maddela et al., 2022).

Regulatory and quality control challenges: The identification and reduction of plastic contamination in agricultural produce pose significant issues in terms of regulatory compliance and quality assurance. The continuous concern for regulatory authorities and the food sector is the development of efficient testing methods and establishment of appropriate standards to address the issue of plastic contamination in food products (Yusuf et al., 2022).

In order to effectively mitigate the problem of crop contamination, it is essential to implement strategies aimed at minimizing the introduction of MNPs into the soil and their subsequent absorption by plants. Implementing strategies such as enhanced waste management, the adoption of sustainable agricultural practices, and the advancement of plastic-free alternatives within the agricultural sector are crucial measures in ensuring the preservation of food safety and quality. Furthermore, it is imperative to conduct further investigations on the enduring consequences of plastic pollution in agricultural produce in order to comprehend fully the possible hazards to human well-being and to formulate remedies that are grounded in empirical data (Yusuf et al., 2022).

8.3.2 Impact on Soil Health and Crop Productivity

Soil health and agricultural productivity may be negatively affected by the presence of MNPs in soil ecosystems. Plastics can change the composition of soil and its microbial community, which can have negative effects on crop health and yield (Pérez-Reverón et al., 2022).

8.3.2.1 Alteration of Soil Properties

Physical changes: Plastics possess the capacity to alter the attributes of soil through the disruption of its physical structure. The presence of plastics in the soil might result in the formation of areas characterized by reduced air circulation, diminished water infiltration, and limited root penetration. The interference described has the potential to impede the development of vegetation and hinder its capacity to assimilate essential nutrients (Zhou et al., 2023).

Chemical changes: Plastic materials possess the capacity to engage in chemical interactions with soil, hence potentially altering the accessibility of nutrients. The absorption of important nutrients by plastic particles can pose challenges for plants in acquiring these essential elements. In contrast, plastics have the potential to release chemicals or additives into the soil, thereby potentially altering its chemical composition (Zhou et al., 2023).

8.3.2.2 Microbial Communities

Microbial diversity: The presence of plastics in soil can influence the diversity and composition of microbial communities inhabiting the soil environment. The growth of specific microbial populations may be inhibited by plastics, whereas the presence of plastics may also hamper the growth of other microbial species. Changes in the microbial diversity of soil can impact nutrient cycling, the decomposition of organic matter in the soil, and the overall soil fertility (Chen et al., 2022).

Nutrient cycling: Microorganisms play a crucial role in nutrient cycling as they facilitate the decomposition of organic materials, so enabling plants to get essential nutrients. Disruptions in microbial communities have the potential to hinder the intricate processes of nutrient cycling, leading to a decrease in the accessibility of essential nutrients for agricultural crops (Pérez-Reverón et al., 2022).

8.3.3 Crop Health and Productivity

Reduced crop yield: The potential occurrence of diminished agricultural productivity may be attributed to a convergence of causes encompassing alterations in soil attributes and microbial populations. The presence of plastics can impede the development of plant roots, hinder their capacity to uptake nutrients, and compromise their overall well-being, resulting in reduced agricultural productivity (Huang et al., 2023).

Food insecurity: The potential exacerbation of food insecurity in regions heavily dependent on agriculture for livelihood may arise due to diminished crop yields resulting from soil degradation caused by the presence of plastics. Consequently, agricultural communities may have heightened economic susceptibility and a growing dependence on imported food sources (Huang et al., 2023).

The maintenance of food security in our rapidly changing global landscape is contingent upon our capacity to mitigate the adverse impacts of plastics on soil health and crop productivity. Examples of sustainable agricultural practices that can mitigate the impact of these issues include the reduction of plastic usage, improved waste management, and the implementation of soil conservation measures. In order to ensure the protection of global food supply, it is imperative to conduct research on the enduring consequences of plastic pollution in soil ecosystems and develop innovative methods to remediate soils impacted by plastic contamination.

8.3.4 Ecological Implications of Microplastics and Nanoplastics Pollution in Soil

The issue of plastic pollution in soil extends beyond the immediate contamination of crops and covers a variety of wider ecological concerns. The aforementioned issues originate from the significant modifications that plastics can impose on soil ecosystems, potentially causing disturbances to intrinsic processes and interdependencies crucial for biodiversity and the provision of ecosystem services that sustain food production (Liu et al., 2021).

8.3.4.1 Soil Ecosystem Disruption

Soil microorganisms are essential to nutrient cycling, breakdown of organic materials, and overall soil health. Microbe populations may be impacted by plastic pollution, altering their structure and function. Alterations in the composition and activity of soil microbes can have far-reaching effects on the cycling of nutrients in the soil and the health of entire ecosystems (Liu et al., 2021). Soil aeration, nitrogen cycling, and organic matter digestion are all processes in which earthworms and arthropods play an important part. There is concern that plastic pollution could have direct or indirect effects on these animals, leading to a reduction in population size or variety (Wang et al., 2023).

8.3.4.2 Biodiversity Impact

Connectivity between soil and plant communities occurs through processes such as nitrogen cycling and plant-soil feedbacks. Plastic pollution can alter soil health and composition, which in turn affects plant and animal distribution and abundance (Rai et al., 2021). Some plant and microbial species may benefit more than others from plastic pollution, which could alter the makeup of plant and soil communities. Impacts on higher trophic levels may rise from lower ones as a result of these shifts (Hurley and Nizzetto, 2018; Rai et al., 2021).

8.3.4.3 Ecosystem Services

Ecosystem services, such as nitrogen cycling and soil fertility, are very important and can only be provided by healthy soil ecosystems. These functions can be jeopardized by plastic-induced disruptions in the nutrient cycle, which in turn reduces ecological and agricultural productivity (Hurley and Nizzetto, 2018). Plastics can change the way soil behaves, which has consequences for water infiltration and runoff. Alterations to water flow patterns, such as the increased transport of toxins into aquatic ecosystems, can have an effect on water quality (Walker and Fequet, 2023).

8.3.4.4 Long-Term Consequences

MNPs, due to their extreme durability, can store many different types of environmental toxins. In soil ecosystems, plastics may soak up and hold contaminants like pesticides and heavy metals, effectively sealing them inside the soil matrix. The potential for these toxins to leak out into the soil and water supply increases the longer plastics are left there due to their persistence in the environment (Bhat et al., 2023). This event is dangerous in more ways than one. To begin with, agricultural output might be negatively impacted by the gradual discharge of accumulated pollutants due to their negative effects on soil health and fertility. Second, these poisons threaten aquatic ecosystems and water sources as they make their way into water bodies through runoff and leaching. Reducing plastic pollution in soils is crucial to larger environmental conservation efforts due to the long-lasting nature of plastics (Bhat et al., 2023).

8.4 CONCLUSION

The issue of MNPs in soil is a multifaceted one with significant consequences for food security. Gaining a comprehensive understanding of the origins, routes, and mechanisms of plastic pollution in soil is necessary in order to formulate efficient measures for its reduction. The urgency of tackling the issue of plastics rests in their impact on crop contamination, soil health, and ecosystems, which in turn poses a threat to food supplies and global food security. Ongoing research, the formulation of policies, and the cultivation of public awareness are essential measures in addressing the potential hazards associated with MNPs present in soil.

REFERENCES

Adhikari, S., Kelkar, V., Kumar, R., & Halden, R. U. (2022). Methods and challenges in the detection of microplastics and nanoplastics: A mini-review. *Polymer International*, *71*(5), 543–551.

Agboola, O. D., & Benson, N. U. (2021). Physisorption and chemisorption mechanisms influencing micro (nano) plastics-organic chemical contaminants interactions: A review. *Frontiers in Environmental Science*, *9*, 167.

Alimi, O. S., Farner Budarz, J., Hernandez, L. M., & Tufenkji, N. (2018). Microplastics and nanoplastics in aquatic environments: Aggregation, deposition, and enhanced contaminant transport. *Environmental science & technology*, *52*(4), 1704–1724.

Allouzi, M. M. A., Tang, D. Y. Y., Chew, K. W., Rinklebe, J., Bolan, N., Allouzi, S. M. A., & Show, P. L. (2021). Micro (nano) plastic pollution: The ecological influence on soil-plant system and human health. *Science of the Total Environment*, *788*, 147815.

Aransiola, S. A., Ikhumetse, A. A., Babaniyi, B. R., Abioye, O. P., Oyedele, O. J., & Falade. N. O. (2023). Phytoaccumulation of micro and nanoplastics: Root Uptake. In N. R. Maddela, K. V. Reddy, & P. Ranjit (Eds.), *Micro and Nanoplastics in Soil*. Springer Cham. https://doi.org/10.10007/978-3-031-21195-9_8.

Aransiola, S. A., Victor-Ekwebelem, M. O., Ajiboye, A. E., Leh-Togi Zobeashia, S. S., Ijah, U. J. J., & Oyedele, O. J. (2023). Ecological impacts and toxicity of micro and nanoplastics in agroecosystem. In N. R. Maddela, K. V. Reddy, & P. Ranjit (Eds.), *Micro and Nanoplastics in Soil*. Springer Cham. https://doi.org/10.10007/978-3-031-21195-9_10.

Astner, A. F., Hayes, D. G., O'Neill, H., Evans, B. R., Pingali, S. V., Urban, V. S., & Young, T. M. (2019). Mechanical formation of micro-and nano-plastic materials for environmental studies in agricultural ecosystems. *Science of the Total Environment*, *685*, 1097–1106.

Auta, S. H., Abioye, O. P., Aransiola, S. A., Bala, J. D., Chukwuemeka, V. I., Hassan, A., Aziz, A., & Fauziah, S. H. (2022). Enhanced microbial degradation of PET and PS microplastics under natural conditions in mangrove environment. *Journal of Environmental Management*, *304*(2022), 114273. https://doi.org/10.1016/j.jenvman.2021.114273

Azeem, I., Adeel, M., Ahmad, M. A., Shakoor, N., Jiangcuo, G. D., Azeem, K., & Rui, Y. (2021). Uptake and accumulation of nano/microplastics in plants: A critical review. *Nanomaterials*, *11*(11), 2935.

Bandini, F., Taskin, E., Bellotti, G., Vaccari, F., Misci, C., Guerrieri, M. C., & Puglisi, E. (2022). The treatment of the organic fraction of municipal solid waste (OFMSW) as a possible source of micro-and nano-plastics and bioplastics in agroecosystems: A review. *Chemical and Biological Technologies in Agriculture*, *9*(1), 1–17.

Bhat, M. A., Gedik, K., & Gaga, E. O. (2023). Atmospheric micro (nano) plastics: Future growing concerns for human health. *Air Quality, Atmosphere & Health*, *16*(2), 233–262.

Bratovcic, A. (2019). Degradation of micro-and nano-plastics by photocatalytic methods. *Journal of Nanoscience and Nanotechnology Applications*, *3*, 206.

Cao, X., Liang, Y., Jiang, J., Mo, A., & He, D. (2023). Organic additives in agricultural plastics and their impacts on soil ecosystems: Compared with conventional and biodegradable plastics. *TrAC Trends in Analytical Chemistry*, 117212.

Chen, G., Li, Y., Liu, S., Junaid, M., & Wang, J. (2022). Effects of micro (nano) plastics on higher plants and the rhizosphere environment. *Science of the Total Environment*, *807*, 150841.

Focker, M., van Asselt, E. D., Berendsen, B. J. A., van de Schans, M. G. M., van Leeuwen, S. P. J., Visser, S. M., & Van der Fels-Klerx, H. J. (2022). Review of food safety hazards in circular food systems in Europe. *Food Research International*, *158*, 111505.

Gong, X., Shi, G., Zou, D., Wu, Z., Qin, P., Yang, Y., . . . Zhou, Y. (2023). Micro-and nano-plastics pollution and its potential remediation pathway by phytoremediation. *Planta*, *257*(2), 35.

Guerrera, M. C., Aragona, M., Porcino, C., Fazio, F., Laurà, R., Levanti, M. & Germanà, A. (2021). Micro and nano plastics distribution in fish as model organisms: Histopathology, blood response and bioaccumulation in different organs. *Applied Sciences*, *11*(13), 5768.

He, D., Zhang, Y., & Gao, W. (2021). Micro (nano) plastic contaminations from soils to plants: Human food risks. *Current Opinion in Food Science*, *41*, 116–121.

Huang, F., Hu, J., Chen, L., Wang, Z., Sun, S., Zhang, W. & Fang, L. (2023). Microplastics may increase the environmental risks of Cd via promoting Cd uptake by plants: A meta-analysis. *Journal of Hazardous Materials*, *448*, 130887.

Hurley, R. R., & Nizzetto, L. (2018). Fate and occurrence of micro (nano) plastics in soils: Knowledge gaps and possible risks. *Current Opinion in Environmental Science & Health*, *1*, 6–11.

Iqbal, S., Xu, J., Allen, S. D., Khan, S., Nadir, S., Arif, M. S., & Yasmeen, T. (2020). Unraveling consequences of soil micro-and nano-plastic pollution on soil-plant system: Implications for nitrogen (N) cycling and soil microbial activity. *Chemosphere*, *260*, 127578.

Kiran, B. R., Kopperi, H., & Venkata Mohan, S. (2022). Micro/nano-plastics occurrence, identification, risk analysis and mitigation: Challenges and perspectives. *Reviews in Environmental Science and Bio/Technology*, *21*(1), 169–203.

Kumar, V., Singh, E., Singh, S., Pandey, A., & Bhargava, P. C. (2023). Micro-and nano-plastics (MNPs) as emerging pollutant in ground water: Environmental impact, potential risks, limitations and way forward towards sustainable management. *Chemical Engineering Journal*, *459*, 141568.

Liu, Y., Huang, Q., Hu, W., Qin, J., Zheng, Y., Wang, J. & Xu, L. (2021). Effects of plastic mulch film residues on soil-microbe-plant systems under different soil pH conditions. *Chemosphere*, *267*, 128901.

Lynch, J. P. (2022). Harnessing root architecture to address global challenges. *The Plant Journal*, *109*(2), 415–431.

Maddela, N. R., Ramakrishnan, B., Kakarla, D., Venkateswarlu, K., & Megharaj, M. (2022). Major contaminants of emerging concern in soils: A perspective on potential health risks. *RSC Advances*, *12*(20), 12396–12415.

Maity, S., Guchhait, R., Sarkar, M. B., & Pramanick, K. (2022). Occurrence and distribution of micro/nanoplastics in soils and their phytotoxic effects: A review. *Plant, Cell & Environment*, *45*(4), 1011–1028.

Musa, I. O., Auta, H. S., Ilyasu, U. S., Aransiola, S. A., Makun, H. A., Adabara, N. U., Abioye, O. P., Aziz, A., Jayanthi, B., Maddela, N. R., & Prasad, R. (2024). Micro and nanoplastics in environment – degradation, detection, ecological impact. *International Journal of Environmental Research*, *18*, 1. https://doi.org/10.1007/s41742-023-00551-9

Ng, E. L., Lwanga, E. H., Eldridge, S. M., Johnston, P., Hu, H. W., Geissen, V., & Chen, D. (2018). An overview of microplastic and nanoplastic pollution in agroecosystems. *Science of the Total Environment*, *627*, 1377–1388.

Peng, L., Mehmood, T., Bao, R., Wang, Z., & Fu, D. (2022). An overview of micro (nano) plastics in the environment: Sampling, identification, risk assessment and control. *Sustainability*, *14*(21), 14338.

Pérez-Reverón, R., Álvarez-Méndez, S. J., Kropp, R. M., Perdomo-González, A., Hernández-Borges, J., & Díaz-Peña, F. J. (2022). Microplastics in agricultural systems: Analytical methodologies and effects on soil quality and crop yield. *Agriculture*, *12*(8), 1162.

Perković, S., Paul, C., Vasić, F., & Helming, K. (2022). Human health and soil health risks from heavy metals, micro (nano) plastics, and antibiotic resistant bacteria in agricultural soils. *Agronomy*, *12*(12), 2945.

Qi, R., Jones, D. L., Li, Z., Liu, Q., & Yan, C. (2020). Behavior of microplastics and plastic film residues in the soil environment: A critical review. *Science of the Total Environment*, *703*, 134722.

Rai, P. K., Lee, J., Brown, R. J., & Kim, K. H. (2021). Micro-and nano-plastic pollution: Behavior, microbial ecology, and remediation technologies. *Journal of Cleaner Production*, *291*, 125240.

Rubio, L., Marcos, R., & Hernández, A. (2020). Potential adverse health effects of ingested micro-and nanoplastics on humans. Lessons learned from in vivo and in vitro mammalian models. *Journal of Toxicology and Environmental Health, Part B*, *23*(2), 51–68.

Song, Y., Cao, C., Qiu, R., Hu, J., Liu, M., Lu, S., & He, D. (2019). Uptake and adverse effects of polyethylene terephthalate microplastics fibers on terrestrial snails (Achatina fulica) after soil exposure. *Environmental Pollution*, *250*, 447–455.

Sorasan, C., Edo, C., González-Pleiter, M., Fernández-Piñas, F., Leganés, F., Rodríguez, A., & Rosal, R. (2022). Ageing and fragmentation of marine microplastics. *Science of the Total Environment*, *827*, 154438.

Tian, W., Song, P., Zhang, H., Duan, X., Wei, Y., Wang, H., & Wang, S. (2022). Microplastic materials in the environment: Problem and strategical solutions. *Progress in Materials Science*, 101035.

Ullah, R., Tsui, M. T. K., Chow, A., Chen, H., Williams, C., & Ligaba-Osena, A. (2023). Micro (nano) plastic pollution in terrestrial ecosystem: Emphasis on impacts of polystyrene on soil biota, plants, animals, and humans. *Environmental Monitoring and Assessment*, *195*(1), 252.

van den Berg, P., Huerta-Lwanga, E., Corradini, F., & Geissen, V. (2020). Sewage sludge application as a vehicle for microplastics in eastern Spanish agricultural soils. *Environmental Pollution*, *261*, 114198.

Walker, T. R., & Fequet, L. (2023). Current trends of unsustainable plastic production and micro (nano) plastic pollution. *TrAC Trends in Analytical Chemistry*, 116984.

Wang, F., Feng, X., Liu, Y., Adams, C. A., Sun, Y., & Zhang, S. (2022). Micro (nano) plastics and terrestrial plants: Up-to-date knowledge on uptake, translocation, and phytotoxicity. *Resources, Conservation and Recycling*, *185*, 106503.

Wang, J., Zhang, X., Li, X., & Wang, Z. (2023). Exposure pathways, environmental processes and risks of micro (nano) plastics to crops and feasible control strategies in agricultural regions. *Journal of Hazardous Materials*, 132269.

Yusuf, A., Sodiq, A., Giwa, A., Eke, J., Pikuda, O., Eniola, J. O., & Bilad, M. R. (2022). Updated review on microplastics in water, their occurrence, detection, measurement, environmental pollution, and the need for regulatory standards. *Environmental Pollution*, *292*, 118421.

Zhou, J., Jia, R., Brown, R. W., Yang, Y., Zeng, Z., Jones, D. L., & Zang, H. (2023). The long-term uncertainty of biodegradable mulch film residues and associated microplastics pollution on plant-soil health. *Journal of Hazardous Materials*, *442*, 130055.

9 Environmental Pollution and the Entrance of Toxic Elements into the Food Chain

Babafemi Raphael Babaniyi, Isaac Gbolahan Olamide, Damilola Eunice Fagbamigbe, Joshua Ibukun Adebomi, and Iyabode Felicia Areo

9.1 INTRODUCTION

The environment is an intricate system encompassing ecosystems, biodiversity, and natural resources, serving as the foundational support for life on Earth. Understanding and studying the environment are imperative for addressing contemporary challenges, including climate change, biodiversity loss, and environmental pollution, that impact the intricate balance of our planet. Environmental science, at its core, delves into ecosystems: complex networks of living organisms interacting with their physical surroundings, influencing climate regulation, nutrient cycles, and providing essential ecosystem services (Daily, 1997; Millennium Ecosystem Assessment, 2005). Biodiversity, encapsulating genetic diversity, species diversity, and ecosystem diversity, assumes a pivotal role in the resilience, adaptation, and sustainability of ecosystems (CBD, 1992; Sala et al., 2000; Cardinale et al., 2012; Babaniyi et al., 2023).

The contemporary environment experiences unprecedented changes primarily driven by anthropogenic activities. Climate change, attributed to the escalating levels of greenhouse gas emissions, presents substantial challenges to ecosystems and global societies (IPCC, 2014; Hoegh-Guldberg et al., 2018). Environmental pollution, a critical global concern resulting from human activities, spans air, water, soil, noise, and light pollution, impacting human health, biodiversity, and ecosystems (UNEP, 2018; Landrigan et al., 2018; Aransiola, Victor-Ekwebelem et al., 2021). In response to these challenges, the concept of sustainable development gains prominence, emphasizing the equilibrium between economic growth, environmental conservation, and social equity for long-term planetary health (Brundtland, 1987; United Nations, 2015). Addressing environmental issues necessitates interdisciplinary collaboration, integrating fields such as environmental science, ecology, climatology, and policy studies. A holistic understanding of the interconnectedness between natural systems and human activities is essential for comprehensive solutions (Leichenko & O'Brien, 2008; Steffen et al., 2015). This introduction offers a glimpse into the diverse facets of the environment and its multifaceted challenges, urging the adoption of a comprehensive and collaborative approach. Integrating scientific research, policy initiatives, and public engagement becomes imperative to foster a sustainable and resilient future.

Environmental pollution is a pervasive and critical issue, and is defined as the introduction of contaminants into the environment, leading to adverse changes that impact ecosystems, biodiversity, and human well-being. This multifaceted problem stems from both natural processes, such as volcanic eruptions and wildfires, and human activities like industrial processes, transportation, agriculture, and waste disposal. Pollution manifests in various forms across air, water, soil, noise, and light, each necessitating specific monitoring and mitigation approaches. Air pollution, primarily caused by the combustion of fossil fuels and industrial emissions, poses significant threats to

DOI: 10.1201/9781032683751-11

human health and the environment (World Health Organization, 2006; Lelieveld et al., 2015). The consequences are far-reaching, contributing to respiratory diseases and cardiovascular problems. Water pollution, arising from industrial discharges, agricultural runoff, and marine activities like oil spills, jeopardizes aquatic ecosystems and human health (UNEP, 2006; Carpenter et al., 1998). The impact extends beyond immediate environmental concerns to include economic challenges affecting industries, agriculture, and healthcare systems (Díaz et al., 2019; Landrigan et al., 2018; World Bank, 2019). Soil pollution, caused by pesticides, heavy metals, and improper waste disposal, affects soil fertility, structure, and the health of plants and animals (Kabata-Pendias & Mukherjee, 2007; Panagos et al., 2013; Abioye et al., 2017; Aransiola et al., 2022). Noise pollution, stemming from vehicle traffic, industrial activities, and recreational events, causes stress and adverse health effects (Basner et al., 2014; Münzel et al., 2014). Furthermore, light pollution, emanating from streetlights, urban development, and outdoor advertising, interferes with natural darkness, impacting ecosystems and human health (Falchi et al., 2016; Gaston et al., 2014).

The impacts of environmental pollution are vast, encompassing the loss of biodiversity, ecological disruption, and adverse health effects on humans. These include not only immediate health concerns but also long-term consequences such as economic burdens on various sectors. Addressing these challenges necessitates a comprehensive, interdisciplinary approach that incorporates regulatory frameworks, technological innovations, and global collaboration (UNEP, 2018; Gao et al., 2016). Sustainable practices, clean technologies, and eco-friendly lifestyles emerge as integral components of preventing further environmental degradation (Leal Filho et al., 2019; IPCC, 2018).

As societies strive for sustainable development, the urgent need to address environmental pollution becomes paramount. A thorough understanding of its sources, types, and impacts is crucial for developing effective strategies to mitigate adverse effects and safeguard the health of ecosystems and human societies. This requires a holistic approach that considers the interconnectedness of environmental, social, and economic systems. Regulatory frameworks play a pivotal role in curbing pollution by establishing standards, monitoring mechanisms, and enforcement protocols. Governments and international organizations need to collaborate to create and enforce stringent regulations that hold industries and individuals accountable for their environmental impact. Additionally, technological innovations can provide viable solutions to mitigate pollution. Advancements in renewable energy, waste management technologies, and green infrastructure can significantly reduce the environmental footprint of human activities (Aransiola et al., 2019).

Global collaboration is essential in tackling pollution, as environmental issues often transcend national boundaries. Initiatives like international agreements, partnerships, and knowledge-sharing platforms can facilitate the exchange of best practices and technologies. The sharing of expertise can help countries with varying levels of development to implement effective pollution control measures. Educating and raising awareness among the public is another crucial aspect of pollution mitigation. Sustainable practices and eco-friendly lifestyles need to be promoted to instill a sense of responsibility among individuals. This includes advocating for reduced consumption, proper waste disposal, and the adoption of environmentally friendly alternatives.

9.2 POLLUTION

Pollution, a pervasive and pressing environmental challenge, emerges from the release of contaminants into the natural environment, presenting formidable threats to ecosystems, human health, and the overall quality of life. The multifaceted nature of pollution is evident in its various forms, encompassing air pollution, water pollution, soil contamination, noise pollution, and light pollution, each characterized by distinct sources and far-reaching consequences (Table 9.1).

Air pollution, arising from diverse origins such as industrial emissions, vehicular exhaust, combustion of fossil fuels, and natural phenomena like wildfires and volcanic activities, casts a shadow of adverse impacts. These include the insidious effects of respiratory and cardiovascular diseases,

TABLE 9.1
Pollution Types and Sources

Air	
	– Industrial discharges
	– Vehicular exhaust
	– Burning of fossil fuels
	– Natural sources (wildfires, volcanic activities)
Water	
	– Industrial discharges
	– Agricultural runoff
	– Untreated sewage
Soil	
	– Agricultural chemicals
	– Industrial waste
	– Improper disposal of hazardous materials
	– Mining activities
Noise	
	– Urbanization
	– Industrial activities
	– Vehicular traffic
	– Loud machinery
Light	
	– Artificial lighting from streetlights, buildings, outdoor fixtures

the overarching concern of global warming, the corrosive impact of acid rain on ecosystems, and widespread damage to flora and fauna (World Health Organization, 2018; Lelieveld et al., 2015).

Water pollution, emanating from sources as varied as industrial discharges, agricultural runoff, untreated sewage, and catastrophic oil spills, engenders myriad challenges. From the contamination of drinking water sources to the disruptive influence on aquatic ecosystems, the consequences extend to the loss of biodiversity and the pervasive threats posed to aquatic life (United Nations Environment Programme, 2016; Carpenter et al., 1998).

Soil contamination, a consequence of agricultural chemicals, industrial waste, improper disposal of hazardous materials, and intensive mining activities, unveils a realm of concerns. These include compromised agricultural productivity due to reduced crop yields, the degradation of soil quality, the perilous contamination of groundwater, and the consequential threats to human health (Alloway, 2013; Khan et al., 2009).

Noise pollution, a by-product of urbanization, industrial activities, vehicular traffic, and the operation of loud machinery, manifests in a spectrum of adverse impacts. Ranging from the threat of hearing loss to the pervasive influence of stress and sleep disturbances, the repercussions extend to the negative effects on wildlife, disrupting ecosystems and threatening biodiversity (Basner et al., 2014).

Light pollution, stemming from the pervasive glow of artificial lighting through streetlights, buildings, and outdoor fixtures, introduces a novel dimension of challenges. These include the disruptive interference with natural ecosystems, the compromising of astronomical observations, and the pervasive adverse effects on human health, particularly related to disrupted circadian rhythms and sleep patterns (Falchi et al., 2016).

Mitigating the complex issue of pollution necessitates a multifaceted approach involving regulatory measures, technological innovations, and concerted public awareness campaigns. The global nature of pollution requires collaborative efforts to develop sustainable solutions and address the interconnected environmental and health challenges posed by pollution. Only through such concerted actions can we hope to create a healthier and more sustainable future for our planet and its inhabitants.

9.3 TOXIC ELEMENTS

Toxic elements, also known as heavy metals, metalloids, and other hazardous substances, pose significant threats to human health and the environment. These elements, often naturally occurring, can become pollutants when released into the air, water, or soil through anthropogenic activities such as industrial processes, mining, and improper waste disposal. This section will explore the sources, effects, and mitigation strategies related to toxic elements, drawing upon well-cited references to provide a comprehensive understanding of this critical environmental and public health concern.

Sources of Toxic Elements

1. *Industrial activities*: Industrial processes are major contributors to the release of toxic elements into the environment. Factories and manufacturing plants can release heavy metals such as lead (Pb), mercury (Hg), and cadmium (Cd) as by-products of production processes (Nriagu, 1988).
2. *Mining and smelting*: Mining activities, especially for metals like Pb, arsenic (As), and chromium (Cr), contribute significantly to the release of toxic elements. The extraction and processing of ores release these elements into the surrounding air, water, and soil (Alloway, 2013).
3. *Agricultural practices*: The use of certain fertilizers and pesticides can introduce toxic elements into the soil. Contamination can occur through the application of phosphate-based fertilizers, leading to the presence of Cd and other heavy metals in agricultural products (McBride, 1994).
4. *Waste disposal*: Improper disposal of electronic waste and other hazardous materials can result in the leaching of toxic elements into the soil and water. Landfills and incineration facilities may release pollutants like Pb, Hg, and As into the environment (Agency for Toxic Substances and Disease Registry, 2007).

Effects of Toxic Elements

1. *Human health*: Exposure to toxic elements poses severe health risks. Exposure to Pb, for example, is associated with cognitive impairment in children (Needleman, 2004), while Hg exposure can lead to neurological and developmental issues (Grandjean et al., 2014). Chronic exposure to As in drinking water is linked to skin, lung, and bladder cancers (International Agency for Research on Cancer, 2012) (Table 9.2).
2. *Environmental impact*: Toxic elements can have detrimental effects on ecosystems. Mercury, for instance, bioaccumulates in aquatic environments, leading to contamination of fish and posing risks to higher trophic levels, including humans (Eagles-Smith et al., 2008). Soil contamination with heavy metals can disrupt plant growth and microbial activity, affecting overall ecosystem health (Alloway, 2013).

Mitigation Strategies

1. *Regulatory measures*: Strengthening and enforcing regulations on industrial emissions, waste disposal, and mining practices are crucial steps in preventing the release of toxic

TABLE 9.2
Toxic Elements and Their Health Implications

Toxic Element	Sources	Health Implications	Environmental Implications
Arsenic	Natural deposits in soil	Neurological effects	Water contamination, posing risks to aquatic life
	Agricultural pesticides	Anemia and developmental issues	Soil contamination, negatively impacting crops
	Industrial processes	Cardiovascular and respiratory issues	Bioaccumulation in aquatic and terrestrial ecosystems
Cadmium	Mining and smelting	Cognitive impairment in children	Air pollution through combustion processes
	Battery manufacturing	Respiratory issues	Contamination of fish, posing risks to humans
Chromium	Industrial discharges	Skin, lung, and bladder cancers	Groundwater contamination, affecting drinking water
	Leather tanning	Minamata disease (methylmercury exposure)	Soil contamination, reducing crop yields
	Stainless steel production	Dermatitis and other skin issues	Water contamination, harmful to aquatic organisms
Lead	Industrial processes	Cardiovascular effects	Bioaccumulation in aquatic ecosystems
	Lead-based paints	Kidney damage and osteoporosis	Soil contamination, harmful to plants and animals
	Battery manufacturing	Neurological effects in adults	Water contamination, posing risks to aquatic life
Mercury	Coal combustion	Respiratory and gastrointestinal issues	Soil contamination, affecting plant growth
	Industrial processes	Cardiovascular and respiratory issues	Water contamination, harmful to aquatic life
	Artisanal gold mining	Lung cancer (hexavalent chromium exposure)	Contamination of fish, posing risks to humans

elements. Stringent standards can limit the permissible levels of these elements in air, water, and soil, safeguarding both human health and the environment (Hopenhayn-Rich et al., 1996).

2. *Waste management*: Implementing proper waste disposal methods, particularly for electronic waste and hazardous materials, is essential. Recycling programs and the promotion of responsible disposal practices can reduce the leaching of toxic elements from landfills (Widmer et al., 2005).
3. *Soil remediation*: Techniques such as phytoremediation, involving the use of plants to extract, stabilize, or degrade toxic elements, can be employed to remediate contaminated soils (Salt et al., 1995). Additionally, soil amendments and microbial processes can be harnessed to immobilize or transform toxic elements, reducing their bioavailability (Kumpiene et al., 2006).
4. *Water treatment*: Implementing effective water treatment processes is crucial for removing toxic elements from drinking water sources. Advanced treatment methods, including coagulation, filtration, and ion exchange, can help mitigate the risks associated with heavy metal contamination (Agency for Toxic Substances and Disease Registry, 2007).

9.4 RELATIONSHIP BETWEEN POLLUTION AND TOXIC ELEMENTS

The classification of toxic elements as pollutants is well supported by a body of literature that highlights their adverse effects on both human health and ecosystems. These elements, encompassing heavy metals and metalloids, are often introduced into the environment through anthropogenic activities such as industrial processes, mining, agriculture, and improper waste disposal. A comprehensive understanding of the environmental significance of toxic elements is elucidated through authoritative references.

Industrial activities play a pivotal role in the release of toxic elements into the environment, with heavy metals like Pb, Hg, and Cd being prominent pollutants (Alloway, 2013). The impact of industrial processes on soil contamination and the bioavailability of these elements is extensively discussed in Alloway's work on heavy metals in soils (Alloway, 2013). Similarly, mining and smelting activities contribute significantly to the global dissemination of toxic elements, including As, Cr, and Cd (Nriagu, 1988). Nriagu's historical analysis sheds light on the extensive metal pollution resulting from such activities (Nriagu, 1988).

Agricultural practices, notably the use of fertilizers and pesticides, introduce toxic elements such as Cd and As into the soil, emphasizing the role of agriculture in environmental contamination (McBride, 1994). McBride's work on environmental chemistry of soils provides valuable insights into the interactions between soil chemistry and agricultural practices (McBride, 1994). Improper waste disposal, including electronic waste and hazardous materials, contributes to the leaching of toxic elements into the environment, with Pb being a common pollutant resulting from such practices (Agency for Toxic Substances and Disease Registry, 2007).

The impacts of toxic elements on human health and ecosystems further underscore their classification as pollutants. Exposure to Pb, primarily from industrial processes and lead-based paints, has been unequivocally linked to cognitive impairment in children, as detailed in Needleman's comprehensive review on Pb poisoning (Needleman, 2004). The profound effects of toxic elements extend to ecosystems, with Hg, for example, accumulating in aquatic environments and posing risks to fish and higher trophic levels (Eagles-Smith et al., 2008). Eagles-Smith's synthesis provides a holistic perspective on the environmental contamination and bioaccumulation of Hg in western North America (Eagles-Smith et al., 2008).

In conclusion, the well-established evidence presented in these authoritative references highlights the intricate relationship between pollution and toxic elements. Recognizing the pervasive nature of these pollutants is crucial for developing effective strategies to mitigate their adverse effects and safeguard both environmental quality and human well-being. The intersection of anthropogenic activities and the release of toxic elements into the environment underscores the urgency of addressing this complex environmental and public health challenge.

9.5 HOW TOXIC ELEMENTS ENTER THE FOOD CHAIN AND THEIR IMPACTS

The ingress of toxic elements into the food chain is an intricate and multifaceted process encompassing both natural and anthropogenic (human-induced) sources. These toxic elements, including heavy metals like Hg, Pb, Cd, and As, have the potential to permeate the environment and subsequently contaminate the food chain, thereby posing significant threats to both human health and ecosystems.

9.5.1 NATURAL PROCESSES

Natural processes play a pivotal role in the complex distribution patterns of toxic elements within the environment. The Earth's crust harbors an array of elements, some endowed with toxic properties. Geological processes act as conduits for the introduction of these elements into the environment, contributing to the dynamics of environmental contamination. The fundamental geological process of rock weathering facilitates the breakdown of minerals, releasing elements into the soil and water. During this weathering process, minerals containing toxic elements, such as As, Pb, and Hg, can undergo dissolution, thereby releasing these elements into the surrounding environment (Alloway, 2013; Plumlee & Morman, 2011; Selinus et al., 2005). Furthermore, volcanic activity stands out as a potent source of toxic elements, with volcanic eruptions releasing substantial quantities of metals and metalloids into the atmosphere, soil, and water. The ensuing deposition of volcanic ash, rich in various elements, settles on the ground, influencing soil composition and subsequently impacting plant uptake (Horwell et al., 2013; Zoller et al., 1974). Erosion, another natural process, significantly contributes to the dispersion of toxic elements. Through the erosive action of water and wind on rocks and soil, elements are transported downstream and deposited in varied locations. This phenomenon is particularly relevant in river systems, where transported sediments may contain heightened concentrations of toxic elements, thereby influencing water quality and ecosystems (Lambert et al., 2019; Sutherland et al., 2019).

Moreover, the bioavailability of toxic elements is intricately linked to environmental factors such as pH, organic matter content, and mineral composition (Alloway, 2013; Kabata-Pendias & Mukherjee, 2007; Scheckel et al., 2018). These environmental parameters exert influence over the mobility and uptake of toxic elements by plants, subsequently impacting the transfer of these elements along the food chain.

9.5.2 Human Activities

Human activities are significant contributors to the introduction of toxic elements into the environment, exerting profound impacts on air, water, and soil. Industrial processes constitute a noteworthy source of toxic element release. Various industries discharge pollutants into the air and water, contributing to the dispersion of heavy metals. Emissions from activities such as metal smelting, coal combustion, and chemical manufacturing release elements like Pb, Hg, and Cd into the atmosphere, resulting in air pollution and subsequent deposition in terrestrial and aquatic environments (Nriagu, 1988; Pacyna et al., 2007). Mining activities, essential for resource extraction, are notorious for releasing heavy metals into the environment. The excavation and processing of ores can result in the release of toxic elements into nearby water bodies, leading to contamination (Plumlee & Morman, 2011; Smedley & Kinniburgh, 2017). Runoff from mining sites introduces metals such as As, copper (Cu), and zinc (Zn) into aquatic ecosystems, posing threats to aquatic life and potentially impacting downstream water quality (Kabata-Pendias & Mukherjee, 2007; Younger, 2001). Agricultural practices, including the application of fertilizers, pesticides, and irrigation water, also contribute to the release of toxic elements. Certain fertilizers contain elements like Cd and As, which can accumulate in soils over time (Alloway, 2013). Additionally, the use of pesticides containing heavy metals may lead to soil contamination. The utilization of contaminated irrigation water further exacerbates the issue, potentially affecting crops and entering the food chain (Kabata-Pendias & Mukherjee, 2007; Scheckel et al., 2018).

The release of heavy metals into the environment due to human activities raises concerns about potential health risks. Elevated exposure to toxic elements can have detrimental effects on human health, including neurological disorders, developmental issues, and various chronic diseases (Grandjean et al., 2014; Nordberg et al., 2015). Furthermore, the ecological impacts on biodiversity, particularly in aquatic ecosystems, are significant, with potential disruptions in food chains and ecosystem functioning (Rainbow, 2002).

9.5.3 Air and Water Pollution

Air and water pollution emerge as critical conduits for the widespread dissemination of toxic elements, presenting substantial concerns for both the environment and human health. Industrial emissions constitute a primary source of airborne pollutants containing heavy metals. Various industrial processes, including metal smelting, chemical manufacturing, and power generation, release notable quantities of toxic elements into the atmosphere (Nriagu, 1988; Pacyna et al., 2007). These emissions encompass metals such as Pb, Hg, and Cd, dispersing over extensive areas before depositing onto land or water surfaces. Waste incineration stands as another anthropogenic activity contributing to air pollution laden with toxic elements. The incineration of municipal and industrial waste releases heavy metals into the air in the form of fine particles and gases (Santillo et al., 2001). The combustion of materials containing metals, such as electronic waste or construction debris, can lead to the emission of hazardous airborne pollutants, fostering long-range transport and environmental contamination. The combustion of fossil fuels, particularly coal and oil, emerges as a significant source of air pollution containing heavy metals. Power plants and vehicles release pollutants like sulfur dioxide and nitrogen oxides, which can react with metals in the atmosphere, forming particulate matter capable of extensive transport (Pacyna et al., 2007). Atmospheric deposition of these particles can contaminate soils and water bodies, thereby impacting ecosystems and human health.

The atmospheric transport of these pollutants, once airborne, is influenced by atmospheric conditions, prevailing wind patterns, and the physical and chemical characteristics of the released particles (Gao et al., 2015). Certain heavy metals, such as Hg, exhibit the capacity for extensive atmospheric transport and can be deposited far from their initial emission sources. Upon settling, these pollutants contribute to water pollution by contaminating water bodies. Rainfall serves as a mechanism to wash deposited particles into rivers, lakes, and oceans, introducing toxic elements into aquatic ecosystems. This contamination holds significant implications for aquatic life and the quality of drinking water supplies (Fuge & Wallschläger, 2010; Smedley & Kinniburgh, 2017).

9.6 IMPACTS

9.6.1 Accumulation in Plants

Plants play a crucial role as key mediators in the transfer of toxic elements into the food chain, serving as conduits for the absorption and accumulation of these elements from the soil or water through their complex root systems. The process of accumulation in plants stands as a pivotal step in the movement of toxic elements through ecosystems, ultimately influencing both wildlife and human populations.

Upon exposure to contaminated soil or water, plants demonstrate the capacity to absorb toxic elements through their roots. The uptake of elements such as As, Cd, and Pb is facilitated by specific transport mechanisms in plant roots (Alloway, 2013; Clemens, 2006). The bioavailability of these elements in the soil, a factor influenced by parameters like pH and organic matter content, significantly dictates their uptake by plants (Kabata-Pendias & Mukherjee, 2007). Once absorbed, toxic elements exhibit the propensity to accumulate in various plant tissues, encompassing roots, stems, leaves, and fruits. The distribution of these elements within the plant is contingent upon factors such as the type of element, plant species, and physiological characteristics (Lombi et al., 2009; Scheckel et al., 2018). For instance, certain elements may predominantly accumulate in the roots, while others are translocated to aerial parts of the plant. Particular plant species, identified as hyperaccumulators, demonstrate an exceptional ability to accumulate high concentrations of specific heavy metals in their tissues (Baker & Brooks, 1989). Hyperaccumulators have evolved unique physiological and biochemical mechanisms enabling them to flourish in environments characterized by elevated concentrations of toxic elements. These plants assume a pivotal role in phytoremediation, a process utilizing plants to extract, stabilize, or detoxify pollutants from the environment (Rascio & Navari-Izzo, 2011).

The accumulation of toxic elements in plants gives rise to concerns regarding potential health risks associated with the consumption of contaminated crops. Food crops cultivated in polluted environments have the propensity to uptake heavy metals, and upon ingestion by humans or animals, these elements can infiltrate the food chain, potentially leading to health issues (Alloway,

2013; Scheckel et al., 2018). The bioavailability and toxicity of these elements are contingent upon various factors, including the chemical form of the element and its concentration in the edible parts of the plant.

9.6.2 Bioaccumulation in Animals

The process of bioaccumulation in animals stands as a profound ecological phenomenon, wherein the concentration of toxic elements amplifies as they traverse distinct trophic levels within a food chain. Herbivores, occupying the primary consumer role, assume a pivotal function in this process by ingesting plants contaminated with toxic elements, thereby accumulating these substances in their tissues. Subsequently, carnivores and omnivores that feed on contaminated herbivores undergo further bioaccumulation, resulting in an escalation of toxic element concentrations at each trophic level. Herbivores assimilate toxic elements from the plants they consume, absorbing these substances through their dietary intake and accumulating them in various tissues, including muscles and organs. The concentration of toxic elements in herbivores is subject to influencing factors such as the plant types consumed, the duration of exposure, and the bioavailability of the elements in the environment (Saloña-Bordas et al., 2017). Heavy metals like Hg, Cd, and Pb are common participants in this bioaccumulation process.

As herbivores become contaminated, carnivores and omnivores that prey upon them inherit the accumulated toxic elements. This transference of toxic substances through successive trophic levels is termed biomagnification. Carnivores and omnivores, over their life span, may consume multiple contaminated herbivores, resulting in higher concentrations of toxic elements in their tissues compared to herbivores (Borgå et al., 2012; Smith et al., 2012). Moreover, the biomagnification process may lead to certain predators exhibiting concentrations of toxic elements several orders of magnitude higher than those found in their prey. The repercussions of bioaccumulation and biomagnification extend beyond individual organisms to impact entire ecosystems. Elevated concentrations of toxic elements in top predators can induce adverse effects on their health, reproductive success, and overall population dynamics. Additionally, this phenomenon poses risks to human health, particularly in communities reliant on fish or wildlife for sustenance.

Documented instances of bioaccumulation of toxic elements in animals span various ecosystems worldwide. Notable examples include extensive research on Hg bioaccumulation in fish from polluted water bodies, underscoring potential risks associated with consuming contaminated seafood (Sunderland, 2007). Similarly, studies on the bioaccumulation of persistent organic pollutants in marine mammals underscore the pervasive nature of this phenomenon (Routti et al., 2014).

9.6.3 Human Consumption

Human exposure to toxic elements is primarily mediated through the consumption of contaminated food, posing potential risks to public health. Myriad food sources, including seafood and crops, can act as conduits for the transfer of toxic elements into the human diet. The bioaccumulation and biomagnification processes, stemming from both natural and anthropogenic sources, converge to yield elevated concentrations of toxic elements in various food items. Seafood exemplifies a significant food source susceptible to contamination, particularly with elements such as Hg. Mercury, introduced into water bodies via natural processes and industrial activities, undergoes biomagnification within aquatic ecosystems. Predatory fish, such as tuna and swordfish, accumulate heightened concentrations of Hg through their diet, posing risks to human consumers (Karimi & Fisher, 2020). Prolonged exposure to elevated Hg levels may result in adverse health effects, particularly impacting the nervous system.

Crops grown in contaminated soil or irrigated with polluted water represent another substantial route for human exposure to toxic elements. Plants, acting as accumulators, absorb elements from their environment, and the concentration of these elements in crops is contingent upon factors such as soil characteristics, irrigation water quality, and plant species (Alloway, 2013; Scheckel et al., 2018). For instance, rice cultivated in regions with elevated levels of As in the soil or water may contain increased concentrations of this toxic element, thereby posing health risks to consumers (Meharg et al., 2009).

The risks associated with human consumption of contaminated food are diverse and hinge on factors such as the type and concentration of toxic elements, the frequency of exposure, and individual susceptibility. Chronic exposure to toxic elements through the diet has been correlated with various health issues, including neurological disorders, developmental problems, and an augmented risk of certain cancers (Grandjean et al., 2014; Nordberg et al., 2015). Vulnerable populations, such as pregnant women, infants, and young children, may exhibit heightened susceptibility to the adverse effects of exposure.

9.7 MEDIATING MEASURES OF THESE TOXIC ELEMENTS

Mediating measures to address the presence and impact of toxic elements in the environment involve a combination of preventive strategies, remediation techniques, and regulatory measures. These efforts aim to reduce the release of toxic elements into the environment, limit exposure pathways, and mitigate the associated risks to ecosystems and human health. The following are some key mediating measures.

9.7.1 Environmental Monitoring and Risk Assessment

Environmental monitoring and risk assessment are integral components of a proactive strategy to identify, quantify, and mitigate the impact of toxic elements on the environment and human health. This multifaceted approach involves continuous surveillance and sophisticated analyses to inform decision-making and regulatory frameworks.

Soil quality monitoring: Regular monitoring and analysis of soil samples represent a fundamental strategy for tracking the presence and concentrations of toxic elements, offering crucial insights into identifying contamination sources (Smedley & Kinniburgh, 2017). This proactive approach allows for the early detection of environmental issues and facilitates informed decision-making in remediation efforts. Techniques such as inductively coupled plasma–mass spectrometry (ICP-MS) and x-ray fluorescence (XRF) spectroscopy are widely employed for soil analysis, enabling precise measurements of diverse elements in soil samples (Kabata-Pendias & Mukherjee, 2007). These sophisticated analytical methods enhance the accuracy of soil quality assessments, supporting effective environmental management practices.

Water quality monitoring: Continuous monitoring of water quality, with a focus on parameters such as heavy metal concentrations, plays a crucial role in early contamination detection and management. Techniques such as atomic absorption spectrometry (AAS) and inductively coupled plasma–optical emission spectrometry (ICP-OES) are integral components of water quality monitoring efforts (Fuge & Wallschläger, 2010). By employing these advanced analytical techniques, researchers and environmental agencies can assess the levels of toxic elements in water bodies, providing essential information for formulating targeted interventions. Early detection through water quality monitoring is essential for preventing adverse effects on aquatic ecosystems and safeguarding human health in communities relying on water sources for various purposes.

Air quality monitoring: Real-time monitoring of air quality is imperative for assessing the levels of particulate matter and gaseous pollutants, contributing to the overall understanding of environmental contamination. Techniques such as high-volume air samplers and inductively coupled plasma–mass spectrometry (ICP-MS) are employed in the analysis of airborne contaminants (Gao et al., 2015). These methods enable the identification and quantification of toxic elements present in the air, aiding in the evaluation of potential health risks associated with air pollution. By continuously monitoring air quality, authorities can implement timely interventions to mitigate the impact of airborne contaminants on both the environment and public health. The integration of advanced analytical technologies in air quality monitoring enhances the precision and reliability of data, supporting comprehensive efforts to manage and reduce environmental pollution.

9.7.2 Regulatory Frameworks

Risk assessment studies play a crucial role in comprehending the pathways through which toxic elements infiltrate ecosystems and human populations. Employing exposure pathway analysis, these studies aim to identify and quantify routes of exposure, including ingestion, inhalation, and dermal contact (ATSDR, 2017). Evaluating the relationship between the amount of exposure to toxic elements and the resulting health effects is essential, and dose-response assessments assist in establishing toxicity thresholds for different contaminants (Grandjean et al., 2014). Furthermore, risk assessments take into account vulnerable populations, such as children and pregnant women, who may exhibit heightened susceptibility to the adverse effects of toxic element exposure. This comprehensive approach ensures a thorough evaluation of potential health risks (Nordberg et al., 2015).

9.7.3 Phytoremediation

Phytoremediation, considered a sustainable and innovative approach, harnesses the unique capabilities of plants, including hyperaccumulators, to remediate environments contaminated with toxic elements. This method holds significant promise for addressing soil and water pollution, providing an eco-friendly alternative to conventional remediation techniques. The ensuing discussion explores the principles and mechanisms of phytoremediation, drawing insights from seminal references in the field.

Utilizing plants, particularly hyperaccumulators, constitutes a fundamental aspect of phytoremediation. Hyperaccumulators exhibit an extraordinary ability to absorb and accumulate high concentrations of specific toxic elements in their tissues, acting as nature's green machines that draw contaminants from the soil into their biomass (Salt et al., 1995; Aransiola et al., 2013).

In the realm of phytoremediation, plants play a dual role by stabilizing toxic elements through immobilization in the soil or facilitating detoxification through mechanisms such as phytoextraction, rhizofiltration, and phytostabilization (Rascio & Navari-Izzo, 2011). This multifaceted approach underscores the potential of phytoremediation as an effective and environmentally friendly strategy for mitigating the impact of toxic elements in contaminated environments.

9.7.4 Waste Management

Waste management stands as a critical component of environmental protection, playing a pivotal role in ensuring the proper disposal and treatment of industrial and municipal waste to prevent the release of toxic elements into ecosystems. This section delves into the significance of waste management in mitigating environmental contamination, drawing insights from key references in the field (Aransiola, Ijah et al., 2021).

Efficient waste management practices concentrate on averting the release of toxic elements from industrial and municipal waste into the environment. This necessitates the implementation of proper containment, treatment, and disposal procedures aimed at minimizing the impact on soil, water, and air quality (Santillo et al., 2001).

An integral aspect of waste management involves the comprehensive understanding of waste composition. Waste is characterized and classified based on its attributes and potential environmental impact. This classification guides decisions on whether specialized treatment is necessary or if safe landfill disposal is an appropriate option (UNEP, 2013). The meticulous characterization and classification of waste contribute to informed decision-making, enhancing the effectiveness of waste management practices in safeguarding environmental integrity.

9.7.5 Global Collaboration

Global collaboration and international agreements play indispensable roles in addressing the intricate and interlinked challenges presented by the transport and deposition of toxic elements. This section delves into the significance of international cooperation in managing transboundary issues, referencing key works in the field.

Toxic elements, frequently released into the environment through diverse human activities, possess the capability to traverse borders and impact neighboring regions. The necessity for international cooperation becomes evident in addressing these transboundary challenges, as pollutants can traverse air and water masses, exerting influences on ecosystems and human health (Foley et al., 2005; Pacyna et al., 2007). Addressing cross-border pollution necessitates collaborative efforts to establish and enforce mechanisms for pollution control. Agreements and protocols that delineate responsibilities, standards, and response strategies are pivotal in minimizing the transboundary movement of toxic elements (Gardiner et al., 2019). This collaborative approach reflects the interconnected nature of environmental challenges and underscores the importance of concerted global efforts to safeguard ecosystems and human well-being.

9.8 CONCLUSION

This chapter discussed the connection between environmental pollution and the infiltration of toxic elements into the food chain. Through an examination of the intricate interplay between pollution and toxic substances, the study underscores the pervasive impact of human activities on the environment and the associated risks to ecosystems and public health. The identification of toxic elements, including heavy metals and persistent organic pollutants, underscores the urgency of addressing this complex issue. The exploration of pathways through which these harmful elements enter the food chain, involving processes like bioaccumulation and biomagnification, sheds light on the depth of the challenge. The ripple effects on different trophic levels lead to significant threats to human well-being, resulting in chronic diseases and developmental disorders. This highlights the necessity for proactive measures to mitigate the consequences of environmental pollution on the food chain. The proposal of mediating measures becomes crucial to counteract and prevent further contamination. Suggestions such as improved waste management, adoption of sustainable agricultural practices, and the integration of eco-friendly technologies provide a tangible framework for intervention. Moreover, the emphasis on robust regulatory frameworks and international collaboration underscores the global nature of environmental pollution and the need for concerted efforts.

In essence, this study offers a comprehensive comprehension of the complex dynamics between pollution and toxic elements, providing insights into the vulnerabilities of the food chain. By delineating practical measures to alleviate these impacts, the research contributes to a collective endeavor focused on preserving ecosystems, ensuring food safety, and protecting human health in the face of escalating environmental challenges. The call for immediate and coordinated action is evident, as the consequences of inaction could result in irreparable harm to our environment and the sustainability of the food chain.

REFERENCES

Abioye, O. P., Ijah, U. J. J., & Aransiola, S. A. (2017). Phytoremediation of Soil Contaminants by Biodiesel Plant Jatropha curcas. In Chapter 4 of K. Bauddh et al. (Eds.), *Phytoremediation potential of bioenergy plants, DOI 10.1007/978-981-10-3084-0_4*. Springer Nature Singapore Pte Ltd. https://link.springer.com/book/10.1007%2F978-981-10-3084-0.

Agency for Toxic Substances and Disease Registry. (2007). *Toxicological profile for lead.* Retrieved from https://www.atsdr.cdc.gov/toxprofiles/tp13.pdf

Agency for Toxic Substances and Disease Registry (ATSDR). (2017). *Principles of environmental health: Lesson 5 – Risk assessment.* Retrieved from https://www.atsdr.cdc.gov/training/principles/riskassessment/

Alloway, B. J. (2013). *Heavy metals in soils: Trace metals and metalloids in soils and their bioavailability.* Springer.

Aransiola, S. A., Ijah, U. J. J., & Abioye, O. P. (2013). Phytoremediation of lead polluted soil by glycine max L. *Applied and Environmental Soil Science*, 2013, Article ID 631619. doi: 10.1155/2013/631619. https://www.hindawi.com/journals/aess/2013/631619/abs/.

Aransiola, S. A., Ijah, U. J. J., Abioye, O. P., & Bala, J. D. (2019). Microbial-aided Phytoremediation of heavy metals contaminated soil: A review. *European Journal of Biological Research*, 9(2), 104–125. http://www.journals.tmkarpinski.com/index.php/ejbr/article/view/157.

Aransiola, S. A., Ijah, U. J. J., Abioye, O. P., & Bala, J. D. (2022). Vermicompost-assisted phytoremediation of toxic trace element-contaminated soil in Madaka, Nigeria using Melissa officinalis L and Sida acuta. *International Journal of Environmental Science and Technology.* http://dx.doi.org/10.1007/s13762-022-04105-y

Aransiola, S. A., Ijah, U. J. J., Abioye, O. P., & Victor-Ekwebelem, M. O. (2021). Anammox in wastewater treatment. In N. R. Maddela, L. C. García Cruzatty, & S. Chakraborty (Eds.), *Advances in the domain of environmental biotechnology. Environmental and microbial biotechnology.* Springer. https://doi.org/10.1007/978-981-15-8999-7_15

Aransiola, S. A., Victor-Ekwebelem, M. O., Ikhumetse, A. A., & Abioye, O. P. (2021). Challenges and future prospects of biotechnology. In N. R. Maddela & L. C. García (Eds.), *Innovations in biotechnology for a sustainable future.* Springer Nature Switzerland AG. https://doi.org/10.1007/978-3-030-80108-3_20.

Babaniyi, B. R., Ogundele, O. D., Bisi-omotosho, A., Babaniyi, E. E., & Aransiola, S. A. (2023). Remediation approaches in environmental sustainability. In Naga Raju Maddela, Lizziane Kretli Winkelstroter Eller, & Ram Prasad (Eds.), *Microbiology for cleaner production and environmental sustainability.* CRC Press. ISBN: 9781032496061. https://www.routledge.com/Microbiology-for-Cleaner-Production-and-Environmental-Sustainability/Maddela-Eller-Prasad/p/book/9781032496061

Baker, A. J., & Brooks, R. R. (1989). Terrestrial higher plants which hyperaccumulate metallic elements – a review of their distribution, ecology and phytochemistry. *Biorecovery*, 1(2), 81–126.

Basner, M., Babisch, W., Davis, A., Brink, M., Clark, C., Janssen, S., & Stansfeld, S. (2014). Auditory and non-auditory effects of noise on health. *The Lancet, 383*(9925), 1325–1332. doi: 10.1016/S0140-6736(13)61613-X

Borgå, K., Kidd, K. A., Muir, D. C., & Berglund, O. (2012). Trophic magnification factors: Considerations of ecology, ecosystems, and study design. *Integrated Environmental Assessment and Management*, 8(1), 64–84.

Brundtland, G. H. (1987). *Our common future: The world commission on environment and development.* Oxford University Press.

Cardinale, B. J., et al. (2012). Biodiversity loss and its impact on humanity. *Nature*, 486(7401), 59–67.

Carpenter, S. R., Caraco, N. F., Correll, D. L., Howarth, R. W., Sharpley, A. N., & Smith, V. H. (1998). Nonpoint pollution of surface waters with phosphorus and nitrogen. *Ecological Applications*, 8(3), 559–568. doi: 10.2307/2641247

CBD (Convention on Biological Diversity). (1992). *Convention on biological diversity.* Retrieved from https://www.cbd.int/convention/text/

Clemens, S. (2006). Toxic metal accumulation, responses to exposure and mechanisms of tolerance in plants. *Biochimie*, 88(11), 1707–1719.

Daily, G. C. (1997). *Nature's services: Societal dependence on natural ecosystems.* Island Press.

Díaz, S., Pascual, U., Stenseke, M., Martín-López, B., Watson, R. T., Molnár, Z., Hill, R., Chan, K. M. A., Baste, I. A., Brauman, K. A., Polasky, S., Church, A., Lonsdale, M., Larigauderie, A., Leadley, P. W., van Oudenhoven, A. P. E., van der Plaat, F., Schröter, M., Lavorel, S., & Shirayama, Y. (2019). *Summary for policymakers of the global assessment report on biodiversity and ecosystem services of the intergovernmental science-policy platform on biodiversity and ecosystem services.* Retrieved from https://www.ipbes.net/news/ipbes-global-assessment-summary-policymakers-pdf

Eagles-Smith, C. A., et al. (2008). Mercury in western North America: A synthesis of environmental contamination, fluxes, bioaccumulation, and risk to fish and wildlife. *Science of the Total Environment*, 407(1), 219–238.

Falchi, F., Cinzano, P., Elvidge, C. D., Keith, D. M., & Haim, A. (2016). Limiting the impact of light pollution on human health, environment and stellar visibility. *Journal of Environmental Management*, 92(10), 2714–2722. https://doi.org/10.1016/j.jenvman.2010.05.012

Foley, J. A., DeFries, R., Asner, G. P., Barford, C., Bonan, G., Carpenter, S. R., . . . Snyder, P. K. (2005). Global consequences of land use. *Science*, 309(5734), 570–574.

Fuge, R., & Wallschläger, D. (2010). Appendix 1: Naturally occurring substances in soils. In *Essentials of medical geology* (pp. 577–621). Springer.

Gao, S., Yang, Y., Zhao, X., & Niu, Z. (2015). Progress of atmospheric environmental monitoring in China. *Environmental Science and Pollution Research*, 22(22), 17476–17488.

Gao, Y., Wu, Y., Liu, Y., Li, Y., Cheng, X., Shen, Z., & Tian, Y. (2016). Global trends of solid waste research from 1997 to 2011 by using bibliometric analysis. *Scientometrics*, 109(1), 231–249. https://doi.org/10.1007/s11192-016-1953-y

Gardiner, N. J., Maas, R., Draper, G., & Richardson, J. (2019). Global atmospheric emissions inventory of hexachlorocyclohexanes (HCHs) and Dichlorodiphenyltrichloroethane (DDT) from 1980 to 2030. *Environmental Science & Technology*, 53(4), 2137–2146.

Gaston, K. J., Bennie, J., Davies, T. W., & Hopkins, J. (2014). The ecological impacts of nighttime light pollution: A mechanistic appraisal. *Biological Reviews*, 88(4), 912–927. https://doi.org/10.1111/brv.12036

Grandjean, P., Bellinger, D., Bergman, Å., Cordier, S., Davey-Smith, G., Eskenazi, B., . . . Heindel, J. J. (2014). The Faroes statement: Human health effects of developmental exposure to chemicals in our environment. *Basic & Clinical Pharmacology & Toxicology*, 115(1), 162–169.

Hoegh-Guldberg, O., et al. (2018). *The Intergovernmental Panel on Climate Change (IPCC) special report on the impacts of global warming of 1.5°C above pre-industrial levels and related global greenhouse gas emission pathways, in the context of strengthening the global response to the threat of climate change*. World Meteorological Organization.

Hopenhayn-Rich, C., et al. (1996). Cerebral dysfunction in children with chronic exposure to manganese from drinking water. *Environmental Research*, 73(2–3), 175–180.

Horwell, C. J., Williamson, B. J., Donaldson, K., Le Blond, J. S., Damby, D. E., Bowen, L., . . . Gosschalk, K. (2013). The structure of volcanic cristobalite in relation to its toxicity; relevance for the variable crystalline silica hazard. *Particle and Fibre Toxicology*, 10(1), 5.

International Agency for Research on Cancer. (2012). *IARC monographs on the evaluation of carcinogenic risks to humans – Arsenic, metals, fibres, and dusts.* Retrieved from https://publications.iarc.fr/121

IPCC (Intergovernmental Panel on Climate Change). (2014). *Climate change 2014: Synthesis report.* Retrieved from https://www.ipcc.ch/report/ar5/syr/

IPCC (Intergovernmental Panel on Climate Change). (2018). *Global warming of 1.5°C.* Retrieved from https://www.ipcc.ch/sr15/

Kabata-Pendias, A., & Mukherjee, A. B. (2007). *Trace elements from soil to human.* Springer.

Karimi, R., & Fisher, N. S. (2020). Mercury and selenium in marine fish: Concentrations, interactions, and sources of variation. *Ecotoxicology and Environmental Safety*, 196, 110541.

Khan, S., Hesham, A. E. L., & Qiao, M. (2009). *Reclamation of heavy metal-contaminated soils.* Springer.

Kumpiene, J., et al. (2006). Assessment of zerovalent iron for stabilization of chromium, copper, and arsenic in soil. *Environmental Pollution*, 144(1), 62–69.

Lambert, M. R., Carey, J. C., Bao, H., & Vance, J. M. (2019). Geochemical stability of deep-sea sediments in the northeast Pacific Ocean. *Chemical Geology*, 523, 43–54.

Landrigan, P. J., Fuller, R., Acosta, N. J. R., Adeyi, O., Arnold, R., Basu, N., Balde, A. B., Bertollini, R., Bose-O'Reilly, S., Boufford, J. I., Breysse, P. N., Chiles, T., Mahidol, C., Coll-Seck, A. M., Cropper, M. L., Fobil, J., Fuster, V., Greenstone, M., Haines, A., . . . Zhong, M. (2018). The lancet commission on pollution and health. *The Lancet*, 391(10119), 462–512. https://doi.org/10.1016/S0140-6736(17)32345-0

Leal Filho, W., Salvia, A. L., & Frankenberger, F. (2019). A critical appraisal of the 'making cities resilient' campaign by local governments and community stakeholders. *Sustainability*, 11(6), 1743. https://doi.org/10.3390/su11061743

Leichenko, R., & O'Brien, K. L. (2008). *Environmental change and globalization: Double exposures*. Oxford University Press.

Lelieveld, J., Evans, J. S., Fnais, M., Giannadaki, D., & Pozzer, A. (2015). The contribution of outdoor air pollution sources to premature mortality on a global scale. *Nature*, 525(7569), 367–371. doi: 10.1038/nature15371

Lombi, E., Scheckel, K. G., & Kempson, I. M. (2009). In situ speciation and bioavailability of trace elements in hyperaccumulators using X-ray absorption spectroscopy. *Plant Physiology*, 151(3), 1254–1263.

McBride, M. B. (1994). *Environmental chemistry of soils*. Oxford University Press.

Meharg, A. A., Williams, P. N., Adomako, E., Lawgali, Y. Y., Deacon, C., Villada, A., . . . Raab, A. (2009). Geographical variation in total and inorganic arsenic content of polished (white) rice. *Environmental Science & Technology*, 43(5), 1612–1617.

Millennium Ecosystem Assessment. (2005). *Ecosystems and human well-being: Synthesis*. Island Press.

Münzel, T., Gori, T., Babisch, W., & Basner, M. (2014). Cardiovascular effects of environmental noise exposure. *European Heart Journal*, 35(13), 829–836. https://doi.org/10.1093/eurheartj/eht565

Needleman, H. L. (2004). Lead poisoning. *Annual Review of Medicine*, 55, 209–222.

Nordberg, G. F., Fowler, B. A., Nordberg, M., & Friberg, L. T. (2015). *Handbook on the toxicology of metals*. Academic Press.

Nriagu, J. O. (1988). A history of global metal pollution. *Science of the Total Environment*, 100, 149–165.

Pacyna, J. M., Pacyna, E. G., Steenhuisen, F., & Wilson, S. (2007). Global anthropogenic mercury emission inventory for 2000. *Atmospheric Environment*, 41(30), 8127–8139.

Panagos, P., Van Liedekerke, M., Yigini, Y., & Montanarella, L. (2013). Contaminated sites in Europe: Review of the current situation based on data collected through a European network. *Journal of Environmental and Public Health*, 2013, 158764. https://doi.org/10.1155/2013/158764

Plumlee, G. S., & Morman, S. A. (2011). The role of airborne mineral dusts in human disease. *Aeolian Research*, 3(1), 93–114.

Rainbow, P. S. (2002). Trace metal concentrations in aquatic invertebrates: Why and so what?. *Environmental Pollution*, 120(3), 497–507.

Rascio, N., & Navari-Izzo, F. (2011). Heavy metal hyperaccumulating plants: How and why do they do it? And what makes them so interesting?. *Plant Science*, 180(2), 169–181.

Routti, H., Letcher, R. J., Born, E. W., & Dietz, R. (2014). Long-term temporal trends of organohalogen contaminants and their biomagnification in ringed seals from East Greenland. *Environmental Science & Technology*, 48(9), 4964–4972.

Sala, O. E., Chapin III, F. S., Armesto, J. J., Berlow, E., Bloomfield, J., Dirzo, R., . . . Wall, D. H. (2000). Global biodiversity scenarios for the year 2100. *Science*, 287(5459), 1770–1774.

Saloña-Bordas, M. I., Iavicoli, I., & Berglund, M. (2017). Bioaccumulation of essential and toxic trace elements in the land snail Helix aspersa: Implications for human health. *Environmental Geochemistry and Health*, 39(3), 699–710.

Salt, D. E., Blaylock, M., Kumar, N. P., Dushenkov, V., Ensley, B. D., Chet, I., & Raskin, I. (1995). Phytoremediation: A novel strategy for the removal of toxic metals from the environment using plants. *Biotechnology*, 13(5), 468–474.

Santillo, D., Johnston, P., & Stringer, R. L. (2001). Human exposure to hexachlorobenzene resulting from the combustion of household waste in barrels. *Journal of Environmental Monitoring*, 3(2), 196–201.

Scheckel, K. G., Jacob, D. L., Catarino, T., Melby, E. S., Fakra, S. C., & Juillot, F. (2018). Nickel speciation in the rhizosphere of hyperaccumulator plants from New Caledonia: A XAS and TEM study. *Environmental Science & Technology*, 52(15), 8548–8556.

Selinus, O., Alloway, B., Centeno, J. A., Finkelman, R. B., Fuge, R., Lindh, U., & Smedley, P. (Eds.). (2005). *Essentials of medical geology: Impacts of the natural environment on public health*. Academic Press.

Smedley, P. L., & Kinniburgh, D. G. (2017). A review of the source, behaviour and distribution of arsenic in natural waters. *Applied Geochemistry*, 17(5), 517–568.

Smith, B. W., Funk, W. H., & Sax, N. I. (2012). Selenium accumulation in benthic invertebrates and fish in mercury-laden waters of the South River, Virginia, USA. *Environmental Toxicology and Chemistry*, 31(12), 2856–2863.

Steffen, W., Richardson, K., Rockström, J., Cornell, S. E., Fetzer, I., Bennett, E. M., . . . Sörlin, S. (2015). Planetary boundaries: Guiding human development on a changing planet. *Science*, 347(6223), 1259855.

Sunderland, E. M. (2007). Mercury exposure from domestic and imported estuarine and marine fish in the US seafood market. *Environmental Health Perspectives*, 115(2), 235–242.

Sutherland, R. A., Rahman, M. M., Baker, A., & Molla, A. H. (2019). Arsenic distribution in different landscape units within an abandoned gold mine, New Zealand. *Science of the Total Environment*, 647, 750–761.

United Nations. (2015). *Transforming our world: The 2030 agenda for sustainable development*. Retrieved from https://sdgs.un.org/2030agenda

UNEP (United Nations Environment Programme). (2006). *Marine pollution in the Mediterranean: A diagnostic of the situation, UNEP/MAP*. Retrieved from https://www.unepmap.org/index.php/en/pam

United Nations Environment Programme (UNEP). (2013). *Guidance manual for the classification of hazardous waste under the Basel convention*. Retrieved from https://wedocs.unep.org/bitstream/handle/20.500.11822/8566/Basel_classification_manual_2013.pdf

United Nations Environment Programme (UNEP). (2016). *Marine plastic debris and microplastics – Global lessons and research to inspire action and guide policy change*. Retrieved from https://wedocs.unep.org/bitstream/handle/20.500.11822/17787/GPA2016_Marine_PlasticDebrisAndMicroplastics.pdf

UNEP (United Nations Environment Programme). (2018). *Towards a pollution-free planet*. Retrieved from https://www.unenvironment.org/resources/report/towards-pollution-free-planet

Widmer, R., et al. (2005). Global perspectives on e-waste. *Environmental Impact Assessment Review*, 25(5), 436–458.

World Bank. (2019). *Pollution: The hidden pandemic*. Retrieved from https://openknowledge.worldbank.org/handle/10986/32380

World Health Organization. (2006). *Air quality guidelines: Global update 2005*. Retrieved from https://www.who.int/phe/health_topics/outdoorair/outdoorair_aqg/en/
World Health Organization (WHO). (2018). *Ambient air pollution: A global assessment of exposure and burden of disease.* Retrieved from https://www.who.int/airpollution/data/AAP_GBD_2018.pdf
Younger, P. L. (2001). *Mine water pollution. Environmental and pollution science* (2nd ed., pp. 197–240). Blackwell Science.
Zoller, W. H., Parrington, J. R., & Kotra, J. M. (1974). Heavy metal aerosols in the Alaskan Arctic. *Environmental Science & Technology*, 8(7), 661–669.

10 Acute and Chronic Effects of Contaminants from Food Crops

Pabbati Ranjit, Kathuroju Harikrishna, P. Gnana Deepu, and Kondakindi Venkateswar Reddy

10.1 INTRODUCTION TO FOOD CROP CONTAMINATION

Food crops are plants that are grown specifically for human consumption. They include fruits, vegetables, grains, and other crops that provide essential nutrients for our bodies. The escalating demand to feed the increasingly mounting world population has put a lot of pressure on our agriculture system (Ashraf et al., 2019). The urge to grow more food made us reliant on the supplementary use of various agricultural practices by compromising the quality of food for quantity.

Contamination of food crops poses significant risks to both the environment as well as to human health resulting in both short-term (acute) and long-term (chronic) effects. These contaminants enter in different ways, such as through exposure to pesticides, heavy metals, or other pollutants. They can also be contaminated through exposure to animal waste, dirty water, or other pathogenic sources like bacteria, fungi, and viruses. The deposition of several contaminants in agricultural soil, air, water, and even into the food chain has disturbed the proper functioning of the earth's ecosystem (Pavlidis and Tsihrintzis, 2018), affecting plants and animals. In the agricultural system, crops are contaminated easily, either by land situated near heavy industries that release contaminants through soil, air, and water (Lebelo et al., 2021), or by heavy usage of fertilizers and insecticides (Bhalla, 2019) that affect the quality and safety of the crops. Globally, heavy metal contamination in food crops has resulted in an outbreak of toxicities and illnesses in both people and animals, when consuming these crops cultivated in soil polluted by heavy metals (Onakpa et al., 2018).

The acute effects of food crop contamination include abdominal pain, vomiting, nausea, diarrhea, headache, dizziness, and allergic reactions that may occur shortly after ingestion of contaminated food. Some contaminants, such as aflatoxins and pesticide residues, can also cause acute poisoning, which can be fatal in severe cases. For example, aflatoxins are potent carcinogens generated by certain fungi that infect crops like maize, cotton, and peanuts. Aflatoxins can cause acute liver failure, hemorrhage, and death in humans and animals (Kumar et al., 2021). Pesticide residues are chemicals that remain in food crops after the application of pesticides to control pests and diseases. Pesticide residues can cause acute toxicity, such as neurotoxicity, respiratory distress, and cardiac arrest, in humans and animals (World Health Organization, 2022).

Chronic effects of food crop contamination are those that occur after long-term or repeated exposure to low levels of contaminants and may include cancer, reproductive disorders, developmental defects, hormonal imbalances, and immune suppression. Some contaminants, such as heavy metals and radioactive substances, can accumulate in the body and cause chronic toxicity, which can damage various organs and systems. For example, heavy metals, such as lead (Pb), cadmium (Cd), mercury (Hg), and arsenic (As), can contaminate food crops through soil, water, and air pollution. Heavy metals can cause chronic effects, such as neurodegeneration, kidney damage, bone loss, and anemia, in humans and animals (Singh et al., 2018). Radioactive substances, such as cesium

DOI: 10.1201/9781032683751-12

(Cs), iodine (I), and strontium (Sr), can contaminate food crops through nuclear accidents, weapons testing, and waste disposal. Radioactive substances can cause chronic effects, such as genetic mutations, leukemia, thyroid cancer, and bone cancer, in humans and animals (Khan et al., 2022).

Food crop contamination can also have adverse effects on the environment, such as water pollution, soil degradation, biodiversity loss, and ecosystem disruption. Some contaminants, like pesticides and mycotoxins, can last in the environment for a long time and affect natural resources and other nonspecific organisms. For example, pesticides can leach into the soil and water and affect soil fertility, microbial activity, and aquatic life (Awuchi et al., 2022). Mycotoxins can spread through the food and affect wildlife, human health, and livestock (Kumar et al., 2021). Some contaminants, like heavy metals and radioactive substances, can alter the physiochemical properties of the environment and affect plant growth, crop yield, and food quality. For example, heavy metals can reduce the photosynthesis, transpiration, and nutrient uptake of plants and decrease crop productivity and nutritional value (Singh et al., 2018). Radioactive substances can induce physiological and morphological changes in plants and reduce crop quality and safety (Khan et al., 2022).

10.2 TYPES OF CONTAMINANTS OF FOOD CROPS

Food contaminants can be referred to as harmful substances that enter into food through various methods and can cause a serious impact on consumers (humans and animals).

The contamination might accidentally enter the food during agricultural production, the environment, storage, transit, sale, and processing. The entry of contaminants can be outside or its formation inside the food crops. The contaminants can be grouped into three categories: biological, chemical, and physical (Abdolshahi and Shokrollahi Yancheshmeh, 2020).

10.2.1 Biological Contaminants

Biological contamination of food crops can be defined as contamination through microorganisms or their toxins: the microorganisms like bacteria, viruses, parasites, and fungi are the main agents which are responsible for biological contamination. They can gain access to crops by various routes like air, water, and various macroscopic organisms like insects and rodents. Biological pollutants can induce human illnesses through various mechanisms like infection, intoxication, and immune responses (Concon, 1988).

10.2.1.1 Bacterial Contamination

Bacteria are microscopic organisms that are considered pathogenic to humans and animals who consume them through food crops. Bacteria contaminate food crops after their harvest (i.e., during transport or storage), which can cause the spoilage of food crops and bring huge economic losses to the farmers. They can gain access to food crops in various ways as mentioned above (Fu et al., 2016).

These bacteria by themselves can cause serious infections, and toxins released by them can cause a serious impact. Various factors like moisture, pH, and nutrients can favor the bacteria in food crops, hence control of them can impair their growth. Bacteria like *Escherichia coli* and *Salmonella* spp. Are examples of direct infections that can cause urinary and intestinal infections, respectively. Bacteria like *Clostridium botulinum* and *Staphylococcus aureus* are examples of intoxications that affect humans and animals by toxins released by them, which are even fatal sometimes if not treated (Machado-Moreira, 2019).

10.2.1.2 Viral Contamination

Viruses are very tiny organisms that depend on a host for their survival and multiplication. Unlike bacteria, viruses cannot survive outside the host; they can only multiply in living cells. The primary cause of viral contamination of food crops is unhygienic handling of the harvest. There are various routes of viral contamination but one major known route of infection is through irrigation. Food crops like vegetables and fruits, during harvest, are contaminated by irrigation water, by washing or spraying water on them to prevent drying. These fruits, when consumed without proper washing, can infect humans and animals. Consuming them raw can increase the risk of contamination (Gerba and Choi, 2006).

Viruses have been identified as a major source of foodborne illness in recent years. Different viruses can contaminate food, and the most frequent foodborne viral infections attack the gastrointestinal system (Seymour and Appleton, 2001).

Viruses, unlike bacteria, do not proliferate or generate poisons in food, therefore food items just serve as vehicles for their transmission. Human norovirus (NoV) and hepatitis A virus (HAV) are the most commonly involved viruses in foodborne infections, but other viruses such as enterovirus (EV), human rotavirus (HRV), hepatitis E virus (HEV), astroviruses (AstV), Aichi virus (AiV), sapoviruses, coronavirus, parvovirus, and human adenovirus (HadV) can also be transmitted (Petrović and D'Agostino, 2016).

10.2.1.3 Fungal Contamination

Fungi are a group of microorganisms that can occur in yeast, mold, or a combination of both. The fungi grow on dead and decaying organic matter under moist conditions. Fungal contamination can occur by various routes through air, soil, water, or improper and unhygienic handling of food crops during pre- and post-harvest, storage, and transport. The food crops that allow this fungus to enter can also be contaminated by injuries caused by insects and nematodes. The invasion of fungi into crops has detrimental effects on both the natural defense mechanisms of crop plants against fungal attacks and the purity of crop seeds, leading to the formation of aflatoxins. Fungal infections significantly impact the growth, productivity, and market value of crops (Kumar et al., 2021).

The most common food pollutants worldwide are filamentous fungi. These grow on a variety of food crops, such as grains, oilseeds, roots, tubers, legumes, fruits, and their by-products, as well as herbs and medicinal plants. Food quality may decline, physical and chemical changes may occur, food may become discolored, and fungi may produce toxins as a result of invasion. The development of fungus and the synthesis of mycotoxins requires a number of abiotic conditions like temperature, humidity, pH, and water (Aasa et al., 2022).

The major concern with fungal contamination is toxins produced by the fungal mold. Contamination of crops with aflatoxin poses a worldwide hazard to food and feed safety, as well as to the agricultural economy and small-scale companies that rely on crops. Certain fungi can generate harmful by-products known as mycotoxins, which can contaminate food under specific conditions. The main genera of molds that produce toxins include *Fusarium, Alternaria, Penicillium*, and *Aspergillus* (Khodaei et al., 2021).

Several varieties of fungi produce mycotoxins of which aflatoxins (Afs) B1, B2, G1, G2, and M1 are considered most significant as they are genotoxic and carcinogens. Humans exposed to long-term AF develop liver cancer, infections, and growth damage; high exposures result in acute symptoms and even death (Eskola et al., 2020).

10.2.1.4 Parasite contamination

Parasites, which include worms and protozoa, are one of the major increasing concerns of food crop conditions worldwide. Contamination of food crops with parasite larvae or eggs can happen in various ways, but one of the most known routes is by cultivation of food crops with improperly treated sewage water (Eraky et al., 2014).

The most common cause of parasitic infection through food contamination is the consumption of raw uncooked or improperly washed fruits and vegetables by parasites such as *Trichuris trichiura, Cryptosporidium* spp., *Ascaris lumbricoides, Hymenolepis* spp., *Entamoeba histolytica, Giardia intestinalis, Enterobius vermicularis*, and *Taenia* spp. (Bekele et al., 2017).

10.2.2 Chemical Contaminants

The chemical contamination of food crops is also an increasing global concern. It is becoming increasingly difficult to protect the public from varying degrees of food contamination. Chemical contamination is a term used to describe situations in which chemicals are found more than what is considered acceptable or in places where they shouldn't be present at all. Chemical contaminants are one of the key contributors to food contamination and outbreaks of foodborne diseases (Faille et al., 2018). Contamination may result from pollutants introduced by humans or from naturally existing toxins in the environment. Food contamination is also significantly influenced by the stages of food preparation, packaging, transportation, and storage (Rather et al., 2017). Chemical pollutants can come from a variety of sources and end up in our food. Foods can be contaminated primarily as a result of the utilization of agrochemicals. Other causes include the usage of unapproved food additives and adulterants, contamination from environmental sources such as water, air, or soil pollution, cross-contamination during food processing, migration from packaging materials, and presence or contamination by natural toxins (Di Stefano and Avellone, 2014).

10.2.3 Physical Contamination

Food tainted by extraneous objects throughout processing stages is referred to as physically contaminated food. Internal sources like bone chips and pits, as well as exterior ones like metal pieces, can both contribute to physical pollutants in food. Inadequate farming techniques can cause them to inadvertently enter into food during pre- and post-harvesting, as well as during manufacture, storage, transit, and sale. Foods can get physically contaminated by foreign items like metal filings, glass, jewelry, stones, or bone chips. Occasionally, these things can also cause biological contamination if they contain harmful microorganisms (Faille et al., 2018). There are several other types of physical contamination, including plastic contamination and wood contamination (Di Stefano and Avellone, 2014).

10.3 ACUTE EFFECTS OF CONTAMINATION OF FOOD CROPS

Upon consumption of contaminated food crops, they cause immediate and often several health consequences; in extreme conditions they cause acute poisoning. They may be caused by the presence of toxins like mycotoxins, aflatoxins, the presence of abnormal levels of heavy metals, or contamination through pesticides, fertilizers, and insecticides. Consumption of these contaminated crops leads to symptoms like poisoning (food poisoning, aflatoxin poisoning, heavy metal poisoning), pesticide intoxication, allergic reactions, nausea, vomiting, and diarrhea.

10.3.1 Poisoning

Poisoning is caused by the presence of harmful substances in the consumed food crops. They are of different types based on contaminants like microbial pathogens (food poisoning), molds like *Aspergillus* (aflatoxin poisoning), and heavy metals (heavy metal poisoning).

10.3.1.1 Food Poisoning

Crops can be contaminated by microbial pathogens due to poor hygiene during cultivation, harvesting, or storage.

Examples include the 2006 spinach *E. coli* outbreak (Sharapov et al., 2016) and the 2014 bean sprout *Salmonella* outbreak (Mohle-Boetani et al., 2009), both in the United States.

10.3.1.2 Aflatoxin Poisoning

Aflatoxin poisoning is caused by the release of aflatoxins by molds like *Aspergillus.* Acute exposure causes symptoms such as fatigue, loss of appetite, and jaundice.

Examples include acute aflatoxins outbreak in Turkey (maize contamination) in 2001 (Şengül, et al., 2016), maize contamination in eastern and central Kenya in 2004 (Lewis, et al., 2015), and peanut contamination in China in 2013 (Liu et al., 2016).

10.3.1.3 Heavy Metal Poisoning

Crops grown in contaminated soil may accumulate heavy metals. The accumulation of these heavy metals leads to Heavy metal poisoning.

Examples include Pb poisoning from contaminated turmeric species in Pakistan (Baig et al., 2019) and Cd poisoning (contamination in rice) leading to itai-itai disease in Japan (Yoshida et al., 1999).

10.3.2 Pesticide Intoxication

Residues of pesticides (e.g., organophosphates, pyrethroids) on crops can cause acute poisoning.

Examples include endosulfan poisoning in Kasaragod, India, in 2001 due to the use of this organochlorine pesticide on cashew plants (Sreekumar and Prathapan, 2021), and methyl parathion poisoning in India in 2001 due to consumption of wheat flour contaminated with this pesticide (Palimar et al., 2012).

10.3.3 Other Symptoms

Other symptoms of poisoning include allergic reactions (caused by the ingestion of contaminated crops, which trigger allergic reactions like hives, itching, or swelling), mycotoxin-induced illness, nausea, vomiting, and diarrhea (when bacterially contaminated food is ingested).

10.4 CHRONIC EFFECTS OF CONTAMINATION OF FOOD CROPS

Exposure to contaminants in food crops for an extended period can lead to various chronic health issues. The impact of these chronic effects depends on the type and level of contaminants, as well as individual susceptibility.

Aflatoxins are a type of secondary metabolite produced by various molds and fungi, with the primary culprits being *Aspergillus parasiticus* and *Aspergillus flavus*. These toxins have been extensively studied, and out of the more than 18 identified aflatoxins, aflatoxin B1 (AFB1) is considered the most potent and well-known. It has been connected to the development of hepatocellular carcinoma (HCC), a kind of liver cancer seen in people and animals. Exposure to low levels of aflatoxins over an extended period can lead to chronic illnesses, with cancer being the most prominent and deadly outcome. While liver malignancies like HCC and bile duct hyperplasia have long been associated with aflatoxin exposure, there have also been connections to other types of cancers such as kidney, bone, pancreas, bladder, and viscera cancers (Dabuo et al., 2022). Furthermore, the consumption of aflatoxin-contaminated food crops can result in gradual health deterioration. This includes liver damage, weakened immune system function, oxidative stress, and abnormalities in fetal growth (Negash and Demissie, 2018). Therefore, it is crucial to address the presence of aflatoxins in food sources to mitigate their detrimental effects on human health.

Persistent organic pollutants (POPs) have been a major cause of concern for food safety due to their enduring nature and toxicity (Gaur et al., 2018). Numerous health issues, including hormone disruption, heart disease, cancer, diabetes, birth deformities, and compromised immunological and reproductive systems, can arise from exposure to these contaminants. POP exposure has been linked to metabolic disorders like obesity and diabetes as well as cardiovascular diseases. In addition, POP exposure has detrimental effects on the endocrine system (Vandelac, 2000). The endocrine-disrupting actions of POPs may have negative impacts on the immunological, neurological, and reproductive systems. They may also raise the risk of hormone-dependent cancer formation and impact the differentiation, growth, and development of sexual behavior. For masculine embryos and infants, certain POPs in food have been linked to carcinomas and hypospadias; for females, food contaminated by POP causes endometriosis, breast cancer, and cystic ovaries (Ruiz-Fernández et al., 2014).

Pesticides carry the risk of being dangerous to human health and can cause both immediate and long-term effects. Continuous exposure to pesticides can lead to lasting health issues, such as reproductive problems and an increased risk of cancer. The individuals most at risk for health complications from pesticide exposure are those who regularly come into contact with them, whether through their occupation, agriculture, or in their own homes and gardens. Chronic ingestion of pesticide residues in food crops has been associated with cancer, hormone disruption, diabetes, heart disease, birth deformities, and compromised reproductive and immunological systems. Pesticides, when consumed over an extended period, may accumulate in organs such as the kidneys and liver, potentially leading to damage or dysfunction. Some pesticides have neurotoxic properties and may contribute to long-term neurological disorders, affecting cognitive function and behavior (Abioye et al., 2021; Dhuldhaj et al., 2023).

Heavy metal pollution is a worldwide problem that puts human health in grave danger and disrupts the ecology (Gudkov et al., 2021). There is evidence that inhaling and consuming heavy metals accumulated in food crops might expose one to health risks (Mamat et al., 2014). Once within the body, these metals impede on enzymes, causing vital physiological processes to slow down or cease entirely. They have the potential to cause major health problems like anemia, renal failure, and brain damage over time (Pappas et al., 2006; Aransiola et al., 2013). Prolonged contact with heavy metals including Pb, Hg, Cd, and As, which are frequently found in polluted soil, can cause kidney damage, neurological abnormalities, and several kinds of cancer (Manwani et al., 2023). Neurological alterations that are sometimes irreversible can result from mercury exposure (Islam et al., 2007). Exposure to Pb can have significant adverse effects on human health in addition to substantially contaminating the environment. It primarily affects the gastrointestinal (GI) tract and central nervous system (CNS) in both adults and children. Through immunological modulation and inflammatory, and oxidative mechanisms, Pb exposure can cause respiratory, urinary, cardiovascular, and neurological diseases (Obeng-Gyasi et al., 2021). Long-term airborne nickel (Ni) exposure is thought to be the cause of chronic inflammation, lung and nasal carcinomas, and chronic bronchitis (Mudgal et al., 2010). Harmful Cd levels lead to major issues in the circulatory system, liver, pancreas, kidneys, and lungs (Alina et al., 2012). Long-term As exposure can result in skin lesions and cancer. Exposure during early childhood has been linked to negative effects on cognitive development and an increase in young adult fatalities. Chronic exposure ultimately results in illness across several body systems, such as peripheral neuropathy (tingling and numbness), enlargement of the liver and spleen, cognitive impairment, diabetes, heart disease, and damage to the portal vein (Hashim, 2023).

Pathogens: Numerous illnesses are being caused by foodborne bacteria, which have a major impact on both human health and the economy. Microorganisms that have the potential to cause foodborne illnesses are referred to as foodborne pathogens, which encompass both bacteria and viruses. Foodborne disease occurs when a toxigenic pathogen colonizes a food item and produces a toxin that is subsequently ingested by a person, or when a pathogen is consumed with food and establishes itself in the human body, often leading to multiplication. Foodborne sickness can be classified into two main categories: foodborne infection and foodborne intoxication (Bintsis, 2017; Amobonye et al., 2023). Bacterial species like *Salmonella, Shigella* spp., and *S. aureus* are most common bacteria involved in food crop contamination and can cause chronic infections like bacteremia, pneumonia, osteomyelitis, cerebritis, and meningitis (Bacon and Sofos, 2003). Foodborne illness can be caused by different types of enteric viruses; the most prevalent ones are the norovirus and hepatitis A viruses, which cause chronic liver and gastric infection (Scallan et al., 2011).

Microplastics: Food crops that have been exposed to contaminated water and soil can accumulate microplastics. Health issues such as elevated inflammation, type 2 diabetes, insulin resistance, and heart disease can result from chronic exposure to microplastics. Numerous investigations have revealed that microplastics, associated with the additives and pollutants, can cause neurotoxicity, oxidative stress, and immunotoxicity (Udovicki et al., 2022; Aransiola et al., 2023).

Genetically modified (GM) food crops: The genetic material of these food crops is modified with the help of R-DNA technology. There are various concerns on the safety of consumption of GM crops. According to a comprehensive analysis, chronic effects associated with GM crop consumption include organ abnormalities, reduced fertility, tumors or cancer, and impaired learning capacity. Further extensive studies are required to evaluate the safety of GM crop consumption (Shen et al., 2022).

Hormones are used to treat certain food crops to accelerate their growth. These hormones have the potential to cause endocrine system disruption and health problems, including cancer, reproductive illnesses, and developmental issues (Zheng et al., 2014).

Radionuclides: Radioactive materials can contaminate food crops, especially after nuclear accidents. Chronic exposure to these radionuclides can lead to numerous forms of cancer and other health issues (Gupta et al., 2017).

Food crop contamination not only impacts human health but also has far-reaching consequences for the environment in terms of the following.

- *Soil degradation*: Contaminants such as heavy metals, pesticides, and industrial chemicals can accumulate in soil due to contaminated irrigation water or improper waste disposal over the period. This causes the quality of soil to deteriorate, affecting nutrient availability and plant growth. Example: Cadmium-contaminated rice fields in Japan led to soil degradation and reduced agricultural productivity.
- *Water pollution*: Runoff from contaminated fields can carry pollutants into nearby water bodies. Pollutants like pesticides, herbicides, and fertilizers can contaminate rivers, lakes, and groundwater. Example: Agricultural runoff containing pesticides adversely impacts aquatic ecosystems and drinking water quality.
- *Biodiversity loss*: Contaminants affect not only crops but also surrounding flora and fauna. Usage of pesticides harms beneficial insects, birds, and soil organisms. Example: Neonicotinoid pesticides have been linked to declining bee populations, impacting pollination and biodiversity.
- *Air pollution*: Volatile organic compounds (VOCs) from pesticides and fertilizers can contribute to air pollution. Example: Herbicides containing VOCs can release harmful gases into the atmosphere during application.
- *Health risks to wildlife*: Contaminated crops, when consumed by wildlife, lead to health issues. Example: Birds feeding on pesticide-treated seeds may suffer adverse effects.
- *Disruption of ecosystem services*: Contaminated crops affect ecosystem functions. Due to pesticide exposure, pollination is reduced and impacts crop yields. Example: A decline in bee populations affects pollination services.
- *Long-term soil contamination*: Persistent organic pollutants and heavy metals can remain in soil for years. Example: DDT (dichlorodiphenyltrichloroethane) persists in soil, impacting future crops.
- *Loss of beneficial microorganisms*: Soil microbes play a crucial role in nutrient cycling and plant health. Contaminants harm these beneficial microorganisms. Example: Pesticides disrupt soil microbial communities.
- *Contaminated runoff into natural habitats*: Rainwater runoff from contaminated fields can reach natural habitats. Wetlands, forests, and aquatic ecosystems are at risk. Example: Pesticide runoff affecting wetland ecosystems.
- *Reduced carbon sequestration*: Soil contamination affects carbon storage capacity. Contaminants alter soil structure and microbial activity. Example: Heavy metal contamination reduces soil's ability to sequester carbon.

10.5 FACTORS AFFECTING THE CONTAMINATION OF FOOD CROPS

Contamination of food crops arises from a complex interaction of agricultural, environmental, biological, and processing factors. Numerous factors can affect the food crop contamination (Garvey, 2019). Some of the crucial factors include:

Environmental factors such as temperature, rainfall, and humidity can impact the growth of bacteria, fungi, and other microorganisms that produce toxins. Water, air, and soil pollution releases toxic substances into the environment, affecting crops through assimilation. For instance, aflatoxins are toxins produced by fungi that thrive on crops such as corn, peanuts, and cottonseed, and are more likely to grow in humid and warm conditions (Kumar et al., 2021).

Soil factors: Soil composition plays a crucial role in determining the accumulation and uptake of contaminants in crops. Soil factors such as organic matter, nutrient availability, and pH can affect the uptake of contaminants by food crops. Contaminated soil, mostly due to improper waste disposal or industrial activities, can also transfer contaminants to food crops. For example, heavy metals such as Hg, Cd, and Pb accumulate in soil and can be taken up by crops, which can then be ingested by animals and humans (Kharazi et al., 2021).

Climate and weather patterns affect the transport and fate of contaminants in the environment, which in turn impact their presence in food crops. Humidity, temperature, wind speed, and rainfall can impact the volatilization, leaching, deposition, and degradation of contaminants in water and soil. Extreme weather events, such as droughts and floods, can also exacerbate contamination by mobilizing contaminants from soil into water bodies or by concentrating pollutants in soil due to reduced water availability. For example, changes in precipitation patterns and increased temperatures can lead to changes in the dispersion and growth of pathogens and pests that can contaminate food crops (Naeem et al., 2023).

Water quality and irrigation practices are significant factors influencing the contamination of food crops, particularly in areas where irrigation is crucial for agriculture. Contaminants present in irrigation water, such as pesticides, heavy metals, and microbial pathogens, can be absorbed by food crops and accumulate in edible plant parts. Additionally, poor water quality resulting from agricultural runoff, industrial discharges, or wastewater reuse can contribute to contamination of both soil and crops. For example, heavy metals such as Cd can leach into groundwater and be taken up by crops, which can then be consumed by animals and humans (Pachepsky et al., 2011; Aransiola et al., 2021).

Agricultural practices such as the use of herbicides, fertilizers, and pesticides introduce chemicals and nutrients into the soil, some of which may become contaminants if applied improperly or in excess. For instances, pesticides can leave residues on food crops that can be harmful to human health if ingested in large quantities. Additionally, irrigation practices, tillage methods, soil management techniques, and crop rotation can impact the fate and transport of contaminants in agricultural ecosystems (Nabil et al., 2021).

Post-harvest handling practices such as transportation, processing, and storage can impact food crop contamination. Improper harvesting can expose food crops to contamination by rodents, insects, birds, or humans. Microbial contamination can result from improper handling and storage sanitation. Food crops can be physically harmed and exposed to environmental contaminants due to inadequate packaging and shipping methods. For example, while food crops are being stored, mycotoxins like aflatoxins might grow due to unfavorable conditions (Arah et al., 2016).

Genetics and plant physiology: The genetics and physiology of food crops plays a major role in determining their susceptibility to contamination. Different crop varieties may exhibit varying levels of tolerance or sensitivity to specific contaminants, depending on their genetic makeup. Moreover, physiological factors such as nutrient uptake mechanisms, root morphology, and metabolic pathways can influence the translocation, uptake, and detoxification of contaminants within plants (Ronald, 2011).

Biological interactions: The uptake and accumulation of contaminants in food crops can be influenced by biological interactions that occur within agricultural ecosystems. Through processes including biodegradation, volatilization, and immobilization, soil microorganisms like bacteria and fungi can affect the mobility and transformation of contaminants (Kuiper et al., 2004). Contaminants and nutrients in the rhizosphere can also be influenced by plant-microbe interactions, such as symbiotic and mycorrhizal interactions.

Industrial activities are main source of contaminants that can affect food crops. Emissions from industrial operations, such as heavy metals, VOCs, and particulate matter can deposit onto plants and soil, leading to contamination. Furthermore, improper industrial waste and by-products disposal can affect surrounding food crops and ecosystems by contaminating soil and water sources (Cabral-Pinto et al., 2020).

Urbanization and land use changes: Food crops are more susceptible to contamination as a result of urbanization and alterations in land use, which introduces new sources of pollutants. The emissions from moving cars, industry, and improper waste disposal methods can all contribute to higher levels of pollutants in urban areas. In addition to causing soil deterioration, the conversion of agricultural land to urban or industrial uses can exacerbate the contamination.

10.6 CHALLENGES AND FUTURE PERSPECTIVES

10.6.1 Challenges

Controlling food crop contamination is a difficult endeavor that requires tackling several difficulties at various stages of the food production and distribution chain. The issues of managing food crop contamination are numerous, involving aspects such as agriculture, processing, transportation, and regulatory structures. Some of the challenges are mentioned below.

- *Complex supply chains*: Modern food supply chains are intricate, involving multiple stakeholders from farm to table. Ensuring food safety at each step is challenging.
- *Diverse pathogens and toxins*: Different pathogens (bacteria, viruses, fungi) contaminate food, leading to varied health risks.
- *Globalization and imports*: Increased international trade exposes consumers to foods from diverse sources, potentially carrying contaminants.
- *Climate change and environmental factors*: Climate variations impact crop growth, fungal proliferation, and toxin production.
- *Consumer behavior and awareness*: Consumer habits influence food handling, storage, and consumption.
- *Resource constraints*: Limited resources hinder implementation of preventive measures.
- *Regulatory compliance and enforcement*: Ensuring compliance with food safety regulations across diverse regions and cultures.
- *Emerging contaminants and unknown risks*: New contaminants may arise due to evolving agricultural practices or environmental changes.
- *Behavioral challenges*: Human behavior affects contamination levels (e.g., improper handwashing, inadequate storage).
- *Balancing production and safety*: Meeting food demand while ensuring safety can be conflicting.

10.6.2 Future Perspectives

Technological breakthroughs, novel agricultural techniques, enhanced regulatory frameworks, and more global collaboration will all play a role in future food crop contamination management strategies. As we look ahead, many major trends and techniques emerge that offer promise for improving food safety and lowering the risks of contamination in food crops.

- *Integrated pest management (IPM)*: Implement IPM practices to manage pests and diseases without excessive reliance on chemical pesticides.
- *Good agricultural practices*: Follow recommended practices for planting, irrigation, fertilization, and pest control.
- *Crop rotation*: Rotate crops to break pest and disease cycles.
- *Post-harvest handling*: Properly handle harvested crops to prevent contamination during storage and transportation.
- *Natural products and biocontrol agents*: Use natural compounds (e.g., neem, garlic, essential oils) and biocontrol agents (e.g., *Trichoderma*) to suppress pests and pathogens.
- *Soil amendments*: Apply lime, farmyard manure, or cereal crop residues to improve soil health and suppress aflatoxin-producing fungi.
- *Education and training*: Educate farmers on safe practices, hygiene, and proper pesticide use.
- *Regular monitoring and testing*: Regularly monitor crops for signs of contamination and test for mycotoxins.

10.7 CONCLUSION

Food crop poisoning threatens not just human health and the environment but also the economy. Short-term and long-term implications of food crop contamination include soil and water pollution, which affects the local environment and fauna, disturbing ecological equilibrium. When contaminated food crops are ingested, short-term or acute exposure causes health dangers such as poisoning, nausea, vomiting, and so on, but long-term or chronic exposure causes health problems such as cancer, reproductive disorders, and developmental disorders.

To summarize, the acute and chronic effects of food crop contamination require a complex and interdisciplinary approach. The potential consequences for human health, the environment, and the economy highlight the importance of coordinated action at the local, national, and international levels. As we work toward resilient and sustainable food systems, the lessons learned from food crop contamination studies must influence legislation, practices, and public awareness campaigns to ensure a safer and more secure future for global populations.

REFERENCES

Aasa, A. O., P. B. Njobeh, and F. F. Fru. "Incidence of filamentous fungi in some food commodities from Ivory Coast." *Journal of Agriculture and Food Research* 8 (2022): 100304.

Abdolshahi, Anna, and B. Shokrollahi Yancheshmeh. "Food contamination." In *Mycotoxins and Food Safety* (pp. 5–16). Intech Open, 2020.

Abioye O. P., U. J. J. Ijah, S. A. Aransiola, S. H. Auta, and M. I. Ojeba. "Bioremediation of toxic pesticides in soil using microbial products." In R. Prasad, S. C. Nayak, R. N. Kharwar, and N. K. Dubey (eds.), *Mycoremediation and Environmental Sustainability. Fungal Biology*. Springer Cham, 2021. https://doi.org/10.1007/978-3-030-54422-5_1

Alina, M., A. Azrina, A. S. Mohd Yunus, S. Mohd Zakiuddin, H. Mohd Izuan Effendi, and R. Muhammad Rizal. "Heavy metals (mercury, arsenic, cadmium, plumbum) in selected marine fish and shellfish along the Straits of Malacca." *International Food Research Journal* 19, no. 1 (2012).

Arah, Isaac Kojo, Gerald K. Ahorbo, Etornam Kosi Anku, Ernest Kodzo Kumah, and Harrison Amaglo. "Postharvest handling practices and treatment methods for tomato handlers in developing countries: A mini review." *Advances in Agriculture* 2016 (2016): 1–8. https://doi.org/10.1155/2016/6436945.

Aransiola, S. A., U. J. J. Ijah, O. P. Abioye, and M. O. Victor-Ekwebelem. "ANAMMOX in Wastewater Treatment." In N. R. Maddela, L. C. García Cruzatty, and S. Chakraborty (eds.), *Advances in the Domain of Environmental Biotechnology. Environmental and Microbial Biotechnology*. Springer Singapore, 2021. https://doi.org/10.1007/978-981-15-8999-7_15

Aransiola, S. A., U. J. J. Ijah, O. P. Abioye. Phytoremediation of lead polluted soil by *Glycine max* L. *Applied and Environmental Soil Science* (2013), Article ID 631619. Doi: 10.1155/2013/631619. https://www.hindawi.com/journals/aess/2013/631619/abs/

Aransiola, S. A., M. O. Victor-Ekwebelem, A. E. Ajiboye, S. S. Leh-Togi Zobeashia, U. J. J. Ijah, and O. J. Oyedele. "Ecological impacts and toxicity of micro and nanoplastics in agroecosystem." In N. R. Maddela, K. V. Reddy, and P. Ranjit (eds.), *Micro and Nanoplastics in Soil*. Springer, 2023. https://doi.org/10.10007/978-3-031-21195-9_10.

Amobonye, A., C. Aruwa, S. A. Aransiola, O. John, T. D. Alabi, and J. Lalung. "Exploring the potential of fungi in the bioremediation of pharmaceutically active compounds." *Front. Microbiology Section Microbiotechnology* 14 (2023): 1207792, https://doi.10.3389/fmicb.2023.1207792

Ashraf, Ifra, Tanzeel Khan, Nowsheeba Rashid, Shazia Ramzan, and Noureen Khurshid. "Role of some emerging agro-chemicals in groundwater contamination." *Contaminants in Agriculture and Environment: Health Risk and Remediation. Agro Environ Media* 1 (2019): 9–20.

Awuchi, Chinaza Godseill, Erick Nyakundi Ondari, Sarah Nwozo, Grace Akinyi Odongo, Ifie Josiah Eseoghene, Hannington Twinomuhwezi, Chukwuka U. Ogbonna, Anjani K. Upadhyay, Ademiku O. Adeleye, and Charles Odilichukwu R. Okpala. "Mycotoxins' toxicological mechanisms involving humans, livestock and their associated health concerns: A review." *Toxins* 14, no. 3 (2022): 167.

Bacon, R. Todd, and John N. Sofos. "Characteristics of biological hazards in foods." *Food Safety Handbook* 10 (2003): 157–195.

Baig, Jameel Ahmed, Saba Bhatti, Tasneem Gul Kazi, and Hassan Imran Afridi. "Evaluation of arsenic, cadmium, nickel and lead in common spices in Pakistan." *Biological Trace Element Research* 187 (2019): 585–595.

Bekele, Fitsum, Tamirat Tefera, Gelila Biresaw, and Tsegaye Yohannes. "Parasitic contamination of raw vegetables and fruits collected from selected local markets in Arba Minch town, Southern Ethiopia." *Infectious Diseases of Poverty* 6, no. 1 (2017): 1–7.

Bhalla, Tek Chand. "International laws and food-borne illness." In *Food Safety and Human Health*, pp. 319–371. Academic Press, 2019.

Bintsis, Thomas. "Foodborne pathogens." *AIMS Microbiology* 3, no. 3 (2017): 529.

Cabral-Pinto, Marina M. S., Manuela Inácio, Orquídia Neves, Agostinho A. Almeida, Edgar Pinto, Bárbara Oliveiros, and Eduardo A. Ferreira da Silva. "Human health risk assessment due to agricultural activities and crop consumption in the surroundings of an industrial area." *Exposure and Health* 12 (2020): 629–640.

Concon, Jose M. *Food Toxicology. Part A: Principles and Concepts; Part B: Contaminants and Additives*. Marcel Dekker Inc., 1988.

Dabuo, Bismark, Emmanuella Wesome Avogo, Gabriel Owusu Koomson, Maxwell Akantibila, and Daniel Ayendo Gbati. "Aflatoxins: Toxicity, occurrences and chronic exposure." In *Aflatoxins – Occurrence, Detection and Novel Detoxification Strategies*. IntechOpen, 2022. https://doi.org/10.5772/intechopen.105723.

Dhuldhaj, Umesh Pravin, Rishikesh Singh, and Vipin Kumar Singh. "Pesticide contamination in agro-ecosystems: Toxicity, impacts, and bio-based management strategies." *Environmental Science and Pollution Research* 30, no. 4 (2023): 9243–9270.

Di Stefano, Vita, and Giuseppe Avellone. "Food contaminants." *Journal of Food Studies* 3, no. 1 (2014): 88–103.

Eraky, Maysa Ahmad, Samia Mostafa Rashed, Mona El-Sayed Nasr, Azza Mohammed Salah El-Hamshary, and Amera Salah El-Ghannam. "Parasitic contamination of commonly consumed fresh leafy vegetables in Benha, Egypt." *Journal of Parasitology Research* 2014 (2014): 7. Article ID 613960. https://doi.org/10.1155/2014/613960.

Eskola, M., G. Kos, C. T. Elliott, J. Hajšlová, S. Mayar, and R. Krska. "Worldwide contamination of food-crops with mycotoxins: Validity of the widely cited 'FAO estimate' of 25%." *Critical Reviews in Food Science and Nutrition* 60, no. 16 (2020): 2773–2789. https://doi.org/10.1080/10408398.2019.1658570

Faille, Christine, Charles Cunault, Thomas Dubois, and Thierry Benezech. "Hygienic design of food processing lines to mitigate the risk of bacterial food contamination with respect to environmental concerns." *Innovative Food Science & Emerging Technologies* 46 (2018): 65–73.

Fu, L., H. R. Valentino, and Y. Wang. "Bacterial contamination in food production." In *Antimicrobial Food Packaging*, pp. 35–43. Academic Press, 2016.

Garvey, Mary. "Food pollution: A comprehensive review of chemical and biological sources of food contamination and impact on human health." *Nutrire* 44 (2019): 1–13.

Gaur, Nisha, Korrapati Narasimhulu, and Y. PydiSetty. "Recent advances in the bio-remediation of persistent organic pollutants and its effect on environment." *Journal of Cleaner Production* 198 (2018): 1602–1631.

Gerba, Charles P., and Christopher Y. Choi. "Role of irrigation water in crop contamination by viruses." In Viruses in Foods, pp. 257–263. Springer, 2006.

Gudkov, Sergey V., Dmitriy E. Burmistrov, Dmitriy A. Serov, Maxim B. Rebezov, Anastasia A. Semenova, and Andrey B. Lisitsyn. "A mini review of antibacterial properties of ZnO nanoparticles." *Frontiers in Physics* 9 (2021): 641481.

Gupta, D. K., S. Chatterjee, S. Datta, A. V. Voronina, and C. Walther. "Radionuclides: Accumulation and transport in plants." *Reviews of Environmental Contamination and Toxicology* 241 (2017): 139–160.

Hashim, A. "Human health effects of chronic arsenic exposure." In *Arsenic Toxicity Remediation: Biotechnological Approaches*, pp. 45–60. Springer Nature, 2023.

Islam, Ejaz Ul, Xiao-E. Yang, Zhen-li He, and Qaisar Mahmood. "Assessing potential dietary toxicity of heavy metals in selected vegetables and food crops." *Journal of Zhejiang University Science B* 8, no. 1 (2007): 1–13.

Khan, Romana, Alireza Noorpoor, and Abdol Ghaffar Ebadi. "Effects of air contamination on agriculture." In *Sustainable Plant Nutrition Under Contaminated Environments*, pp. 1–16. Springer International Publishing, 2022.

Kharazi, Ava, Mostafa Leili, Mohammad Khazaei, Mohammad Yusef Alikhani, and Reza Shokoohi. "Human health risk assessment of heavy metals in agricultural soil and food crops in Hamadan, Iran." *Journal of Food Composition and Analysis* 100 (2021): 103890.

Khodaei, Diako, Fardin Javanmardi, and Amin Mousavi Khaneghah. "The global overview of the occurrence of mycotoxins in cereals: A three-year survey." *Current Opinion in Food Science* 39 (2021): 36–42.

Kuiper, Irene, Ellen L. Lagendijk, Guido V. Bloemberg, and Ben J. J. Lugtenberg. "Rhizoremediation: A beneficial plant-microbe interaction." *Molecular Plant-Microbe Interactions* 17, no. 1 (2004): 6–15.

Kumar, Abhishek, Hardik Pathak, Seema Bhadauria, and Jebi Sudan. "Aflatoxin contamination in food crops: Causes, detection, and management: A review." *Food Production, Processing and Nutrition* 3 (2021): 1–9.

Lebelo, Kgomotso, Ntsoaki Malebo, Mokgaotsa Jonas Mochane, and Muthoni Masinde. "Chemical contamination pathways and the food safety implications along the various stages of food production: A review." *International Journal of Environmental Research and Public Health* no. 11 (May 28, 2021): 5795. https://doi.org/10.3390/ijerph18115795.

Lewis, L., M. Onsongo, H. Njapau, H., Schurz-Rogers, G. Luber, S. Kieszak, J. Nyamongo, et al. "Aflatoxin contamination of commercial maize products during an outbreak of acute aflatoxicosis in eastern and central Kenya." *Environmental Health Perspectives* 113, no. 12 (2015): 1763–1767.

Liu, Jie: Lvhui Sun, Jiacai Zhang, Jiao Guo, Lei Chen, Desheng Qi, and Niya Zhang. ""Aflatoxin B1, zearalenone and deoxynivalenol in feed ingredients and complete feed from central China." *Food Additives & Contaminants: Part B* 9, no. 2 (2016): 91–97.

Machado-Moreira, Bernardino, Karl Richards, Fiona Brennan, Florence Abram, and Catherine M. Burgess. "Microbial contamination of fresh produce: What, where, and how?." *Comprehensive Reviews in Food Science and Food Safety* 18, no. 6 (2019): 1727–1750.

Mamat, Zulpiya, Hamid Yimit, Rou Zi A. Ji, and Mamattursun Eziz. "Source identification and hazardous risk delineation of heavy metal contamination in Yanqi basin, northwest China." *Science of the Total Environment* 493 (2014): 1098–1111.

Manwani, Seema, Pooja Devi, Tanvi Singh, Chandra Shekhar Yadav, Kumud Kant Awasthi, Narain Bhoot, and Garima Awasthi. "Heavy metals in vegetables: A review of status, human health concerns, and management options." *Environmental Science and Pollution Research* 30, no. 28 (2023): 71940–71956.

Mohle-Boetani, J. C., J. Farrar, P. Bradley, Barak, J. D. P., M Miller, Mandrell, R, P. Mead, et al. "Salmonella infections associated with mung bean sprouts: Epidemiological and environmental investigations." *Epidemiology & Infection* 137, no. 3 (2009): 357–366.

Mudgal, Varsha, Nidhi Madaan, Anurag Mudgal, R. B. Singh, and Sanjay Mishra. "Effect of toxic metals on human health." *The Open Nutraceuticals Journal* 3, no. 1 (2010).

Nabil, Shariful Kibria, Nadira Mustari, and Mohidus Samad Khan. "Sources and health impacts of chemical contaminants in foods." *Techniques to Measure Food Safety and Quality: Microbial, Chemical, and Sensory* (2021): 31–68.

Naeem, M., Ritu Gill, Sarvajeet Singh Gill, Kashmir Singh, Adriano Sofo, and Narendra Tuteja. "Emerging contaminants and their effect on agricultural crops." *Frontiers in Plant Science* 14 (2023): 1296252.

Negash, Demissie. "A review of aflatoxin: Occurrence, prevention, and gaps in both food and feed safety." *Journal of Applied Microbiological Research* 1, no. 1 (2018): 35–43.

Obeng-Gyasi, Emmanuel, Javad Roostaei, and Jacqueline MacDonald Gibson. "Lead distribution in urban soil in a medium-sized city: Household-scale analysis." Environmental Science & Technology 55, no. 6 (2021): 3696–3705.

Onakpa, Michael Monday, Anoka Ayembe Njan, and Ogbureke Chidiebere Kalu. "A review of heavy metal contamination of food crops in Nigeria." *Ann Glob Health* 84, no. 3 (2018): 488–494.

Pachepsky, Yakov, Daniel R. Shelton, Jean ET McLain, Jitendra Patel, and Robert E. Mandrell. "Irrigation waters as a source of pathogenic microorganisms in produce: A review." *Advances in Agronomy* 113 (2011): 75–141.

Palimar, V., K. M. Saralaya, M. Arun, M. K. Mohanty, and B. Singh. "Profile of methyl parathion poisoning in Manipal, India." *Infection* 12 (2012): 23–25.

Pappas, R. S., G. M. Polzin, L. Zhang, C. H. Watson, D. C. Paschal, and D. L. Ashley. "Cadmium, lead, and thallium in mainstream tobacco smoke particulate." *Food and Chemical Toxicology* 44, no. 5 (2006): 714–723.

Pavlidis, G., and V. A. Tsihrintzis. "Environmental benefits and control of pollution to surface water and groundwater by agroforestry systems: A review." *Water Resources Management* 32 (2018): 1–29.

Petrović, T., and M. D'Agostino. "Viral contamination of food." In *Antimicrobial Food Packaging*, pp. 65–79. Academic Press, 2016.

Rather, Irfan A., Wee Yin Koh, Woon K. Paek, and Jeongheui Lim. "The sources of chemical contaminants in food and their health implications." *Frontiers in Pharmacology* 8 (2017): 830.

Ronald, Pamela. "Plant genetics, sustainable agriculture and global food security." *Genetics* 188, no. 1 (2011): 11–20.

Ruiz-Fernández, Ana Carolina, Jorge Feliciano Ontiveros-Cuadras, José L. Sericano, Joan-Albert Sanchez-Cabeza, Laval Liong Wee Kwong, Robert B. Dunbar, David A. Mucciarone, Libia Hascibe Pérez-Bernal, and Federico Páez-Osuna. "Long-range atmospheric transport of persistent organic pollutants to remote lacustrine environments." *Science of the Total Environment* 493 (2014): 505–520.

Scallan, Elaine, Robert M. Hoekstra, Frederick J. Angulo, Robert V. Tauxe, Marc-Alain Widdowson, Sharon L. Roy, Jeffery L. Jones, and Patricia M. Griffin. "Foodborne illness acquired in the United States – major pathogens." *Emerging Infectious Diseases* 17, no. 1 (2011): 7.

Şengül, Ü., E. Yalçın, B. Şengü, and K. Çavuşoğlu. "Investigation of aflatoxin contamination in maize flour consumed in Giresun, Turkey." *Quality Assurance and Safety of Crops & Foods* 8, no. 3 (2016): 385–391.

Seymour, I. J., and H. Appleton. "Foodborne viruses and fresh produce." *Journal of applied microbiology* 91, no. 5 (2001): 759–773.

Sharapov, U. M., A. M. Wendel, J. P. Davis, W. E. Keene, J. Farrar, S. Sodha, E. Hyytia-Trees, et al. "Multistate outbreak of Escherichia coli O157: H7 infections associated with consumption of fresh spinach: United States, 2006." *Journal of Food Protection* 79, no. 12 (2016): 2024–2030.

Shen, Chen, Xiang-Chang Yin, Bo-Yang Jiao, Jing Li, Peng Jia, Xiao-Wen Zhang, Xue-Hao Cheng et al. "Evaluation of adverse effects/events of genetically modified food consumption: A systematic review of animal and human studies." *Environmental Sciences Europe* 34, no. 1 (2022): 1–33.

Singh, Ngangbam Sarat, Ranju Sharma, Talat Parween, and P. K. Patanjali. "Pesticide contamination and human health risk factor." *Modern Age Environmental Problems and Their Remediation* (2018): 49–68.

Sreekumar, K. M., and K. D. Prathapan. "An evidence-based inquiry into the endosulfan tragedy in Kasaragod, Kerala." *Economic Political Wkly* 56, no. 41 (2021): 45–53.

Udovicki, Bozidar, Mirjana Andjelkovic, Tanja Cirkovic-Velickovic, and Andreja Rajkovic. "Microplastics in food: Scoping review on health effects, occurrence, and human exposure." *International Journal of Food Contamination* 9, no. 1 (2022): 1–16.

Vandelac, L. "Endocrine disruption agents: Environment, health, public policies, and the precautionary principle." *Bulletin de L'academie Nationale de Medecine* 184, no. 7 (2000): 1477–86.

World Health Organization. "World Health Organization." 2022, September 15. https://www.who.int/news-room/fact-sheets/detail/pesticide-residues-in-food.

Yoshida, Fumikazu, Akio Hata, and Haruo Tonegawa. "Itai-Itai disease and the countermeasures against cadmium pollution by the Kamioka mine." *Environmental Economics and Policy Studies* 2 (1999): 215–229.

Zheng, Wei, Kelsey N. Wiles, Nancy Holm, Nathan A. Deppe, and Clinton R. Shipley. "Uptake, translocation, and accumulation of pharmaceutical and hormone contaminants in vegetables." In *Retention, Uptake, and Translocation of Agrochemicals in Plants*, pp. 167–181. American Chemical Society, 2014.

11 Plant Uptake of Emerging Contaminants in Foods

The Extent of Its Effects

Srinivasan Kameswaran, Ramesh B. Kasetti, and Bellamkonda Ramesh

11.1 INTRODUCTION

A class of metallic elements known as heavy metals has large densities, atomic weights, and/or numbers. Common examples of these elements are nickel (Ni), copper (Cu), zinc (Zn), cadmium (Cd), mercury (Hg), chromium (Cr), and lead (Pb). When present in very high densities (>5 g/cm^3), they mostly represent inorganic chemical pollutants (Wuana and Okieimen, 2011). Because of their propensity to dissolve in water, metal ions can bioaccumulate in aquatic plants and animals and have a negative impact on the health of a larger ecosystem by contaminating the food chain and water sources. According to Hossain et al. (2008), heavy metals in the soil ecosystem are absorbed by living cells and tissues through polluted subterranean water through chemical redox reactions, leaching, or physical weathering processes. All things considered, the contamination of sediments, soils, and water with heavy metals is a growing problem. Metal contamination is definitely a product of the fast industrialization and urbanization that has occurred (Micó et al., 2006; Oladoye, 2022). Anthropogenic activity is growing in emerging nations, which is proportionately causing heavy metal pollution to become an unmanageable issue (Yan et al., 2018; Babaniyi et al., 2023).

Heavy metals from various anthropogenic activities, such as fertilizer application, hard water irrigation, petrochemical spillage, atmospheric deposition, and animal manure, as well as residues from the combustion of coal and tires, are sucked into the soil (Adeyi and Oladoye, 2020). The extensive distribution of heavy metals in agricultural areas has caused considerable stress to terrestrial ecosystems, and their aftereffects also can result in marked dietary exposure through plant–food chain transfer (Alloway, 2013; Grimm et al., 2008). Along with parallel issues with land tenure, the pollution of agricultural crops and groundwater by heavy metals has had a significant influence on food quality (McLaughlin et al., 2000a, 2000b). Thus, it is crucial to reduce the inputs or sources of heavy metal pollution and, more crucially, to increase the efficacy of remediation techniques used to remediate contaminated soils (Derakhshan Nejad et al., 2018). In addition to prevention, there is a need for improved advancements in the cleanup technologies used to deal with contaminated environments, depending on the kind of pollutant, degree of contamination, and characteristics of the contaminated area. Laws protecting the environment and public health change depending on the kind of contaminated soil or water, and various cleanup procedures are needed (Kabata-Pendias and Pendias, 2001; Aransiola et al., 2013). Therefore, it would be possible to recover soil from heavy metal contamination once the bioavailability and speciation of the metal ions in that soil have been thoroughly characterized. This will provide information on the environmental and health effects of the contamination, as well as its source and basic chemistry. Furthermore, risk assessment serves as an affordable tool that protects the environment and public health, allowing policymakers and/or decision makers to manage efficiently regions polluted by heavy metals (Zhao and Kaluarachchi, 2002).

 DOI: 10.1201/9781032683751-13

In order to develop affordable and appropriate solutions for the restoration of heavy metal-polluted soils, the scientific community has been more interested in finding a solution to the issue (Bolan et al., 2008). The best solutions needed for the restoration of metal-polluted areas include soil washing, phytoremediation, and contaminant immobilization, according to several reports (Wuana and Okieimen, 2011). Though these technologies are inexpensive and environmentally friendly, their main practical uses have reportedly been observed in developed nations; in developing nations, however, their lack of awareness and knowledge has prevented them from being widely applied. Furthermore, it is critical to have accessible simple, affordable, and sustainable remediation methods to lower hazards connected with contaminated soil sites, enhance food security, avoid land tenure–related problems, and make contaminated soils available for farming, particularly in developing nations.

This chapter compiles disparate data from the literature on the harmful effects of heavy metals and the causes of soil pollution, both urban and agricultural. The effects of toxic metals on food security, their potential for bioaccumulation and biotoxicity, and potential uses of phytoremediation by-products for disposal or utilization are all well described. Treatment options for heavy metal-contaminated soil, including in situ and ex situ remediation approaches with a focus on phytoremediation techniques, are also critically examined.

11.2 HEAVY METAL SOURCES IN CONTAMINATED SOILS

Natural and human sources are the two main ways that heavy metals enter the soil.

11.2.1 Natural Sources

One naturally occurring source of heavy metal(loids) in contaminated areas is the parent soil; this is referred to as a lithogenic source. Many heavy metals are found in chemical complexes rather than alone, which makes them easier for live cells and tissues to absorb directly (Derakh shan Nejad et al., 2018). Common heavy metals include Zn, Pb(II), Hg, Ni, Cu, arsenic (As), Cd, and Cr. Heavy metals can also occur in soil as silicates, hydroxides, sulfides, oxides, organic phosphates, and hydroxide compounds. Although they are present in trace levels, they are thought to be major contributors to both environmental and human health issues. According to Kabata-Pendias and Pendias (2001), the weathering of parent soil components through pedogenic processes also explains the trace metal release (<1000 mg/kg) of heavy metals into the soil environment. In addition, some environmental factors as windborne soil particles, sea salt sprays, volcanic eruptions, forest fires, biogenic sources, and rock weathering can raise the concentrations of heavy metals (Wuana and Okieimen, 2011).

11.2.2 Anthropogenic Sources

The everyday production of commodities and mechanical activities to satisfy the high needs of a large human population has resulted in the simultaneous emission of garbage or effluents, which has increased environmental pollution. Sources of heavy metals include anthropogenic activities such as chemical runoffs, sewage, industrial and agricultural processes, and metallurgical and mining operations. Chemical effluent wastes have been found to contribute to soil pollution in industrial urban areas by containing high quantities of heavy metals such as Pb, Cu, Cd, and Zn (Tchounwou et al., 2012; Todorović et al., 2014). Certain industrial activities have been linked to specific heavy metal problems; these include the use of Pb in automobiles; As in insecticides; Zn, Cu, and As in smelting processes; and the combustion of fossil fuels, which releases vanadium (V), Ni, selenium (Se), Hg, and tin (Sn; Vhahangwele and Mugera, 2015).

11.2.3 Possible Harmful Effects of Heavy Metals

A number of physiological and biochemical processes in animals and plants depend on essential nutrients like cobalt (Co), Cr, Ni, Cu, Se, iron (Fe), manganese (Mn), magnesium (Mg), Zn, and molybdenum (Mo; WHO, 1996). Non-essential heavy metals include Pb, Cd, astatine (At), and Hg. They are also important players in a number of oxidation-reduction processes and are substantial components of numerous prominent enzymes. The nucleus, lysosome, cell membrane, endoplasmic reticulum, mitochondria, nuclei, and certain enzymes connected to metabolic, detoxification, and damage repair processes are among the organelles and/or cell components in biological systems that are impacted by heavy metals (Wang and Shi, 2001). However, the health of humans may be seriously harmed by excesses of these micronutrients at levels higher than the maximum allowable intake (Tchounwou and colleagues (2012); Asemoloye and colleagues, 2021).

The absence of any one of these vital micronutrients can result in the expression of many deficient syndromes and/or illnesses. Copper, for example, is a necessary micronutrient required for a number of metalloenzyme related processes, including the production of hemoglobin, the biosynthesis of catecholamines, the metabolism of carbohydrates, and the cross-linking of amino acids and proteins (elastin, collagen, and keratin; Harvey and McArdle, 2008). In redox processes, Cu can alternate between the +2 and +1 oxidation states. The creation of hydroxyl and superoxide radicals, which deteriorate human health, is explained by this transition feature between two oxidation states of Cu (Tchounwou et al., 2008). For this reason, consuming too much Cu has been linked to cell damage and Wilson disease (Tchounwou et al., 2008). While an excess of some important micronutrient elements can cause tissue and cellular damage as well as illnesses, they are needed for the appropriate physiological and biochemical operation of the human system. For some micronutrient metals, like Cu and Cr, there is a relatively small range of concentrations between harmful and beneficial effects.

Exogenous metal(loids) are those that are thought to be non-essential and have no recognized biological role(s). Even at extremely small concentrations, non-essential heavy metals are harmful to people, animals, and vegetation. According to Ali et al. (2019), some examples of these elements are uranium (U), vanadium (V), Sn, titanium (Ti), Ni, Hg, lithium (Li), Pb, indium (In), gold (Au), germanium (Ge), gallium (Ga), Cd, bismuth (Bi), beryllium (Be), barium (Ba), As, antimony (Sb), aluminum (Al), and tellurium (Te). Among these, the most harmful and important to the environment are As, Hg, Pb, Cd, Zn, Cu, Ni, and Cr (Barakat, 2011).

Because these heavy metals can harm both people and wildlife, they are hazardous to the environment. If a chemical compound has the ability to disrupt an organism's ability to grow, survive, or reproduce, it is considered poisonous. Numerous mechanisms, some of which are yet unknown, underlie their interference with various organs, tissues, and systems in various species (Ali et al., 2019). The distinct physicochemical characteristics and attributes of every metal provide its specific mechanisms of action in toxicology. Oxidative stress and reactive oxygen species (ROS) play a significant role in the carcinogenicity and toxicity of metals such as Hg (Sutton et al., 2002; Sutton and Tchounwou, 2007), Pb (Tchounwou et al., 2004b; Yedjou and Tchounwou, 2007a), As (Tchounwou et al., 2004a; Yedjou and Tchounwou, 2007b), and Cd (Tchounwou et al., 2001). The severe toxicity of heavy metals (As, Cr, Cd, Hg, and Pb) has led to their classification as a public health problem. This pollution is becoming increasingly severe because of the quick expansion of industry, urbanization, and greater reliance on synthetic agrochemicals. According to Pan et al. (2016), there is a detrimental impact of the high concentration of heavy metals in soil on human health, food quality, soil structure, function, nutrients, and biological activity. According to Ali et al. (2006), humans can come into contact with heavy metals through skin contact, ingestion, and inhalation from the soil. For instance, renal failure and cardiovascular and cerebrovascular illness are associated with cadmium (Bernard, 2008). The hematological system, the kidneys, and neurons can all suffer long-term damage from exposure to Pb-contaminated settings. A thorough evaluation of their toxicities has been done by Tchounwou et al. (2012), Ali et al. (2019), and Wuana and Okieimen (2011).

11.3 EFFECTS OF HEAVY METAL POLLUTION OF AGRICULTURAL SOILS ON THE AVAILABILITY OF FOOD

11.3.1 Heavy Metals and Agricultural Soils

There has been evidence over the past 10 years of an increase in the pollution of agricultural soils with heavy metals. This has been linked to the complexity and rise of a number of human activities, including mechanization, chemical application, and excessive industrialization, to name a few (Ashraf et al., 2019; Suman et al., 2018). Heavy metal pollution of agricultural soils is a major environmental issue and a danger to agricultural productivity because of the metals' biological half-lives, non-biodegradable nature, toxicity, persistence, and bioaccumulation in the food chain (Lü et al., 2018). Chemical chemicals and heavy metals may cause irreversible damage and loss of some agricultural functions (Streets et al., 2017). Topographical features, regional climate, and geology are the primary natural elements thought to be responsible for the environmental dispersion of heavy metals. The lengthy residence duration of heavy metals is caused by their interactions with certain components of soil. Chemical forms and metal speciation affect the presence and movement of heavy metals in soils, which can lower the quality of an aquifer (Barbulescu, 2020; Al-Taani et al., 2021). The most well-known anthropogenic causes include waste management, transportation, industrial emissions, and fuel combustion (El-Radaideh and Al-Taani, 2018). Numerous writers have emphasized the importance that mining operations play in soil degradation, especially when it comes to heavy metals (Khlifi et al., 2014; Vimercati et al., 2017). The main sources of heavy metals in agricultural soils include liming, irrigation, sewage sludge, and inorganic fertilizers. Different quantities of Cr, Cd, Zn, Pb, and Ni have been added to soil by other sources, such as fungicides and inorganic fertilizers (Tóth et al., 2016). Repeated application of phosphate fertilizer raises levels of key heavy metals in agricultural soil that are normally low (Carnelo et al., 1997). Cr, Cu, Zn, Pb, Ni, and Cd are added to the soil by sewage sludge, whereas animal manure contributes Co, Cu, Mn, and Zn (Srivastava et al., 2017; Sharma et al., 2017). Pesticides also introduce heavy metal pollution to agricultural crops. In a similar vein, irrigation with wastewater plays a major role in heavy metal contamination. Therefore, the makeup and rate of application of pesticides, inorganic fertilizers, wastewater, and/or sewage sludge, as well as soil characteristics, all affect the concentrations and amounts of heavy metals in agricultural soil.

Risk assessment is a practical tool used by decision makers to manage contaminated fields and protect human health and the environment (Zhao and Kaluarachchi, 2002). Geochemical distribution of heavy metals in soil is greatly influenced by pedological processes, environmental features, and variations in the parent rock (Rezapour et al., 2014). Hazardous concentrations of heavy metals, especially Pb, Zn, As, Cr, and Cu, can seriously damage human health and lead to a host of additional negative outcomes (Brevik, 2009). Major causes of heavy metal pollution in agricultural regions include the introduction of pesticides and fertilizers, low-quality manure and compost, industrial wastes, and several other results of anthropogenic activity (Keshavarzi and Kumar, 2019). The ecological risks index (RI), geo-accumulation index, contamination factor (CF), and enrichment factor (EF) are some of the techniques that researchers have used to quantify the amount of heavy metal pollution (Tian et al., 2017; Elbehiry et al., 2019).

11.3.2 Effects of Heavy Metals on Food Security

Both the second and third Sustainable Development Goals (SDGs), which were approved by the UN General Assembly for the year 2030, emphasize how important it is to reduce food insecurity (United Nations, 2015). Recent reports claim that heavy metals in food have impacted the attempts of the SDGs to encourage sustainable living for present and future generations, leading to poor health and well-being, particularly in developing and third-world nations where little is known about their effects (Anani et al., 2020). The main topics surrounding heavy metals, which are now hotly debated in the realm of environmental toxicity, include farming practices and lithogenic origin (Anani and Olomukoro, 2017). Food gathered from contaminated soil has been shown to have higher levels of metals than are allowed by law, which might be harmful to human health or cause cancer. It can result in serious disorders of the liver, lung, nervous system, and immune system, as well as bone fractures and malformations, cardiovascular problems, kidney dysfunction, hypertension, and other major diseases of these systems (El-Kady and Abdel-Wahhab, 2018). Furthermore, consumption of contaminated water and seafood has been linked to the detection of some complex compounds based on heavy metal xenobiotics (Anani and Olomukoro, 2018a, 2018b). The terrestrial and aquatic animals and vegetation are killed by these complexes when the rate of accumulation exceeds their absorptive potentials. On the other hand, their suitability for ingesting is called into doubt when they are found to contain metal concentrations (bioaccumulation).

Numerous approaches have been used to study the impact of toxic metal(s) on food security. In their study, Zhang et al. (2015) examined the effects of heavy metal accumulation on food safety in China. They discovered that Cd had the greatest contamination rate at 7.75%, followed by Hg, Cu, Ni, and Zn, while Pb and Cr had the lowest rates at less than 1%. The scientists found that the main heavy metals responsible for 10.18% of the overall contamination in Chinese farming soil were Cd, Hg, Cu, and Ni. Aside from harming agricultural soil, human activities including smelting and mining, industrialization, sewage irrigation, urbanization, and fertilizer application have also discharged trace quantities of heavy metals into the soil. A probable geographical trajectory in a soil-food crop subsystem as well as health problems for humans were discussed by Rai et al. (2019) in relation to heavy metal pollution. Food crops contaminated with heavy metals have also been thoroughly investigated with regard to the physiological and molecular translocation processes involved. The effects of heavy metal contamination on human health were thoroughly reviewed by Anani et al. (2020), with a particular emphasis on food contamination brought on by environmental and agricultural practices like pesticide use, which lead to high levels of toxic metal element accumulation and adsorption. The scientists also described the interactions that several dangerous heavy metals have with the physiological and molecular translocation mechanisms in food crops. Pandiyan et al. (2021) measured heavy metals in another investigation using a double-beam atomic absorption spectrophotometer. Sediment included greater quantities of Hg, Co, and Cr, whereas mollusks and polychaetes had higher concentrations of Ni and Cd, respectively. They also showed that fish had greater Zn levels while crabs had higher amounts of Pb and Cu.

11.4 USING PHYTOREMEDIATION METHODS TO REMEDIATE SOIL POLLUTED WITH HEAVY METALS

11.4.1 Modes of Phytoremediation

Understanding the methods a plant utilizes, its response to a specific metal, its concentration, and its toxicity level is essential for the effective implementation of phytoremediation technology. The ability to plan and execute a sustainable phytoremediation method successfully requires these elements. Additionally, it is critical to comprehend how heavy metals are absorbed or taken up, moved throughout the cell of the plant, and detoxified. Heavy metal detoxification is a crucial process that plants employ, as several experts have highlighted (Figure 11.1).

11.4.1.1 Phytoextraction

The process by which plants absorb metals from the soil or water is known as phytoextraction of heavy metals. It deals with how heavy metals are absorbed, relocated, and accumulated in the shoots and other tissues (Salt et al., 1995; Jacob et al., 2018). According to Ali et al. (2013) and Sarwar et al. (2017), plants are believed to be most successful in removing heavy metals from soil and water by this process. It appears to have a lasting effect on removing heavy metal contaminants from the environment, which makes it potentially useful and widely applicable in the business

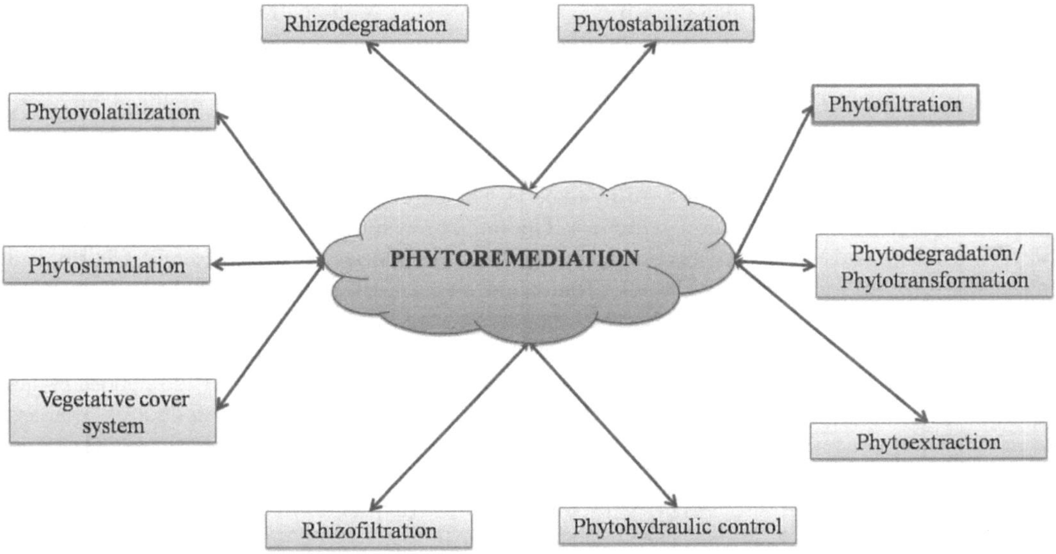

FIGURE 11.1 The different processes of phytoremediation.

sector. However, according to Li et al. (2010), this process is mostly dependent on the biomass of the shoots, the type and concentration of heavy metals, and the root structure. Many researchers have listed the steps involved in the phytoextraction mechanism, which include the mobilization of heavy metals towards the rhizosphere, the absorption/uptake of heavy metals by the plant's roots, the translocation of heavy metals ions through the aerial parts, and the sequestration and compartmentation of these ions in the tissue (Milić et al., 2012; Ali et al., 2012).

11.4.1.2 Rhizofiltration

In contrast to shoots (caulofiltration) or seedlings (blastofiltration), this process primarily deals with plant roots (rhizomes). According to Mesjasz-Przybyłowicz et al. (2004), it is also known as phytofiltration of environmental contaminants. Heavy metals are either absorbed inside the root (through absorption) or adsorb onto its surface (through adsorption) in the rhizofiltration mechanism. Numerous variables have been linked to this process; the functions of root exudates have also been extensively documented. According to Javed et al. (2019), the pH of the rhizosphere is greatly influenced by root exudates, which also have a significant impact on the precipitation of heavy metals on the root system. Typically, plants with long, hairy roots that are capable of accumulating heavy metals are employed for this process (Rezania et al., 2016; Dhanwal et al., 2017). In order to initiate a large root system for the rhizofiltration mechanism, these plants are typically grown hydroponically in clean water to start. After a few specific days, the clean water is replaced with polluted water or soil to acclimatize the plant, and it is then moved to contaminated soil for in situ remediation (Wuana and Okieimen, 2011). Rhizofiltration plants often have a thick root system and a high tolerance to heavy metals before being eliminated after a certain amount of time. They comprise both aquatic and terrestrial plants. Popular aquatic plants that are frequently utilized for rhizofiltration are poplar, azolla, cattail, duckweed, and water hyacinth (Hooda, 2007). Indian mustard (*Brassica juncea*) and sunflower (*Helianthus annuus*) are two common terrestrial plants employed for this process.

11.4.1.3 Phytostabilization

Strongly tolerant plants that can immobilize heavy metals in the soil, increase their bioavailability, and stop the metals from migrating into the ecosystem are known to use this mechanism, which lowers the likelihood that the metals will enter an ecosystem's food chain (Wong, 2003; Marques et al., 2009). The plant may typically absorb the heavy metals by adsorbing them onto the cell wall of the root, precipitating them, or decreasing their valence in the rhizosphere (Ginn et al., 2008; Gerhardt et al., 2017). The development of these plants improves the maintenance of the health of the soil in areas contaminated by heavy metals and the creation of vegetation cover to reduce the amount of heavy metals that leak into subsurface water or are dispersed by wind together with soil particles (Vangronsveld et al., 2009). According to certain research, phytostabilizing plants like Syrian bean-caper (*Zygophyllum fabago* L.), wiregrass (*Gentiana pennelliana* Fernald), red fescue (*Festuca rubra* L. [Griff.]), and hippo grass (*Vossia cuspidata* Roxb.) can effectively stabilize Cr, Cu, Pb, and Zn in contaminated soils (Galal et al., 2017; Radziemska et al., 2017; Aransiola et al., 2019).

In contrast to phytoextraction and rhizofiltration, phytostabilization does not call for the removal of treated plants (Wuana and Okieimen, 2011). However, for this mechanism to work and immobilize heavy metals, stabilize the soil structure, and stop soil erosion, it is necessary to choose plants that are highly tolerant of heavy metals and have an efficient root system. With enough biomass and a rapid growth rate, they ought to be simple to maintain under field conditions (Berti and Cunningham, 2000; Marques et al., 2009). A thorough study of plants with the capacity to stabilize heavy metals was conducted by Burges et al. (2018), while Wong (2003) and Marques et al. (2009) demonstrated how this mechanism works to stop heavy metals from entering the food chain and potentially migrating into the environment.

11.4.1.4 Phytovolatilization

By absorbing pollutants from the soil and converting them through intricate metabolic pathways from harmful to less toxic or non-toxic compounds, plants release or excrete these pollutants into the atmosphere. This is the mechanism at work. *B. juncea* was reported to be an excellent volatilizer of Se, whereas certain important members of the Brassicaceae family of plants have been shown to be good volatilizers of heavy metals (Terry et al., 1992; Banuelos et al., 1993). Mahar et al. (2016) have confirmed that this method works for a number of heavy metals, including As, Se, and Hg. Depending on the type of plant, this technique is assumed to entail distinct phases of growth. The majority of the time, assimilation processes transform heavy metals into complexes. For example, Se is assimilated into selenium-amino acids such as selenocysteine (SeCys) and selenomethionine (SeMet). *B. juncea* uses a biomethylation mechanism to further transform the selenoamino acid (SeMe) into dimethylselenide (DMSe). In comparison with inorganic Se, DMSe is a less harmful volatile molecule that is more readily released into the atmosphere (de Souza et al., 2000; Terry et al., 2000). At room temperature, Hg exists naturally as a liquid; it is very reactive and readily volatilizes. According to Marques et al. (2009), it therefore mostly resides in the environment as the divalent cation molecule Hg_2C. When a plant absorbs mercury from a leaf's root, it transforms it into an ionic form that is then released into the environment as methylmercury, a volatile chemical that is comparatively less harmful (Bizily et al., 2000).

Compared to other phytoremediation methods, phytovolatilization has the advantage of being effective for metalloid contaminants. It can remove these pollutants from polluted sites and disperse them in a gaseous form without having any bioaccumulation effect, so there is no need to dispose of plants. Nevertheless, it may not remove all of the pollutant because some of it may still be released into the environment (Vangronsveld et al., 2009). Therefore, before this mechanism is applied in situ, greater risk evaluation is needed.

11.4.2 Plants for Phytoremediation

Phytoremediation techniques are the most effective way to treat heavy metal pollution (Singh and Prasad, 2015; Gouda et al., 2018). Plants have developed defense mechanisms against heavy metal poisoning, and their abundant and renewable biomass greatly increases the bioaccumulation of poisons in soils and waterways (Singh and Prasad, 2015). Nevertheless, plant growth will be hindered when cultivated on soil that has significant concentrations of heavy metals because of the negative impact of heavy metals on plants. Plants used in the phytoremediation process are commonly referred to as metallophytes. According to Zalewska and Danowska (2017), there are three types of metallophytes: metal hyperaccumulators, excluders, and indicators.

11.4.2.1 Metal Hyperaccumulator

A group of plants from various families known as metal hyperaccumulators are those that can grow on soils rich in metalliferous elements and build up substantially higher concentrations of heavy metals in their aerial organs than are found in most other species while remaining completely unaffected (Ali et al., 2012; Rascio and Navari-Izzo, 2011; Salt et al., 1998). Three discernible traits set hyperaccumulators apart from similar non-hyperaccumulating taxa: a notably elevated rate of heavy metal absorption, a quicker root-to-shoot translocation, and an enhanced capacity to detoxify and accumulate heavy metals in leaves. Particularly for phytoremediation of soils polluted by heavy metals, the hunt for suitable hyperaccumulators has intensified. It has been determined thus far that about 400 plant species are hyperaccumulators (Sheoran et al., 2010; Liu et al., 2018). These plants, which belong to the classes Brassicaceae, Fabaceae, Euphorbiaceae, Asteraceae, Lamiaceae, and Scrophulariaceae, range in size from shrubs to trees (Salt et al., 1998). Furthermore, several grasses including spear grass (Ali et al., 2012), *Panicum maximum* (Asemoloye, Jonathan et al., 2017; Asemoloye et al., 2017a, 2017b, 2020a, 2020b), and *Trifolium alexandrinum* (Malik et al., 2010). The plants' vast root systems, ability to withstand metal toxicity, elevated biomass, and ability to multi-happen throughout a single growth phase (Ali et al., 2012). According to Sachs (1865), *Thlaspi caerulescens* was the first known hyperaccumulator due to its exceptional tolerance to Zn. In their 2004 study, Yang et al. discovered that *Sedum alfredii* was an effective co-hyperaccumulator of Cd and Zn, accumulating high concentrations of both metals (thousands of mg/kg for Zn and several hundreds of mg/kg for Cd). More than 5000 mg/kg may be stored by *Pteris vittata*. As with its pinna leaf, Chen et al. (2002) report that the accumulation coefficient of pinna leaf can approach 80, which is the ratio of the metal concentration in plants to that in the surrounding soil. As of right now, several pot and field trials using hyperaccumulators have been carried out worldwide to remediate soils contaminated with heavy metals (Abid et al., 2019; Huang et al., 2020). Certain high-biomass crops, such as *Cannabis sativa, Nicotiana tabacum*, and *Zea mays*, have been utilized in phytoextraction processes to remove heavy metals (Herzig et al., 2014). Similarly, reports of some woody plant species exist (Suman et al., 2018; Burges et al., 2018). They have been compared to them due to their inedibility, high biomass output, deep root systems, erosion resilience, and ability to stop contaminated soil from spreading.

11.4.2.2 Metal Excluder

Reduced metal concentrations in the aerial portions of plants are the outcome of the metal excluder strategy, which limits the quantity of metals that are translocated from roots to shoots. High concentrations of metals have been found in the roots of excluder species (Baker and Walker, 1990). The discovery of plants that reject heavy metals is crucial because, although metal excluders may flourish in highly contaminated soils, their heavy metal absorption is minimal even in the presence of high concentrations in soils (Suman et al., 2018). Examples of excluder species include *Commelina communis, Agrostis stolonifera* L., *Silene maritima*, and woody plants including *Oenothera biennis, Populus, Salix*, and *Pinus radiata* (Maestri et al., 2010). The growth performance of propagules and 6-month-old *Bruguiera gymnorhiza* seedlings in the presence of contaminating Cu and Cd was examined. It was found that older mangrove seedlings are more able to tolerate metal than younger seedlings because of their more effective exclusion mechanism (Wang et al., 2013). By developing many adaptation strategies, such as ion exclusion and actively avoiding metal absorption, *Avicennia marina* has demonstrated increased resistance to heavy metals (Burchett et al., 2003; Nikalje and Suprasanna, 2018). Furthermore, Pb is claimed to accumulate less in roots than in other metals in this species, and it was mostly eliminated from leaf tissue (at 400 μg/g Pb sediment concentrations) with no detrimental effects on development (MacFarlane and Burchett, 2002). *Atriplex* roots and shoots exhibited greater amounts of Ni, Cu, Pb, and Zn than plants grown in uncontaminated soils when heavy metals were present in the soil at higher concentrations (Walker et al., 2014). According to Kachout et al. (2012), metals that *Atriplex* consumes are mostly distributed in root tissues, suggesting that they have a broad metal tolerance exclusion mechanism.

11.4.2.3 Metal Indicator

Similar to accumulators, metal indicators also accumulate metals in their aerial tissue; however, the concentration of metals in their above-ground tissue often reflects the concentration of metals in the surrounding environment (Baker and Walker, 1990). If the plants continue to take in metals, they will eventually perish. Since these plants absorb pollutants like accumulators and are pollution indicators, they are significant for both biology and ecology (Mganga et al., 2011; Balaferj et al., 2020). To remove heavy metals from the soil, research has been done employing metal indicators. To determine if *Phragmites australis* organs are appropriate for contaminants biomonitoring in soil and water, Bonanno and Giudice (2010) examined the organs' varying degrees of accumulation. Phillips et al. (2015) examined the capacity of *Spartina maritima, Typha capensis*, and *P. australis* to accumulate metals in the urbanized Swartkops Estuary. According to Eid et al. (2020), four aquatic macrophytes were evaluated for their capacity to collect Cd, Ni, and Pb, as well as their use in phytoremediation and signaling of these metals in polluted wetlands. These macrophytes were *Eichhornia crassipes*, *Ludwigia stolonifera*, *Echinochloa stagnina*, and *P. australis.*

11.5 IN SITU, EX SITU, AND MIXED METHODS TO PHYTOREMEDIATION TECHNOLOGY USE IN CONTAMINATED SOILS

During the past 10 years, there has been a lot of support for the use of plant-based phytoremediation (treatment of contaminated environments). Based on the modalities of application, this method is divided into two categories: ex situ and in situ phytoremediation. Applying phytoremediation therapy to soil or water distant from the original site area is part of the ex situ approach. In order to remediate soil or water that has been contaminated by heavy metals, the affected area is often excavated or moved away. This kind of remediation is often carried out by mechanical or physical excavation, detoxification, and demolition of polluted areas in order to produce restored sites. Compared to in situ phytoremediation, this application technique is less popular.

The process of treating contaminated areas on the property is known as in situ phytoremediation. According to Nejad et al. (2018), it entails either the removal or detoxification of hazardous chemicals from polluted areas. Compared to the ex situ system, this application technique is believed to be more economical and environmentally benign. Additionally, it encourages the application of restoration across a sizable area, necessitates less disruption, and does not call for the use of huge machinery for the extraction of soil or water (Mahmood et al., 2015; Khalid et al., 2017). When it comes to in situ remediation of contaminated groundwater, the rhizofiltration mechanism has received more attention than phytoextraction and phytostabilization processes (Lorestani et al., 2013). This method, which uses macrophytes and macroalgae in particular, has been effectively used to clean groundwater polluted with metals and chemical effluents (Mazej and Germ, 2009). *Lemna gibba* L. and *Hydrilla verticillata* (L.f.) Royle are two common examples that have also been noted as effective bioindicators (Mkandawire and Dudel, 2005). According to Robinson et al. (2006), three aquatic macrophytes – *Ceratophyllum demersum* L., *L. gibba* L., and *H. verticillata* – were used to remediate a site polluted with heavy metals. In a different study, Upadhyay et al. (2017) and Srivastava et al. (2014) reported the capacities of five microphages (*Lemna minor* L., *C. demersum, E. crassipes* (Mart.) Solms, *Wolffia globosa* (Roxb.) Hartog and Plas, and *H. verticillata*) in addition to three macrophages (*C. demersum, L. gibba*, and *H. verticillata*) in cleaning up metal-polluted sources.

In order to obtain the best outcomes while reducing the ecological dangers, it has been widely advised that phytoremediation – regardless of the processes employed – be combined with other techniques. Rehabilitating abandoned mining sites in situ has been suggested to be accomplished by the use of phytostabilization with chemical additives such as phosphates, lime, and compost tea (Chaney and Baklanov, 2017). Rhizosphere plant growth–promoting microorganisms are a great way to support plant performance during phytoremediation. Numerous studies have indicated that plant-associated microbes can facilitate phytoremediation. Beneficial microorganisms in the rhizosphere are assumed to promote root growth and improve the plant's reaction to the chemical makeup of the soil through a plant-microbe association process (Gupta et al., 2013; Fasani et al., 2018). Plant growth–promoting rhizobacteria/rhizofungi (PGPR) are a group of rhizobacteria and rhizosphere fungi that are known to aid plants in their defense against soil pathogens; enhance growth, fitness, and nutrient absorption; and stimulate their tolerance response to heavy metals (Ma et al., 2011).

Supplementing these microbes with different chemicals, organic acids, siderophores, enzymes, or phytohormones might improve their activity (Ma et al., 2011). Certain PGPR have been shown to create 1-aminicyclopropane-1-carboxylate (ACC) deaminase, which is capable of degrading the ethylene precursor ACC (Arshad et al., 2007; Glick, 2014). By producing ACC deaminase, these PGPR can lower the amount of ethylene that plants produce, which will encourage plant development. According to certain scientists (Huang et al., 2004; Arshad et al., 2007), cultivating plants with ACC deaminase-producing PGPR showed improved biomass, root densities, and "phytoremediation potentials as observed on heavy metal absorption."

Additionally, according to reports from other studies, some PGPR have the ability to create auxin, also known as indole acetic acid (IAA). This may function as a phytohormone to encourage the growth of root hairs and the beginning of lateral roots in plants, hence advancing phytoremediation (Glick, 2010; DalCorso et al., 2019). While certain siderophore-secreting microorganisms were described by Schalk et al. (2011) and Braud et al. (2009), Becerra-Castro et al. (2011) identified several endophytes that promote plants. Arbuscular mycorrhizal fungi (AMF) have received a lot of attention for their significant roles in promoting plant development and their potential for phytoremediation. Chen et al. (2003) observed that when compared to the control, *Trifolium pratense* L. infected with AMF generated higher biomass and zinc accumulation. Due to their vast hypha networks, AMF have been demonstrated to increase the absorptive surfaces in plant roots, which improve the absorption of nutrients and water as well as the bioavailability of heavy metals (Göhre and Paszkowski, 2006; Olowe et al., 2020). According to Vamerali et al. (2010) and Sarwar et al. (2017), some AMFs can also generate plant hormones to promote phytoremediation and plant development.

Enhancing phytoextraction can also be achieved by raising the heavy metal bioavailability. According to Blaylock and Huang (2000), heavy metals in soil are not always easily assimilated by plants; in most situations, only small percentages are accessible in soluble form. Cr and Pb are the least accessible metals, whereas Co, Fe, and Mn are moderately available (Prasad, 2003). Examples of widely available metals/metalloids are Cd, Cu, Ni, Se, and Zn. Phytoextraction is less effective against heavy metals that are less bioavailable. The solubility, soil profile, and metal's capacity to form complexes with soil particles are the main factors that determine the bioavailability of heavy metals in soils. To improve the phytoextraction process, soil parameters such as pH, chelating agents, chemical profile, and soil microorganisms can be observed (Wang et al., 2006; Thangavel and Subbhuraam, 2004). The solubility, mobility, and bioavailability of heavy metals all can be improved by supplementing plant rhizospheres with metal-mobilizing compounds (such as siderophores and carboxylates) and organic acids (Padmavathiamma and Li, 2012).

The use of chelating agents has been demonstrated to increase the bioavailability of heavy metals (Wuana and Okieimen, 2011). These agents frequently combine with heavy metals in soil to form water-soluble complexes that prevent precipitation, increase mobility, facilitate desorption, and increase bioavailability (Salt et al., 1995; Ali et al., 2012). The two types of common heavy metal chelators are synthetic and organic chelators. To improve phytoextraction, several synthetic chelators have been employed, including EDTA (ethylene diamine tetraacetic acid), EGTA (ethylene glycol tetraacetic acid), and DTPA (diethylene triamine pentaacetic acid; Gupta and Sinha,

2008; Sarwar et al., 2017). However, the use of these artificial chelators has been criticized due to their low biodegradability and persistence in soil, which are thought to encourage heavy metal leaching (Smolińska and Król, 2012; Lee and Sung, 2014). It has been recommended to employ organic chelators such as malic, oxalic, citric, and acetic acids (Sarwar et al., 2017). These have a low environmental risk due to their natural origins and ease of biodegradation (Souza et al., 2013).

A number of intriguing findings have also been provided about the use of organic compost to improve phytoremediation. According to reports (Epelde et al., 2009; Burges et al., 2018; Asemoloye et al., 2020b), adding organic amendments to the polluted soil can improve this technology by altering speciation, reducing the solubility of heavy metals, and increasing their bioavailability through changes in pH and redox status. The biological qualities of the soil's organic matter can be improved by adding organic amendments like animal dung, spent mushroom compost, and composted plant biomass. These properties are necessary for the proper colonization of the soil by microorganisms that promote plant growth, aid in phytostabilization, and increase the soil's capacity to retain water (Mastretta et al., 2009; Ma et al., 2011). According to reports, a large number of these advantageous microorganisms can aid in root initiation for improved phytostabilization of heavy metal pollution, either directly or indirectly (Göhre and Paszkowski, 2006).

11.6 USING AND/OR DISCARDING THE BY-PRODUCTS OF PHYTOREMEDIATION

11.6.1 Pre-Treatment Methods

Plants that are employed for phytoremediation of heavy metal(loid)-contaminated soil frequently become proportionately contaminated, either directly or indirectly. Plants become poisonous as a result, and they need to be treated to prevent further environmental damage. Many researchers have described in literatures various techniques for disposing of and/or using by-products of phytoremediation (Blaylock et al., 1997; Flathman and Lanza, 1998; Garbisu and Alkorta, 2001; Kruger et al., 1997; Kumar et al., 2021; Mulligan et al., 2001; Raskin et al., 1997; Salt et al., 1995, 1998). These findings are based on the works of Sas-Nowosielska et al. (2004) and Kovacs and Szemmelveisz (2017). Three main methods need to be utilized in order to minimize the volume of the plants used for phytoremediation so they may be more easily moved to the ultimate treatment facility: composting, compaction, and pyrolysis.

11.6.1.1 Composting

This technique for biomass volume reduction has been proposed by several researchers on contaminated biomass volume reduction by composting (Garbisu and Alkorta, 2001; Raskin et al., 1997). Zhao et al. (2000) demonstrated "the water solubility of bio-accumulated zinc and other metallic substances" using *Arabidopsis halleri*. Consequently, it should be noted that this volume-reduction approach necessitates close monitoring to prevent the leaching of toxic metal ions. The main advantages of using composting as a pre-treatment method for contaminated biomass are weight and volume reduction (Sas-Nowosielska et al., 2004). This lowers transportation expenses, either directly or indirectly, to the location where hazardous waste disposal is done in the end. Compost produced from contaminated plants frequently has significant concentrations of hazardous metal ions, rendering it unsuitable for use as manure in agriculture (Kovacs and Szemmelveisz, 2017). Further research on leaching tests on biomass polluted with Pb has shown the emergence of soluble organic compounds that have the potential to improve the solubility of Pb. Two scholars with a foundation in the field have suggested strategies to use in order to avoid these occurrences:

- To stop metal ions from leaching, lime can be added to water hyacinth compost. Singh and Kalamdhad (2013) report that after around 30 days, there was a decrease in the amount of water-soluble metals (e.g., Cr, Cu, Zn) and pentetic acid-soluble metals (like Cd, Ni, Cr, Fe, and Zn).

The composition of the compost was 45 kg of cattle dung, 90 kg of contaminated water hyacinth biomass, and 0%, 1%, 2%, or 3% calcium oxide. It was shown that the addition of calcium oxide had no influence on the levels of the metals.

- Combining contaminated compost with non-contaminated biomass materials: A study has been done to investigate the feasibility of combining high percentages of biodegradable non-contaminated biomass with contaminated compost (Guangwei et al., 2009). By doing this, the concentration of heavy metals would have been decreased, making it appropriate for application as agricultural manure; nevertheless, this method necessitates careful monitoring of the mixture's composition.

Additionally, chelating agents can effectively improve phytoextraction from polluted soil, according to certain research. According to a *Phaseolus vulgaris* investigation, the Zn-EDTA and Pb-EDTA complexes in the shoots were completely and partially dissociated for each complex (Sarret et al., 2001). It may be inferred that plant biomass obtained by induced phytoextraction might include mobile and leachable metal-chelate complexes. Therefore, it is critical to ascertain the effect(s) of the chelating agents on the composting of contaminated plant biomass in further investigations.

11.6.1.2 Compaction

This compaction method was proposed by Salt et al. (1995). After pressure is applied to generate the compacted particle product, leachate must be collected separately during the compaction or "palletization process of contaminated plant biomass" (Kovacs and Szemmelveisz, 2017). As a result, this offers an alternate strategy for resolving the contaminated plant disposal issue. Similar to this, the particle end-products from this compaction should be handled as toxic waste that needs further treatment in a facility designated for disposing of hazardous waste due to their high concentration of heavy metal(loids). Research on the effects of biomass forms, sizes, chemical compositions, and particle densities on compaction processes is necessary, as evidenced by the paucity of experimental data and results on compaction found in the literature (Mani et al., 2006). Furthermore, an evaluation of the methods for recovering heavy metals from the leachate formed from compaction of contaminated biomass and their concentration would be quite beneficial.

11.6.1.3 Pyrolysis

Pyrolysis is another method for treating polluted plant biomass (Bridgwater et al., 1999). According to Stals et al. (2010), the process entails breaking down polluted biomass at a moderate temperature without the presence of air, producing gaseous, liquid, and solid phases. The proportion of the fraction with the highest value depends on the operating temperature. When quickly pyrolyzing contaminated plant biomass, a large volume of liquid and little gaseous fraction are produced (Al Chami et al., 2014; Hornung et al., 2011). An intermediate pyrolysis strategy gives a solid fraction of exceptional quality and vapors without tar. Fast pyrolysis produces a higher-quality liquid fraction, which is why it is typically used for experimental research, even if the ultimate goal of pyrolyzing contaminated biomass is to produce a large volume of solid fraction with a high heavy metal concentration (Bridgwater et al., 1999).

In contrast to combustion or gasification techniques, this method frequently operates at lower temperatures, which results in the production of less polluted vapor or gaseous pollutants (Kuppens et al., 2015). Pyrolytic gas and coke breeze, which include heavy metals, are the final products obtained from the pyrolysis of contaminated biomass. Therefore, further treatment at a hazardous waste treatment plant is also suggested or necessary for the coke breeze. If Pb or Zn cannot be recovered during the smelting procedures, coke breeze should be utilized instead of coke for the smelting of Pb and Zn. To run a typical municipal pyrolysis plant at full capacity, the cost would be around €20–30/t. So, in any plant that is currently in place, the contaminated biomass and solid municipal wastes may be pyrolyzed to prevent this.

A typical municipal unit has a 40-ton daily capacity and costs €20–30/t. The plant material can be treated with municipal garbage in existing facilities to save the high costs associated with construction and operation.

11.6.2 Solid-Liquid Extraction

To extract the amount of heavy metals from contaminated plant biomass, this method applies several solvents. The presence of harmful heavy metals is typically the primary reason that contaminated biomass is handled as hazardous waste. Leached biomass (plant biomass) with a low heavy metal concentration can be disposed of as non-hazardous waste after being leached using different solvents (Kovacs et al., 2013; Sas-Nowosielska et al., 2004). To determine if this approach is suitable for use, a cost-benefit analysis must be conducted, given the significant investment in equipment needed to collect the heavy metal components of the leachate (Mulligan et al., 2001).

11.6.3 Direct Disposal

Another method for plants that have been employed for phytoremediation is the direct disposal of contaminated plant biomass. Although it requires less work than other methods, the hazardous plant biomass must be transported to disposal locations at a considerable expense. While this direct disposal strategy appears to be feasible, it is unwelcoming because of the significant quantity of toxic metals that would be released back into the environment when the contaminated plant biomass at the disposal locations decomposes (Šyc et al., 2012). Kovacs et al. (2013) state that doing several leaching studies using conventional methods, such as the following, might help overcome the drawbacks of the direct disposal technique with regard to toxins being returned to the environment:

- Using either nitric acid or acetic acid to leach the biomass, the leaching extraction process (LEP) tests whether contaminated biomass can release its harmful components into the environment.
- Using the US EPA Toxicity Characterization Leaching Procedure (US EPA TCLP) (METHOD 1311), the contaminated biomass is mixed with extraction fluid (20:1), stirred for 18 hours at 30 rpm, filtered, and its metal content is analyzed (US EPA TCLP, 1992).
- Method 1312 of the US EPA Synthetic Precipitation Leaching Procedure (US EPA SPLP): Similar to this, the process includes filtering and analyzing the leachate for metal content after combining the contaminated biomass with extraction fluid (ratio 20:1) and agitating it for 18 hours at 30 rpm. To replicate acidic rain conditions, extraction fluid in this process is acidified de-ionized water (US EPA Method 1312, 1994).

Ultimately, burning contaminated biomass is becoming more and more popular as a result of these processes and other drawbacks with the direct disposal approach.

11.6.4 Ashing

It is also possible to reduce the weight and volume of contaminated biomass by ashing it. There is not much experimental data accessible, despite the approach being mentioned by multiple investigators (Garbisu and Alkorta, 2001; Kumar et al., 2021; Raskin et al., 1997). Zhang et al.'s study from 2021 examined the effects of pyrolysis and ashing treatments on *Sedum plumbizincicola* plants that hyperaccumulate Zn and Cd. The study found that Cd and Zn in the contaminated *S. plumbizincicola* biomass could be converted from bioavailable form to stable form using both pyrolysis and ashing treatments, and that the stability increased with temperature. By converting Zn into the residual fraction (34.3%) and Zn into an oxidizable state (32.5%), the ashing method was found to reduce the toxic metals Cd and Zn by approximately 42%. In contrast, pyrolysis was able to convert 70.3% of the Zn and 80.0% of the Cd portion of the contaminated *S. plumbizincicola* biomass into a stable state. By using leaching tests (DTPA extraction and TCLP test), which demonstrated that the bioavailable portion of Zn and Cd in the char generated was significantly lower than the solid char generated by ashing of the contaminated *S. plumbizincicola* biomass, the applicability of pyrolysis compared to the ashing method was confirmed. These results point to pyrolysis as a viable approach for contaminated biomass ashing. Future research ought to take the two techniques' operational costs into account.

11.7 LACK OF RESEARCH ON PHYTOREMEDIATION

When choosing plants to clean up contaminated soil, there are a few fundamental things to take into account, including the plant's growth rate, root architecture and features, adaptability, reaction, and tolerance to different environmental circumstances, as well as the type of soil, rhizosphere physicochemical qualities, bioavailability, and toxicity of heavy metals (Sarwar et al., 2017). These elements are equally fundamental in selecting a plant for effective phytoremediation. While many researchers so far have focused on the screening of individual plants for phytoremediation potentials or mechanisms, very little have been focused on plant improvement or engineering for improved or speedy remediation of polluted sites. Hence, the main limitation facing phytoremediation technology is not the lack of knowledge or unavailability of potential plants but the inadequacies to improve the known strategies. Modification or enhancement of different phytoremediation mechanisms can play a good role of sustainable remediation while the inherent phytoremediation plants need to be improved for better performance.

Because many heavy metals are extremely reactive, they can produce large amounts of ROS, which frequently result in oxidative stress. One important consideration in the remediation of heavy metal-polluted soils is the phytoremediating plants' tolerance to heavy metals. According to Kozmínska et al. (2018), boosting a plant's antioxidant capacity is a genius method of preparing it for phytoremediation of heavy metals. The only way this is possible is by overexpressing the genes that regulate antioxidant activity; these are the genes that improve ion absorption, metal translocation, and sequestration, among other important processes (Mani and Kumar, 2014; Das et al., 2016). For improved phytoremediation, genes encoding metal or metalloid transporter mechanisms can be found and inserted into remediating plants. The past few years have seen the identification of numerous heavy metal transporter genes. As metal chelators, these genes can also function as metal binding ligands to improve metal bioavailability, uptake, translocation, and sequestration inside the cell (Wu et al., 2010). Some examples of these genes are members of the HMA, MATE, MTP, and ZIP families.

Some features that boost phytoaccumulation can be added. Native breeding techniques, such as hybridization and the genetic development of transgenic plants, can be used to either enhance plant growth or biomass output for hyperaccumulators (DalCorso et al., 2019). In order to get better results, Brewer et al. (1999) used electrofusion to fuse the protoplasts of two Zn hyperaccumulator plants, *Thlaspi caerulescens* and *Brassica napus*. Brewer et al. (1999) observed that the somatic hybrid selection of *T. caerulescens* produced from *B. napus* exhibited greater capacity for biomass production, hyperaccumulation potential, and resistance to heavy metals (Zn and Cd) in comparison to the wild type. Nehnevajova et al. (2007) described the application of chemical mutagenesis with ethyl methanesulfonate (EMS) to produce a sunflower giant mutant with increased biomass and improved heavy metal phytoextraction. Over time, advances in molecular biology, computer science, and genetic engineering have provided fresh insights on how to apply these fields to improve phytoremediation technologies, particularly for soils contaminated with heavy metals. These days, a plant of interest might have a foreign gene added to its genome to improve its remediation ability. These organisms can be bacteria, plants, or even other animals. In order to do this, it may be possible to create genetically engineered plants by DNA recombination, which would give the necessary traits for a specific phytoremediation process.

Compared to conventional breeding techniques, genetic engineering offers more promise for improving phytoremediation capacity in plants. Plants can be genetically altered to exhibit desired traits, such as high biomass, quick growth rate, and heavy metal accumulation/tolerance rate, resulting in improved performance in a shorter amount of time (Berken et al., 2002; Marques et al., 2009). For best results, these genes must be chosen and designed with a thorough understanding of the kinds and types of heavy metal contaminants present in the specific location. The primary drawback of genetic engineering is the intricacy of heavy metal uptake or detoxification by plants. Due to the vast number of genes involved, it is necessary to manipulate many genes in order to get the desired outcome. This calls for a lot of work that is intricate and time-consuming. The potential threats to food security and the environment posed by genetically modified plants can make it challenging to get field testing and public approval for these plants.

11.8 CONCLUSIONS

Pollution from heavy metals is a serious problem that poses serious health risks. Therefore, it needs to be regarded as an essential component of the key management elements for both food safety and sustainable agriculture. This is required because heavy metals are biogenic, meaning that they may collect in living cells and tissues and spread to other trophic levels. Strict laws and regulations must be followed in order to adequately avoid or mitigate the problem of heavy metal pollution. Additionally, less expensive and ecologically friendly technologies must be used for cleanup. Because of this, phytoremediation has received positive reviews and widespread acceptance. While there are other excellent findings on various phytoremediation methods, as this chapter explains, the use of hyperaccumulators has been demonstrated by several researchers to be a successful strategy. There are hundreds of hyperaccumulator plants reported each year, but as this chapter highlights, there are still a few constraints, including long duration, sluggish development rate, biomass, tolerance, and on-site application of many hyperaccumulators in heavily contaminated sites.

To improve phytoremediation mechanisms, we support combining several approaches. In summary, this chapter highlights the use of many techniques, including the introduction of microorganisms that promote plant development, the application of biological composts, and the use of phytohormones to stimulate plant growth during phytoremediation application. Thanks to advancements in genetic engineering, plants may now be modified to exhibit desirable qualities such stress adaption, high biomass output, increased metal tolerance, fast development, and hormone coordination. To create sustainable phytoremediation systems and learn how to improve them even further for optimal performance, it will thus be crucial to identify plants, as well as understand their heavy metal detoxification processes, signaling pathways, and reactions to specific metals. For instance, heavy metal absorption, translocation, sequestration, and tolerance genes may be strengthened by genetic engineering techniques, and during their practical implementation on sites, they can be supplemented with chelating agents, promoting chemicals, and microbes.

Due to issues with phytoaccumulation, it is crucial to avoid using food crops for phytoremediation of soils polluted with heavy metals. The use of inedible phytoaccumulators to the remediation of heavy metal contamination needs more research. To overcome current constraints, use of genome, proteome, and metabolic research should be taken into consideration for the evaluation of phytoremediation technology, while taking into account the combination of multiple remediation processes.

REFERENCES

Abid, R., Manzoor, M., De Oliveira, L.M., da Silva, E., Rathinasabapathi, B., Rensing, C., Mahmood, S., Liu, X. and Ma, L.Q. 2019. Interactive effects of As, Cd and Zn on their uptake and oxidative stress in As-hyperaccumulator *Pteris Vittata*. *Environ. Pollution*, 248, 756–762.

Adeyi, A.A. and Oladoye, P.O. 2020. Assessment of heavy metals in simulated leachates and the ashes of end-of-life tyres. *Int. J. Environ. Waste Manag.*25(4), 474–486.

Al Chami, Z., Amer, N., Smets, K., Yperman, J., Carleer, R., Dumontet, S. and Vangronsveld, J. 2014. Evaluation of flash and slow pyrolysis applied on heavy metal contaminated *Sorghum bicolor* shoots resulting from phytoremediation. *Biomass and Bioenergy* 63, 268–279.

Ali, B., Schwartz, C., Bitton, G. and Morel, J.L. 2006. Heavy metal contamination from mining sites in South Morocco: 1. Use of a biotest to assess metal toxicity of tailings and soils. *Chemosphere* 63(5), 802–810.

Ali, H., Khan, E. and Ilahi, I. 2019. Environmental chemistry and ecotoxicology of hazardous heavy metals: Environmental persistence, toxicity, and bioaccumulation. *Journal of Chemistry*. 2019, 6730305.

Ali, H., Khan, E. and Sajad, M.A. 2013. Phytoremediation of heavy metals—concepts and applications. *Chemosphere*. 91(7), 869–881.

Ali, H., Naseer, M. and Sajad, M.A. 2012. Phytoremediation of heavy metals by *Trifolium alexandrinum*. *Int. J. Environ. Sci.* 2(3), 1459–1469.

Alloway, B.J. 2013. Sources of heavy metals and metalloids in soils. In: Alloway, B. (Ed.), *Heavy Metals in Soils. Environmental Pollution*. Springer, Dordrecht, vol. 22. https://doi.org/10.1007/978-94-007-4470-7_2.

Al-Taani, A.A., Nazzal, Y., Howari, F.M., Iqbal, J., Bou Orm, N., Xavier, C.M., Bărbulescu, A., Sharma, M. and Dumitriu, C.S. 2021. Contamination assessment of heavy metals in agricultural soil, in the Liwa area (UAE). *Toxics* 9(3), 53.

Anani, O.A. and Olomukoro, J.O. 2017. The evaluation of heavy metal load in benthic sediment using some pollution indices in Ossiomo River, Benin City, Nigeria. *FUNAI Asia Pac J Sci Technol*3(2), 103–119.

Anani, O.A. and Olomukoro, J.O. 2018a. Health risk from the consumption of freshwater prawn and crab exposed to heavy metals in a Tropical River, southern Nigeria. *J Heavy Metal Toxicity Dis* 3(2), 5.

Anani, O.A. and Olomukoro, J.O. 2018b. Trace metal residues in a tropical watercourse sediment in Nigeria: Health risk implications. *IOP Conf. Ser. Earth Environ. Sci* 210, 012005.

Anani, O.A., Mishra, R.R., Mishra, P., Olomukoro, J.O., Imoobe, T.O.T. and Adetunji, C.O. 2020. Influence of heavy metal on food security: Recent advances. In *Innovations in Food Technology: Current Perspectives and Future Goals*. Vancouver, pp. 257–267. Springer, Singapore.

Aransiola, S.A., Ijah, U.J.J. and Abioye, O.P. 2013. Phytoremediation of Lead Polluted Soil by Glycine max L. *Appl. Environ. Soil Sci.*, Article ID 631619. Doi: 10.1155/2013/631619. https://www.hindawi.com/journals/aess/2013/631619/abs/.

Aransiola, S.A., Ijah, U.J.J., Abioye, O.P. and Bala, J.D. 2019. Microbial-aided Phytoremediation of Heavy metals Contaminated Soil: A review. *Eur. J. Biol. Res.* 9(2), 104–125. http://www.journals.tmkarpinski.com/index.php/ejbr/article/view/157.

Arshad, M., Saleem, M. and Hussain, S. 2007. Perspectives of bacterial ACC deaminase in phytoremediation. *Trends Biotechnol.* 25, 356–362.

Asemoloye, M.D., Ahmad, R. and Jonathan, S.G. 2017a. Synergistic action of rhizospheric fungi with *Megathyrsus maximus* root speeds up hydrocarbon degradation kinetics in oil polluted soil. *Chemosphere* 184, 1–12.

Asemoloye, M.D., Ahmad, R. and Jonathan, S.G. 2017b. Transcriptomic responses of catalase, peroxidase and laccase encoding genes and enzymatic activities of oil spill inhabiting rhizospheric fungal strains. *Environ. Pollut.* 235, 55–64.

Asemoloye, M.D., Chukwuka, K.S. and Jonathan, S.G. 2020b. Spent mushroom compost enhances plant response and phytoremediation of heavy metal polluted soil. *J. Plant Nutr. Soil Sci.* 183, 492–499.

Asemoloye, M.D., Jonathan, S.G., Jayeola, A.A. and Ahmad, R. 2017. Mediational influence of spent mushroom compost on phytoremediation of black-oil hydrocarbon polluted soil and response of *Megathyrsus maximus* Jacq. *J. Environ. Manag.* 200, 253e262.

Asemoloye, M.D., Marchisio, M.A., Gupta, V.K. and Pecoraro, L. 2021. Genome-based engineering of ligninolytic enzymes in fungi. *Microb. Cell Factories* 20, 20.

Asemoloye, M.D., Tosi, S., Dacco, 'C., Wang, X., Xu, S., Marchisio, M.A., Gao, W., Jonathan, S.G. and Pecoraro, L. 2020a. Hydrocarbon degradation and enzyme activities of *Aspergillus oryzae* and *Mucor irregularis* isolated from Nigerian crude oil-polluted sites. *Microorganisms* 8(12), 1912.

Ashraf, S., Ali, Q., Zahir, Z.A., Ashraf, S. and Asghar, H.N. 2019. Phytoremediation: Environmentally sustainable way for reclamation of heavy metal polluted soils. *Ecotoxicol. Environ. Saf.* 174, 714–727.

Babaniyi, B.R., Ogundele, O.D., Bisi-omotosho, A., Babaniyi, E.E. and Aransiola, S.A. 2023. Remediation approaches in environmental sustainability. In: Maddela, N.R., Kretli, L., Eller, W. and Prasad, R. (Eds.), *Microbiology for Cleaner Production and Environmental Sustainability*. CRC Press. ISBN: 9781032496061. https://www.routledge.com/Microbiology-for-Cleaner-Production-and-Environmental-Sustainability/Maddela-Eller-Prasad/p/book/9781032496061

Baker, A.J.M. and Walker, P.L. 1990. Ecophysiology of metal uptake by tolerant plants: Heavy metal uptake by tolerant plants. In: Shaw, A.J. (Ed.), *Evolutionary Aspects*. CRC, Boca Raton.

Balaferj, H., Bogusz, D., Triqui, Z.A., Guedira, A., Bendaou, N., Smouni, A. and Fahr, M. 2020. Zinc hyperaccumulation in plants: A review. *Plants* 9, 562.

Bañuelos, G.S., Cardon, G., Mackey, B., Ben-Asher, J., Wu, L., Beuselinck, P., Akohoue, S. and Zambrzuski, S. 1993. Boron and selenium removal in boron-laden soils by four sprinkler irrigated plant species. *J. Environ. Qual.* 22(4), 786–792. https://doi.org/10.2134/jeq1993.00472425002200040021x.

Barakat, M.A. 2011. New trends in removing heavy metals from industrial wastewater. *Arab. J. Chem.* 4, 361–377.

Barbulescu, A. 2020. Assessing groundwater vulnerability: DRASTIC and DRASTIC-like methods: A review. *Water* 12(5), 1356.

Becerra-Castro, C., Prieto-Fernández, A., Álvarez-Lopez, V., Monterroso, C., Cabello-Conejo, M.I., Acea, M.J. and Kidd, P.S. 2011. Nickel solubilizing capacity and characterization of rhizobacteria isolated from hyperaccumulating and non-hyperaccumulating subspecies of *Alyssum serpyllifolium*. *Int. J. Phytoremediation* 13(1), 229–244.

Berken, A., Mulholland, M.M., Leduc, D.L. and Terry, N. 2002. Genetic engineering of plants to enhance selenium phytoremediation. *Crit. Rev. Plant Sci.* 21, 567–582.

Bernard, A. 2008. Cadmium & its adverse effects on human health. *Indian J. Med. Res.* 128(4), 557.

Berti, W.R. and Cunningham, S.D. 2000. Phytostabilization of metals. In: Raskin, I. and Ensley, B.D. (Eds.), *Phytoremediation of Toxic Metals: Using Plants to Clean-Up the Environment*. John Wiley& Sons, Inc., New York, NY, pp. 71–88.

Bizily, S.P., Rugh, C.L. and Meagher, R.B. 2000. Phytodetoxification of hazardous organomercurials by genetically engineered plants. *Nat. Biotechnol.* 18, 213–217.

Blaylock, M.J. and Huang, J. 2000. Phytoextraction of metals. In: Raskin, I. and Ensley, B.D. (Eds.), *Phytoremediation of Toxic Metals: Using Plants to Clean-Up the Environment*. JohnWiley & Sons, Inc., New York, NY, p. 303.

Blaylock, M.J., Salt, D.E., Dushenkov, S., Zakharova, O., Gussman, C., Kapulnik, Y., Ensley, B.D. and Raskin, I. 1997. Enhanced accumulation of Pb in Indian mustard by soil-applied chelating agents. *Environ. Sci. Technol.* 31, 860–865.

Bolan, N., Ko, B., Anderson, C.W. and Vogeler, I. 2008. Solute interactions in soils in relation to bioavailability and remediation of the environment. *Rev. la Cienc. Del suelo y Nutr. Veg.* 8, 1–5.

Bonanno, G. and Giudice, R.L. 2010. Heavy metal bioaccumulation by the organs of *Phragmites australis* (common reed) and their potential use as contamination indicators. *Ecol. Indicat.* 10(3), 639–645.

Braud, A., Jézéquel, K., Bazot, S. and Lebeau, T. 2009. Enhanced phytoextraction of an agricultural Cr- and Pb-contaminated soil by bioaugmentation with siderophoreproducing bacteria. *Chemosphere* 74, 280–286.

Brevik, E.C. 2009. Soil, food security, and human health. In: Verheye, W. (Ed.), *Soils, Plant Growth, and Crop Production*. EOLSS Publishers, Oxford, UK.

Brewer, E.P., Saunders, J.A., Angle, J.S., Chaney, R.L. and Mcintosh, M.S. 1999. Somatic hybridization between the zinc accumulator *Thlaspi caerulescens* and *Brassica napus*. *Theor. Appl. Genet.* 99, 761–771.

Bridgwater, A.V., Meier, D. and Radlein, D. 1999. An overview of fast pyrolysis of biomass. *Org. Geochem.* 30, 1479–1493.

Burchett, M.D., MacFarlane, G.R. and Pulkownik, A. 2003. Accumulation and distribution of heavy metals in the grey mangrove *Avicennia marina* (Forsk.) Vierh: Biological indication potential. *Environ. Pollut.* 123, 139–151.

Burges, A., Alkorta, I., Epelde, L. and Garbisu, C., 2018. From phytoremediation of soil contaminants to phytomanagement of ecosystem services in metal contaminated sites. *Int. J. Phytoremediation* 20, 384–397.

Carnelo, L.G.L., de Miguez, S.R. and Marbán, L. 1997. Heavy metals input with phosphate fertilizers used in Argentina. *Sci. Total Environ.* 204(3), 245–250.

Chaney, R.L. and Baklanov, I.A. 2017. Phytoremediation and phytomining: Status and promise. *Adv. Bot. Res.* 83, 189–221.

Chen, B.D., Li, X.L., Tao, H.Q., Christie, P. and Wong, M.H. 2003. The role of arbuscular mycorrhiza in zinc uptake by red clover growing in a calcareous soil spiked with various quantities of zinc. *Chemosphere* 50, 839–846.

Chen, T., Wei, C., Huang, Z., Huang, Q., Lu, Q. and Fan, Z. 2002. Arsenic hyperaccumulator *Pteris vittata* L. and its arsenic accumulation. *Chin. Sci. Bull.* 47, 207–210.

DalCorso, G., Fasani, E., Manara, A., Visioli, G. and Furini, A. 2019. Heavy metal pollution: State of the art and innovation in phytoremediation. *Int. J. Mol. Sci.* 20, 3412.

Das, N., Bhattacharya, S. and Maiti, M.K. 2016. Enhanced cadmium accumulation and tolerance in transgenic tobacco overexpressing rice metal tolerance protein gene OsMTP1 is promising for phytoremediation. *Plant Physiol. Biochem.* 105, 297–309.

De Souza, M.P., Lytle, C.M., Mulholland, M.M., Otte, M.L. and Terry, N. 2000. Selenium assimilation and volatilization from dimethylselenoniopropionate by Indian mustard. *Plant Physiol.* 122, 1281–1288.

Derakhshan Nejad, Z., Jung, M.C. and Kim, K.H. 2018. Remediation of soils contaminated with heavy metals with an emphasis on immobilization technology. *Environ. Geochem. Health* 40, 927–953.

Dhanwal, P., Kumar, A., Dudeja, S., Chhokar, V. and Beniwal, V. 2017. Recent advances in phytoremediation technology. In: Kumar, R., Sharma, A.K. and Ahluwalia, S.S. (Eds.), *Advances in Environmental Biotechnology*. Springer, Singapore, pp. 227–241.

Eid, E.M., Galal, T.M., Sewelam, N.A., Talha, N.I. and Abdallah, S.M. 2020. Phytoremediation of heavy metals by four aquatic macrophytes and their potential use as contamination indicators: A comparative assessment. *Environ Sci Pollut Res.* 27, 12138–12151.

Elbehiry, F., Elbasiouny, H., El-Ramady, H. and Brevik, E.C. 2019. Mobility, distribution, and potential risk assessment of selected trace elements in soils of the Nile Delta. *Egypt. Environ. Monit. Assess.* 191, 713.

El-Kady, A.A. and Abdel-Wahhab, M.A. 2018. Occurrence of trace metals in foodstuffs and their health impact. *Trends Food Sci. Technol.* 75, 36–45.

El-Radaideh, N. and Al-Taani, A. 2018. Geo-environmental study of heavy metals of the agricultural highway soils, NW Jordan. *Arab. J. Geosci.* 11.

Epelde, L., Becerril, J.M., Mijangos, I. and Garbisu, C. 2009. Evaluation of the efficiency of a phytostabilization process with biological indicators of soil health. *J. Environ. Qual.* 38, 2041–2049.

Fasani, E., Manara, A., Martini, F., Furini, A. and DalCorso, G. 2018. The potential of genetic engineering of plants for the remediation of soils contaminated with heavy metals. *Plant Cell Environ.* 41, 1201–1232.

Flathman, P.E. and Lanza, G.R. 1998. Phytoremediation: Current views on an emerging green technology. *J. Soil Contam.* 7, 415–432.

Galal, T.M., Gharib, F.A., Ghazi, S.M. and Mansour, K.H. 2017. Phytostabilization of heavy metals by the emergent macrophyte *Vossia cuspidate* (Roxb.) Griff: A phytoremediation approach. *Int J Phytorem.* 19(11), 992–999.

Garbisu, C. and Alkorta, I. 2001. Phytoextraction: A cost-effective plant-based technology for the removal of metals from the environment. *Bioresour. Technol.* 77, 229–236.

Gerhardt, K.E., Gerwing, P.D. and Greenberg, B.M. 2017. Opinion: Taking phytoremediation from proven technology to accepted practice. *Plant Sci.* 256, 170–185.

Ginn, B.R., Szymanowski, J.S. and Fein, J.B. 2008. Metal and proton binding onto the roots of *Fescue rubra*. *Chem. Geol.* 253, 130–135.

Glick, B.R. 2010. Using soil bacteria to facilitate phytoremediation. *Biotechnol. Adv.* 28, 367–374.

Glick, B.R. 2014. Bacteria with ACC deaminase can promote plant growth and help to feed the world. *Microbiol. Res.* 169, 30–39.

Göhre, V. and Paszkowski, U. 2006. Contribution of the arbuscular mycorrhizal symbiosis to heavy metal phytoremediation. *Planta* 223, 1115–1122.

Gouda, S., Kerry, R.G., Das, G., Paramithiotis, S., Shin, H.S. and Patra, J.K. 2018. Revitalization of plant growth-promoting rhizobacteria for sustainable development in agriculture. *Microbiol. Res.* 206, 131–140.

Grimm, N.B., Foster, D., Groffman, P., Grove, J.M., Hopkinson, C.S., Nadelhoffer, K.J., Pataki, D.E. and Peters, D.P.C. 2008. The changing landscape: Ecosystem responses to urbanization and pollution across climatic and societal gradients. *Front. Ecol. Environ.* 6, 264–272.

Guangwei, Y., Hengyi, L., Tao, B., Zhong, L., Qiang, Y. and Xianqiang, S. 2009. In-situ stabilisation followed by ex-situ composting for treatment and disposal of heavy metals polluted sediments. *J. Environ. Sci.* 21, 877–883.

Gupta, A.K. and Sinha, S. 2008. Decontamination and/or revegetation of fly ash dykes through naturally growing plants. *J. Hazard. Mater.* 153(3), 1078–1087.

Gupta, D.K., Huang, H.G. and Corpas, F.J. 2013. Lead tolerance in plants: Strategies for phytoremediation. *Environ. Sci. Pollut. Res.* 20, 2150–2161.

Harvey, L.J. and McArdle, H.J. 2008. Biomarkers of copper status: A brief update. *Br. J. Nutr.* 99, 17–20.

Herzig, R., Nehnevajova, E., Pfistner, C., Schwitzguebel, J.-P., Ricci, A. and Keller, C. 2014. Feasibility of labile Zn phytoextraction using enhanced tobacco and sunflower: Results of five-and one-year field-scale experiments in Switzerland. *Int. J. Phytoremediation* 16, 735–754.

Hooda, V. 2007. Phytoremediation of toxic metals from soil and waste water. *J. Environ. Biol.* 28 (2 Suppl. L), 367–376.

Hornung, A., Apfelbacher, A. and Sagi, S. 2011. Intermediate pyrolysis: A sustainable biomass-to-energy concept-biothermal valorisation of biomass (BtVB) process. *J. Sci. Ind. Res. (India)* 70, 664–667.

Hossain, M.B., Jahiruddin, M., Panaullah, G.M., Loeppert, R.H., Islam, M.R. and Duxbury, J.M. 2008. Spatial variability of arsenic concentration in soils and plants, and its relationship with iron, manganese and phosphorus. *Environ. Pollut.* 156, 739–744.

Huang, R., Dong, M., Mao, P., Zhuang, P., Paz-Ferreiro, J., Li, Y., Li, Y., Hu, X., Netherway, P. and Li, Z. 2020. Evaluation of phytoremediation potential of five Cd (hyper)accumulators in two Cd contaminated soils. *Sci. Total Environ.* 721, 137581.

Huang, X.D., El-Alawi, Y., Penrose, D.M., Glick, B.R. and Greenberg, B.M. 2004. Responses of three grass species to creosote during phytoremediation. *Environ. Pollut.* 130, 453–463.

Jacob, J.M., Karthik, C., Saratale, R.G., Kumar, S.S., Prabakar, D., Kadirvelu, K. and Pugazhendhi, A. 2018. Biological approaches to tackle heavy metal pollution: A survey of literature. *J. Environ. Manag.* 217, 56–70.

Javed, M.T., Tanwir, K., Akram, M.S., Shahid, M., Niazi, N.K. and Lindberg, S. 2019. Chapter 20 – phytoremediation of cadmium-polluted water/sediment by aquatic macrophytes: Role of plant-induced pH changes. In: Hasanuzzaman, M., Prasad, M.N.V. and Fujita, M. (Eds.), *Cadmium Toxicity and Tolerance in Plants*. Academic Press, London, pp. 495–529.

Kabata-Pendias, A. and Pendias, H. 2001. *Trace Metals in Soils and Plants*, 2nd ed. CRC Press, Boca Raton, FL.

Kachout, S.S., Mansoura, A.B., Mechergui, R., Leclerc, J.C., Rejeb, M.N. and Ouerghi, Z. 2012. Accumulation of Cu, Pb, Ni and Zn in the halophyte plant *Atriplex* grown on polluted soil. *J. Sci. Food Agric.* 92, 336–342.

Keshavarzi, A. and Kumar, V. 2019. Ecological risk assessment and source apportionment of heavy metal contamination in agricultural soils of northeastern Iran. *Int. J. Environ. Health Res.* 29(5), 544–560.

Khalid, S., Shahid, M., Niazi, N.K., Murtaza, B., Bibi, I. and Dumat, C. 2017. A comparison of technologies for remediation of heavy metal contaminated soils. *J. Geochem. Explor.* 182, 247–268.

Khlifi, R., Olmedo, P., Gil, F., Feki-Tounsi, M., Hammami, B., Rebai, A. and HamzaChaffai, A. 2014. Biomonitoring of cadmium, chromium, nickel and arsenic in general population living near mining and active industrial areas in Southern Tunisia. *Environ. Monit. Assess.* 186, 761–779.

Kovacs, H. and Szemmelveisz, K. 2017. Disposal options for polluted plants grown on heavy metal contaminated brownfield lands – a review. *Chemosphere* 166, 8–20.

Kovacs, H., Szemmelveisz, K. and Palotas, A.B. 2013. Solubility analysis and disposal options of combustion residues from plants grown on contaminated mining area. *Environ. Sci. Pollut. Res.* 20, 7917–7925.

Kozmínska, A., Wiszniewska, A., Hanus-Fajerska, E. and Muszýnska, E. 2018. Recent strategies of increasing metal tolerance and phytoremediation potential using genetic transformation of plants. *Plant Biotechnol. Rep.* 12, 1–14.

Kruger, E.L., Anderson, T.A. and Coats, J.R. 1997. *Phytoremediation of Soil and Water Contaminants*. American Chemical Society, Washington, DC, vol. 664, p. 328.

Kumar, V., Pandita, S., Sharma, A., Bakshi, P., Sharma, P., Karaouzas, I., Bhardwaj, R., Thukral, A.K. and Cerda, A. 2021. Ecological and human health risks appraisal of metal (loid) s in agricultural soils: A review. *Geol. Ecol. Lands.* 5(3), 173–185.

Kuppens, T., Van Dael, M., Vanreppelen, K., Thewys, T., Yperman, J., Carleer, R., Schreurs, S. and Van Passel, S. 2015. Techno-economic assessment of fast pyrolysis for the valorization of short rotation coppice cultivated for phytoextraction. *J. Clean. Prod.* 88, 336–344.

Lee, J. and Sung, K. 2014. Effects of chelates on soil microbial properties, plant growth and heavy metal accumulation in plants. *Ecol. Eng.* 73, 386–394.

Li, J.T., Liao, B., Lan, C.Y., Ye, Z.H., Baker, A.J.M. and Shu, W.S. 2010. Cadmium tolerance and accumulation in cultivars of a high-biomass tropical tree (*Averrhoa carambola*) and its potential for phytoextraction. *J. Environ. Qual.* 39 (4), 1262–1268.

Liu, Y., He, Y., Wang, Z., Wan, K., Xia, J., Liu, J. and Cen, K. 2018. Multi-point LIBS measurement and kinetics modeling of sodium release from a burning Zhundong coal particle. *Combust. Flame* 189, 77–86.

Lorestani, B., Yousefi, N., Cheraghi, M. and Farmany, A. 2013. Phytoextraction and phytostabilization potential of plants grown in the vicinity of heavy metalcontaminated soils: A case study at an industrial town site. *Environ. Monit. Assess.* 185(12), 10217–10223.

Lü, J., Jiao, W.-B., Qiu, H.-Y., Chen, B., Huang, X.-X. and Kang, B., 2018. Origin and spatial distribution of heavy metals and carcinogenic risk assessment in mining areas at You'xi County southeast China. *Geoderma* 310, 99–106.

Ma, Y., Prasad, M., Rajkumar, M. and Freitas, H. 2011. Plant growth promoting rhizobacteria and endophytes accelerate phytoremediation of metalliferous soils. *Biotechnol. Adv.* 29, 248–258.

MacFarlane, G.R. and Burchett, M.D. 2002. Toxicity, growth and accumulation relationships of copper, lead and zinc in the grey mangrove *Avicennia marina* (Forsk.) Vierh. *Mar. Environ. Res.* 54, 65–84.

Maestri, E., Marmiroli, M., Visioli, G., Marmiroli, N., 2010. Metal tolerance and hyperaccumulation: Costs and trade-offs between traits and environment. *Environ. Exp. Bot.* 68, 1–13.

Mahar, A., Wang, P., Ali, A., Awasthi, M.K., Lahori, A.H., Wang, Q. and Zhang, Z. 2016. Challenges and opportunities in the phytoremediation of heavy metals contaminated soils: A review. *Ecotoxicol. Environ. Saf.* 126, 111–121.

Mahmood, Q., Mirza, N. and Shaheen, S. 2015. Phytoremediation using algae and macrophytes: I. In: Ansari, A., Gill, S., Gill, R., Lanza, G. and Newma, L. (Eds.), *Phytoremediation*. Springer, Cham, pp. 265–289.

Malik, R.N., Husain, S.Z. and Nazir, I. 2010. Heavy metal contamination and accumulation in soil and wild plant species from industrial area of Islamabad, Pakistan. *Pak J. Bot.* 42(1), 291–301.

Mani, D. and Kumar, C. 2014. Biotechnological advances in bioremediation of heavy metals contaminated ecosystems: An overview with special reference to phytoremediation. *Int. J. Environ. Sci. Technol.* 11(3), 843–872.

Mani, S., Tabil, L.G. and Sokhansanj, S. 2006. Effects of compressive force, particle size and moisture content on mechanical properties of biomass pellets from grasses. *Biomass Bioenergy* 30, 648–654.

Marques, A.P., Rangel, A.O. and Castro, P.M. 2009. Remediation of heavy metal contaminated soils: Phytoremediation as a potentially promising clean-up technology. *Crit. Rev. Environ. Sci. Technol.* 39, 622–654.

Mastretta, C., Taghavi, S., van der Lelie, D., Mengoni, A., Galardi, F., Gonnelli, C., Barac, T., Boulet, J., Weyens, N. and Vangronsveld, J. 2009. Endophytic bacteria from seeds of *Nicotiana tabacum* can reduce cadmium phytotoxicity. *Int. J. Phytoremediation* 11(3), 251–267.

Mazej, Z. and Germ, M. 2009. Trace element accumulation and distribution in four aquatic macrophytes. *Chemosphere* 74(5), 642–647.

McLaughlin, M.J., Hamon, R.E., McLaren, R.G., Speir, T.W. and Rogers, L.S. 2000a. Review: A bioavailability-based rationale for controlling metal and metalloid contamination of agricultural land in Australia and New Zealand. *Aust. J. Soil Res.* 38, 1037–1086.

Mclaughlin, M.J., Zarcinas, B.A. and Cook, N. 2000b. Soil testing for heavy metals. *Commun. Soil Sci. Plant Anal.* 31(11–14), 1661–1700.

Mesjasz-Przybyłowicz, J., Nakonieczny, M., Migula, P., Augustyniak, M., Tarnawska, M., Reimold, W., Koeberl, C., Przybyłowicz, W. and Głowacka, E. 2004. Uptake of cadmium, lead nickel and zinc from soil and water solutions by the nickel hyperaccumulator *Berkheya coddii*. *Acta Biol. Cracoviensia Ser. Bot.* 46, 75–85.

Mganga, N., Manoko, M.L.K. and Rulangaranga, Z.K. 2011. Classification of plants according to their heavy metal content around north mara gold mine, Tanzania: Implication for phytoremediation. *Tanzan. J. Sci.* 37, 109–119.

Micó, C., Recatalá, L., Peris, M. and Sánchez, J. 2006. Assessing heavy metal sources in agricultural soils of an European Mediterranean area by multivariate analysis. *Chemosphere* 65(5), 863–872.

Milić, D., Luković, J., Ninkov, J., Zeremski-Škorić, T., Zorić, L., Vasin, J. and Milić, S. 2012. Heavy metal content in halophytic plants from inland and maritime saline areas. *Cent. Eur. J. Biol.* 7, 307–317.

Mkandawire, M. and Dudel, E.G. 2005. Accumulation of arsenic in *Lemna gibba* L. (duckweed) in tailing waters of two abandoned uranium mining sites in Saxony, Germany. *Sci. Total Environ.* 336(1–3), 81–89.

Mulligan, N.M., Yong, R.N. and Gibbs, B.F. 2001. Remediation technologies for soils and groundwater. *Remediat. Technol. Soils Groundw.* 60, 1–449.

Nehnevajova, E., Herzig, R., Federer, G., Erismann, K.-H. and Schwitzguébel, J.P. 2007. Chemical mutagenesis – a promising technique to increase metal concentration and extraction in sunflowers. *Int. J. Phytoremediation* 9, 149–165.

Nejad, Z.D., Jung, M.C. and Kim, K.H. 2018. Remediation of soils contaminated with heavy metals with an emphasis on immobilization technology. *Environ. Geochem. Health* 40(3), 927–953.

Nikalje, G.C. and Suprasanna, P. 2018. Coping with metal toxicity – cues from halophytes. *Front. Plant Sci.* 9, 777.

Oladoye, P.O. 2022. Natural, low-cost adsorbents for toxic Pb(II) ion sequestration from (waste) water: A state-of-the-art review. *Chemosphere* 287, 132130.

Olowe, O.M., Akanmu, A.O. and Asemoloye, M.D. 2020. Exploration of microbial stimulants for induction of systemic resistance in plant disease management. *Ann. Appl. Biol.* 177(3), 282–293.

Padmavathiamma, P.K. and Li, L.Y. 2012. Rhizosphere influence and seasonal impact on phytostabilisation of metals – a field study. *Water Air Soil Pollut.* 223, 107–124.

Pan, X.D., Wu, P.G. and Jiang, X.G. 2016. Levels and potential health risk of heavy metals in marketed vegetables in Zhejiang, China. *Sci. Rep.* 6(1), 20317.

Pandiyan, J., Mahboob, S., Govindarajan, M., Al-Ghanim, K.A., Ahmed, Z., Al-Mulhm, N.J. and Krishnappa, R. 2021. An assessment of the level of heavy metals pollution in the water, sediment, and aquatic organisms: A perspective of tackling environmental threats for food security. *Saudi J. Biol. Sci.* 28(2), 1218–1225.

Phillips, D.P., Human, L.R.D. and Adams, J.B. 2015. Wetland plants as indicators of heavy metal contamination. *Mar. Pollut. Bull.* 92(1–2), 227–232.

Prasad, M.N.V. 2003. Phytoremediation of metal-polluted ecosystems: Hype for commercialization. *Russ. J. Plant Physiol.* 50, 686–701.

Radziemska, M., Vaverková, M.D. and Baryła, A. 2017. Phytostabilization – management strategy for stabilizing trace elements in contaminated soils. *Int. J. Environ. Res. Public Health* 14(9), 958.

Rai, P.K., Lee, S.S., Zhang, M., Tsang, Y.F. and Kim, K. 2019. Heavy metals in food crops: Health risks, fate, mechanisms, and management. *Environ. Int.* 125(2019), 365–385.

Rascio, N. and Navari-Izzo, F. 2011. Heavy metal hyperaccumulating plants: How and why do they do it? And what makes them so interesting? *Plant Sci.* 180, 169–181.

Raskin, I., Smith, R.D. and Salt, D.E. 1997. Phytoremediation of metals: Using plants to remove pollutants from the environment. *Curr. Opin. Biotechnol.* 8, 221–226.

Rezania, S., Taib, S.M., Md Din, M.F., Dahalan, F.A. and Kamyab, H. 2016. Comprehensive review on phytotechnology: Heavy metals removal by diverse aquatic plants species from wastewater. *J. Hazard Mater.* 318, 587–599.

Rezapour, S., Golmohammad, H. and Ramezanpour, H. 2014. Impact of parent rock and topography aspect on the distribution of soil trace metals in natural ecosystems. *Int. J. Environ. Sci. Technol.* 11(7), 2075–2086.

Robinson, B., Kim, N., Marchetti, M., Moni, C., Schroeter, L., van den Dijssel, C. and Clothier, B. 2006. Arsenic hyperaccumulation by aquatic macrophytes in the Taupo volcanic zone, New Zealand. *Environ. Exp. Bot.* 58(1–3), 206–215.

Sachs, J. 1865. *Handbuch der Experimental-Physiologie der Pflanzen.* Wilhelm Engelmann Verlag, Leipzig, Germany, pp. 153–154.

Salt, D.E., Blaylock, M., Kumar, N.P., Dushenkov, V., Ensley, B.D., Chet, I. and Raskin, I. 1995. Phytoremediation: A novel strategy for the removal of toxic metals from the environment using plants. *Biotechnology* 13(5), 468–474.

Salt, D.E., Smith, R.D. and Raskin, I. 1998. Phytoremediation. *Annu. Rev. Plant Physiol.* 49, 643–668.

Sarret, G., Vangronsveld, J., Manceau, A., Musso, M., D'Haen, J., Menthonnex, J.J. and Hazemann, J.L. 2001. Accumulation forms of Zn and Pb in *Phaseolus vulgaris* in the presence and absence of EDTA. *Environ. Sci. Technol.* 35(13), 2854–2859.

Sarwar, N., Imran, M., Shaheen, M.R., Ishaque, W., Kamran, M.A., Matloob, A., Rehim, A. and Hussain, S. 2017. Phytoremediation strategies for soils contaminated with heavy metals: Modifications and future perspectives. *Chemosphere* 171, 710–721.

Sas-Nowosielska, A., Kucharski, R., Małkowski, E., Pogrzeba, M., Kuperberg, J.M. and Kryński, K.J.E.P. 2004. Phytoextraction crop disposal – an unsolved problem. *Environ. Pollution.* 128(3), 373–379.

Schalk, I.J., Hannauer, M. and Braud, A. 2011. New roles for bacterial siderophores in metal transport and tolerance. *Environ. Microbiol.* 13, 2844–2854.

Sharma, B., Sarkar, A., Singh, P. and Singh, R.P. 2017. Agricultural utilization of biosolids: A review on potential effects on soil and plant grown. *Waste Manag.* 64, 117–132.

Sheoran, V., Sheoran, A.S. and Poonia, P. 2010. Role of hyperaccumulators in phytoextraction of metals from contaminated mining sites: A review. *Crit. Rev. Environ. Sci. Technol.* 41, 168–214.

Singh, A. and Prasad, S.M. 2015. Remediation of heavy metal contaminated ecosystem: An overview on technology advancement. *Int. J. Environ. Sci. Technol.* 12(1), 353–366.

Singh, J. and Kalamdhad, A.S. 2013. Effects of lime on bioavailability and leachability of heavy metals during agitated pile composting of water hyacinth. *Bioresour. Technol.* 138, 148–155.

Smolińska, B. and Król, K. 2012. Leaching of mercury during phytoextraction assisted by EDTA, KI and citric acid. *J. Chem. Technol. Biotechnol.* 87(9), 1360–1365.

Souza, L.A., Piotto, F.A., Nogueirol, R.C. and Azevedo, R.A. 2013. Use of nonhyperaccumulator plant species for the phytoextraction of heavy metals using chelating agents. *Sci. Agric.* 70, 290–295.

Srivastava, S., Sounderajan, S., Udas, A. and Suprasanna, P. 2014. Effect of combinations of aquatic plants (*Hydrilla, Ceratophyllum, Eichhornia, Lemna* and *Wolffia*) on arsenic removal in field conditions. *Ecol. Eng.* 73, 297–301.

Srivastava, V., Sarkar, A., Singh, S., Singh, P., de Araujo, A.S.F. and Singh, R.P. 2017. Agroecological responses of heavy metal pollution with special emphasis on soil health and plant performances. *Front. Environ. Sci.* 5, 64.

Stals, M., Thijssen, E., Vangronsveld, J., Carleer, R., Schreurs, S. and Yperman, J. 2010. Flash pyrolysis of heavy metal contaminated biomass from phytoremediation: Influence of temperature, entrained flow and wood/leaves blended pyrolysis on the behaviour of heavy metals. *J. Anal. Appl. Pyrolysis* 87, 1–7.

Streets, D., Lu, Z., Levin, L. and Ter schure, A. 2017. 2017 Sunderland, E. Historical releases of mercury to air, land, and water from coal combustion. *Sci. Total Environ.* 615, 131–140.

Suman, J., Uhlik, O., Viktorova, J. and Macek, T. 2018. Phytoextraction of heavy metals: A promising tool for clean-up of polluted environment? Front. *Plant Sci.* 9, 1476.

Sutton, D.J. and Tchounwou, P.B. 2007. Mercury induces the externalization of phosphatidyl-serine in human renal proximal tubule (HK-2) cells. *Int. J. Environ. Res. Publ. Health* 4, 138–144.

Sutton, D.J., Tchounwou, P.B., Ninashvili, N. and Shen, E. 2002. Mercury induces cytotoxicity and transcriptionally activates stress genesin human liver carcinoma (HepG2) cells. *Int. J. Mol. Sci.* 3, 965–984.

Šyc, M., Pohořelý, M., Kameníková, P., Habart, J., Svoboda, K. and Punčochář, M. 2012. Willow trees from heavy metals phytoextraction as energy crops. *Biomass and Bioenergy* 37, 106–113.

Tchounwou, P.B., Centeno, J.A. and Patlolla, A.K. 2004a. Arsenic toxicity, mutagenesis, and carcinogenesis – a health risk assessment and management approach. *Mol. Cell. Biochem.* 255, 47–55.

Tchounwou, P.B., Ishaque, A.B. and Schneider, J. 2001. Cytotoxicity and transcriptional activation of stress genes in human liver carcinoma cells (HepG2) exposed to cadmium chloride. *Mol. Cell. Biochem.* 222, 21–28.

Tchounwou, P.B., Newsome, C., Williams, J. and Glass, K. 2008. Copper-induced cytotoxicity and transcriptional activation of stress genes in human liver carcinoma (HepG(2)) cells. *Met. Ions Biol. Med.* 10, 285–290.

Tchounwou, P.B., Yedjou, C.G., Foxx, D.N., Ishaque, A.B. and Shen, E. 2004b. Lead-induced cytotoxicity and transcriptional activation of stress genes in human liver carcinoma (HepG2) cells. *Mol. Cell. Biochem.* 255, 161–170.

Tchounwou, P.B., Yedjou, C.G., Patlolla, A.K. and Sutton, D.J. 2012. Heavy metal toxicity and the environment. In *Molecular, Clinical and Environmental Toxicology: Volume 3: Environmental Toxicology.* pp. 133–164. Springer, Basel.

Terry, N., Carlson, C., Raab, T. and Zayed, A.M. 1992. Rates of selenium volatilization among crop species. *J. Environ. Qual.* 21, 341–344.

Terry, N., Zayed, A.M., De Souza, M.P. and Tarun, A.S. 2000. Selenium in higher plants. *Annu. Rev. Plant Physiol.* 51, 401–432.

Thangavel, P. and Subbhuraam, C. 2004. Phytoextraction: Role of hyperaccumulators in metal contaminated soils. *Proc. Indian Nat. Sci. Acad. B* 70, 109–130.

Tian, K., Huang, B., Xing, Z. and Hu, W. 2017. Geochemical baseline establishment and ecological risk evaluation of heavy metals in greenhouse soils from Dongtai, China. *Ecol. Indicat.* 72, 510–520.

Todorović, Z.B., Ranđelović, L.M., Marjanović, J.Z., Todorović, V., Cakić, M. and Cvetković, O.G. 2014. The assessment and distribution of heavy metals in surface sediments from the reservoir 'Barje': Serbia. *Savremene Tehnologije* 3(2), 85–95.

Tóth, G., Hermann, T., Da Silva, M.R. and Montanarella, L.J.E.I. 2016. Heavy metals in agricultural soils of the European Union with implications for food safety. *Environ. Int.* 88, 299–309.

United Nations. 2015. Transforming Our World: The 2030 Agenda for Sustainable Development. March 25, 2017. https://sustainabledevelopment.un.org/content/documents/21252030%20Agenda%20for%20Sustainable%20Development%20web.pdf.

Upadhyay, M.K., Gautam, A., Mallick, S. and Srivastava, S. 2017. A successive application approach for effective utilization of three aquatic plants in arsenic removal. *Water Air Soil Pollut.* 228(2), 54.

US EPA. 1992. *Method 1311, Toxicity Characteristic Leaching Procedure (TCLP).* Publication SW) 846: Test Methods for Evaluating Solid Waste, Physical/Chemical Methods.

US EPA Method 1312; U.S. Environmental Protection Agency, 1986; 1994 update.

Vangronsveld, J., Herzig, R., Weyens, N., Boulet, J., Adriaensen, K., Ruttens, A. and van der Lelie, D. 2009. Phytoremediation of contaminated soils and groundwater: Lessons from the field. *Environ. Sci. Pollut. Res.* 16(7), 765–794.

Vhahangwele, M. and Mugera, G.W. 2015. Journal of environmental chemical engineering the potential of ball-milled South African bentonite clay for attenuation of heavy metals from acidic wastewaters: Simultaneous. *Biochem. Pharmacol.* 3, 2416–2425.

Vimercati, L., Gatti, M.F., Gagliardi, T., Cuccaro, F., De Maria, L., Caputi, A., Quarato, M. and Baldassarre, A. 2017. Environmental exposure to arsenic and chromium in an industrial area. *Environ. Sci. Pollut. Res. Int.* 11528–11535.

Walker, D.J., Lutts, S., Sanchez-Garcia, M. and Correal, E. 2014. *Atriplex halimus* L.: Its biology and uses. *J. Arid Environ.* 100–101, 111–121.

Wang, A.S., Angle, J.S., Chaney, R.L., Delorme, T.A. and Reeves, R.D. 2006. Soil pH effects on uptake of Cd and Zn by *Thlaspi caerulescens*. *Plant Soil* 281, 325–337.

Wang, S. and Shi, X. 2001. Molecular mechanisms of metal toxicity and carcinogenesis. *Mol. Cell. Biochem.* 222, 3–9.

Wang, Y., Qiu, Q., Xin, G., Yang, Z., Zheng, J., Ye, Z. and Li, S. 2013. Heavy metal contamination in a vulnerable mangrove swamp in South China. *Environ. Monit. Assess.* 185, 5775–5787.

WHO. 1996. *Trace Elements in Human Nutrition and Health*. World Health Organization, Geneva.

Wong, M.H. 2003. Ecological restoration of mine degraded soils, with emphasis on metal contaminated soils. *Chemosphere* 50, 775–780.

Wu, G., Kang, H., Zhang, X., Shao, H., Chu, L. and Ruan, C. 2010. A critical review on the bio-removal of hazardous heavy metals from contaminated soils: Issues, progress, eco-environmental concerns and opportunities. *J. Hazard Mater.* 174, 1–8.

Wuana, R.A. and Okieimen, F.E. 2011. Heavy metals in contaminated soils: A review of sources, chemistry, risks and best available strategies for remediation. *ISRN Ecol* 1–20.

Yan, X., Liu, M., Zhong, J., Guo, J. and Wu, W. 2018. How human activities affect heavy metal contamination of soil and sediment in a long-term reclaimed area of the Liaohe River Delta, North China. Sustain. *Times* 10, 1–19.

Yedjou, C.G. and Tchounwou, P.B. 2007a. N-acetyl-L-cysteine affords protection against lead-induced cytotoxicity and oxidative stress in human liver carcinoma (HepG2) cells. *Int. J. Environ. Res. Publ. Health* 4, 132–137.

Yedjou, C.G. and Tchounwou, P.B. 2007b. In-vitro cytotoxic and genotoxic effects of arsenic trioxide on human leukemia (HL-60) cells using the MTT and alkaline single cell gel electrophoresis (Comet) assays. *Mol. Cell. Biochem.* 301, 123–130.

Zalewska, T. and Danowska, B. 2017. Marine environment status assessment based on microphytobenthos plants as bioindicators of heavy metals pollution. *Mar. Pollut. Bull.* 118, 281–288.

Zhang, X., Zhong, T., Lei, L. and Liu, X.O. 2015. Impact of soil heavy metal pollution on food safety in China. *PloS One* 10(8), e0135182.

Zhao, F.J., Lombi, E., Breedon, T. and McGrath, S.P. 2000. Zinc hyperaccumulation and cellular distribution in Arabidopsis halleri. *Plant Cell Environ.* 23, 507–514.

Zhao, Q. and Kaluarachchi, J.J. 2002. Risk assessment at hazardous waste-contaminated sites with variability of population characteristics. *Environ. Int.* 28, 41–53.

12 Impact of Contamination of Agricultural Soil on Humans and the Environment

Saheed Ibrahim Musa, Parisa Ebrahimbabaie, Nathan Moses, Kingsley Erhons Enerijiofi, Comfort Okoji, Unoka Edith Chinyere, and Iboyi Nathaniel Onuche

12.1 INTRODUCTION

Over the years, soil contamination has been considered a major environmental pollution possibly due to the multiple roles of soil in supporting life on earth. A substance or chemical that initially finds its ways into the soil, either through natural processes or human activities, and has the capacity of influencing soil nature and properties is known as a contaminant (Enerijiofi and Ekhaise, 2019a). Contaminants based on their properties are generally grouped as either chemical, biological, or physical. They can exist in sufficient or less sufficient concentrations but do not adversely affect organisms and their environments. Soil contamination has raised serious concerns due to the fear that it may lead to soil pollution. Soil pollution is generally seen as a pressing issue in agricultural production because when the soil is polluted, the healthy status of the soil is compromised, thereby affecting the life of other organisms that depend on it. This situation over the years has resulted in loss of biodiversity and has affected food security and safety all over the world. Furthermore, pollutants in the soil have been observed to infiltrate the food chain, threatening food system sustainability (Musa and Ikhajiagbe, 2021).

The devastating effects of these contaminants in agricultural soil are not only limited to the plants but also extend to the microbiome of the arable land as well as any other forms of life found in the agricultural soil. Beyond the environmental impact, the consequences of agricultural contamination can be felt in national economies through decreased crop productivity and quality. Furthermore, there are some plants that are considered as bioaccumulators. These plants may survive the presence of certain contaminants but will pass the contaminants as the food chain progresses which may bring about serious health implication and also affect food security and safety (Ikhajiagbe et al., 2021).

12.2 CONTAMINATION OF AGRICULTURAL SOIL

The contamination of agricultural soils can arise from various sources and pose significant challenges to crop growth and food safety. Some common causes of soil contamination in agriculture include the following.

12.2.1 Pesticides and Herbicides

Based on the acceptable code of conduct on the management of pesticide, it is a substance or mixture of substances either of biological or chemical nature used for repelling, destroying, and controlling pests. Excessive use or improper application of pesticides, herbicides, and insecticide can

DOI: 10.1201/9781032683751-14

gradually build up in the soil, affecting the health of plants and potentially entering the food chain. Pesticides are widely used in controlling vectors of human or animal disease, and globally they can impact wildlife, some non-target plant species, and sometimes human health either directly or indirectly (Kim et al., 2017). Pesticide and herbicides have been documented to have some concentration of heavy metals that, when released into the environment, can contaminate or pollute the soil. Most of these heavy metals that are found in soil have affected microbial and enzymatic actions or activities, which has drastically inhibited the microbial roles in the soil (Enerijiofi and Obade, 2018). Most degradation in the soil is done by enzymes, but because of the high heavy metals in the soil, the quality and quantity of the soil affect fertilization. Most of these heavy metals are absorbed mostly by plant roots and accumulate in the softer tissues of most plants (Soni Kumari and Amarnath Mishra, 2021)

12.2.2 Improper Use and Mismanagement of Fertilizer

Overuse or improper fertilizer application can create serious imbalances in the nutrient and cause the accumulation of nitrates and other chemicals in the soil. This can affect both plant health and water quality. Most plants absorb water and nutrients that are present in the soil through the roots. These fertilizers include synthetics, either organic or inorganic, and minerals meant to provide sufficient nutrients to the growth of the crop. In some countries, overuse of this fertilizer has led to serious environmental hazards, in some cases leaching of the groundwater through runoff to surface water leading to pollution of drinking water. Some of these fertilizers are volatile, especially nitrogen (N) when it is converted by microbes to nitrous oxide which is greenhouse gases, and these gases are responsible for global warming and have contributed to decrease or depletion of the stratospheric ozone layer (Reta et al., 2018). During the phosphate mining process, a lot of dust is generated, sources of fluoride and radon are emitted that are trace/minor elements, and radionuclides are mobilized. (Azzi et al., 2017; Reta et al., 2018). When applying phosphate fertilizer to the soil, this can orchestrate/oversee to an addition of harmful chemicals, like fluorine (F), arsenic (As), chromium (Cr), lead (Pb), and radionuclides such as thorium (Th) and uranium (U), which may cause bioaccumulation in plants and enter the food chain. Excess use or continuous applications of fertilizers can lead to a huge bioaccumulation of potentially toxic elements in the soil, and there will be uptake by plants (Jiao et al., 2012).

There are several other causes of contamination of agricultural soils, such as continuous utilization of wastewater irrigation, urban runoff of polluted water through arable lands, and excessive and uncontrolled use of genetically modified organisms (GMO). These causes of soil pollution become dangerous because of components of the chemicals that are released. These substances can bring about soil contamination; however, this situation definitely depends on the concentrations the chemicals present in the soil to the organism or environment.

12.2.3 Manure and Animal Wastes

Animal manure, although organic, can also cause contamination of agricultural soils. Manure and animal waste are usually added to soils used in agriculture in order to enhance soil properties and improve plant yield. However, if not properly treated before application, it may lead to contamination of agricultural soils. For example, animal waste can be converted to compost through decomposition in the absence of oxygen. Methane produced by anaerobic digestion can be used on a farm for heating or power production (Demirbas and Ozturk, 2005). Animal feed may contain certain heavy metals such as Cr and Pb, which are poisonous depending on the concentration, as well as vital elements like copper (Cu) and zinc (Zn0. A common strategy to increase production efficiency is the addition of necessary trace elements to animal feed (Wang et al., 2017). Between 1990 and 2010 a precarious increase in elemental concentration of manure was observed, according to research (Wang et al., 2014). According to Xiong et al. (2010), the Cu and Zn that are observed in animal feeds produced have potential adverse health implications for soil and the food chain. Depending on the needs, several trace elements are used in the manufacture of animals. For instance, copper sulfate is used in pig and poultry systems as a growth booster (Poulsen, 1998).

12.2.4 Compost

One of the elements may be administered excessively if manures and other organic fertilizers are the only sources used. Each has a unique ratio of nitrogen, phosphorus, and potassium (NPK) concentration. Continuous application of phosphorus (P) may bring about eutrophication on freshwater surfaces (Cameron et al., 2013). To prevent these consequences, it is crucial to use the proper NPK ratio when using fertilizer, whether it is organic or inorganic (FAO, 2019). The fermentation and composition of biological wastes, such as food and domestic waste, can act as fertilizers. Some researchers have indicated that fermented biological waste may also contain microplastics, especially from food packaging before compost, which may be very difficult for remediation technologies due to the persistence and non-degradable properties of the plastic (Gionfra, 2018; Weithmann et al., 2018). It has been discovered that compost made from municipal solid waste can be a source of trace elements when applied as an organic fertilizer in agricultural soils (Madrid et al., 2007).

12.2.5 Wastewater for Irrigation

Many times, companies discharge their wastewater and effluents directly onto farmlands without treating the effluents. This unfortunate practice has been among the main causes of agricultural soil contamination (Enerijiofi et al., 2018). In some cases, local farmers divert untreated industrial effluents onto their farmlands with the aim of irrigation, which in turn negatively affects the soil physical and chemical properties, as well as the plant yield within the region (Scott et al., 2004). Recent information indicated that over 50% of global municipal wastewater is used for irrigation. Sometimes, this municipal wastewater would be discharged into arable regions even before treating or with minimal treatment (Musa and Ikhajiagbe, 2021). The utilization of wastewater for irrigation leads to the pollution of agricultural soils (Qadir et al., 2010).

12.2.6 Agricultural Plastic Waste

In recent decades, plastics have become increasingly used in agriculture. Seedling trays and pots are other items made of plastic which may lead to plastic contamination of agricultural soil and limit productivity due to the persistent nature of plastic. Many of these items are single use only. Agricultural plastics are a substantial cause of soil pollution and environmental harm without efficient recovery strategies that involve identifying the pollutant, picking the pollutant, and sustainable recycling and disposal strategies (Vox et al., 2016). Authorities and the scientific community have based their full concern on this type of emerging contaminant.

12.2.7 Mulching Films

Mulching films have been documented to have the highest possibility of polluting the soil because of their close contact with the ground and the possibility that they could be incorporated into it during post-harvest management or as they deteriorate. Plastic mulch is used to protect seedlings by acting as insulation, to prevent evaporation, and to lessen weed and pest pressure in order to boost yield. It frequently works in tandem with drip irrigation tubing made of plastic. Even though mulching films are used for a single planting season, the area covered by mulching is thought to have expanded further, even though more current data have not been available, given the higher trends shown in earlier years (Jansen et al., 2019).

12.2.8 Greenhouse Film and Protective Netting

To maintain a year-round supply of food, greenhouses are used extensively on all continents. In wealthy nations, high-tech glass greenhouses are replacing plastic ones. However, plastic is still utilized inside greenhouses for other purposes and has a relatively poor recycling rate, creating plastic waste. Plastic films continue to be the most popular low-cost choice in less developed nations (Gruda and Popsimonova, 2017). Although protective netting and greenhouse films are made to last through several harvests, eventually they must be replaced due to weathering and photodeterioration, and the old materials must be properly recycled.

12.3 IMPACTS OF CONTAMINATION OF AGRICULTURAL SOIL ON HUMANS, PLANTS, THE ENVIRONMENT, AND BIODIVERSITY

The contamination of agricultural soils harbors a great negative impact that affect plants, animals, humans, and their habitats. Due to the increased vulnerability of children to diseases, they are at a higher risk when contaminated agricultural soils enter their food chain. Contamination of arable land, especially due to hazardous human-influenced activities, has turned into a worldwide concern. This has caused fear of soil pollution, which has a serious detrimental effect on organisms, their habitat, and their biodiversity. For example, mercury (Hg) pollution has been documented to have reduced the population of earthworms due to its inability to break down the heavy meals. One of the direct impacts of soil pollution is that it could lead to climate change that can lead to the extinction of animal species. For instance, the extinction of organisms such as golden toads can be linked to climate change, which disrupts the normal nature of habitats. Polar bears are at risk due to a decrease in the ice extent of the Arctic region. The North Atlantic right whale is also at risk of extinction because its primary food source, plankton, is declining as a result of climate change. While the exact effects of climate change on India's natural resources are still under thorough investigation, initial studies indicate that certain mammals such as the Nilgiri tahr are at a higher risk of becoming extinct. In addition, temperature as a result of weather change has led to a decrease in some oceanic coral reef ecosystems. Coastal areas could be rapidly inundated as a result of the swift rise in sea levels, which are calculated to have increased by 0.1 to 0.2 m over the past century. This could have catastrophic effects on certain organisms. Previous climate fluctuations have resulted in ecosystems with different species compositions, owing to the varying abilities of species to adjust to the changes in climate. Ensuring the world's biodiversity is preserved would help in maintaining a stable climate.

Furthermore, soil contamination may bring about habitat degradation. Habitat degradation or loss refers to the environmental change that makes a natural habitat unable to support the species living there. This can occur naturally or as a result of disruption of habitat fragmentation, climate change, leading to encroachment of dangerous species and depleting ecosystem nutrients (Enerijiofi et al., 2022). During habitat destruction, the organisms that once occupied the area are displaced or eliminated, leading to a decrease in biodiversity. Human-induced habitat destruction has significantly increased since the mid-twentieth century. Natural habitats are frequently destroyed due to human activities aimed at extracting natural resources for industrial production and urban development. For instance, clearing land for agriculture is a major contributor to habitat destruction (Enerijiofi and Ekhaise, 2022).

Contamination of agricultural soils are known to negatively affect plant health and development by interfering with the normal physiological functioning of plants. Some plants may be hyperaccumulators. But in the absence of contaminants, they may not have any contaminant to accumulate, but if the contaminant is present, the plant will take it up, which will affect the way the plant functions and result in slow development and productivity, or death. Another impact of agricultural soil contamination is acidification of the soil. Soil acidification is reported to be the primary concern for plant ecosystems in developing countries, while heightened ozone and N deposition poses significant issues in Europe and North America. Exposure to acidic soil can also lead to diminished stem growth, alterations in wood density, and the accumulation of pollutants within the stems.

Furthermore, contaminants that are found in arable lands can enter the human body through various pathways, including direct skin contact or inhalation of polluted soil dust. The immediate impact of contaminants in an individual's system may bring about queasiness, vomiting, coughing, thoracic discomfort, wheezing, and headache. Eye and skin irritation, exhaustion, and debility may also be observed. In addition, prolonged health issues are associated with agricultural soil contamination; for example, elevated Pb levels likely bring about irreversible alterations to the nervous system, especially among kids (Devi et al., 2022). It has earlier been reported that heavy metals reach humans when contaminated crops are consumed as food or medicine, leading to serious health challenges that affect both male and female reproductive organs (Enerijiofi and Ajuzie, 2012)

12.4 STRATEGIES TO PREVENT CONTAMINATION OF AGRICULTURAL SOILS

For ecosystems and people to remain healthy and for food to be produced sustainably, agricultural soils must not be contaminated. In agricultural situations, the following precautions can be taken to help prevent soil contamination.

12.4.1 Avoid Overuse of Agrochemicals

The widespread application of chemicals on arable land can have effect and impair the community of organisms, especially microbial community in the applied region. When these chemicals are applied in excess, it becomes chemical misuse, thereby leading to accumulation of these dangerous substance in soils. While some research has shown that pesticides have negative impacts on soil microbial community and biology, other studies proved that microbial community of a region may under quorum sensing coordinate the decomposition of chemicals or sometimes absorb the residue (Devi et al., 2022). It is still unclear how exactly pesticides affect soil microorganisms. Pesticides' effects on soil microorganisms depend on several environmental factors, including their toxicity, persistence, and concentration (Abdel-Mallek et al., 1994). It is challenging to come to firm conclusions on how pesticides interact with the soil ecology because of this complicated web of elements. The biochemical processes of nutrient cycling can generally be disturbed by long-term pesticide treatment (Devi et al., 2022). Agricultural soil pollution that occurs as a result of variety of pesticide chemicals can persist for many years and have a detrimental effect on organisms and the environment (Enerijiofi and Obade, 2018). The application of pesticides has been known to limit the soil properties and thereby influencing the soil microbiota and total diversity. Reducing the use of chemicals leads to higher-quality soil, increased soil mineral content also enhances the absorption of water (CDC, 2007). According to research by Musa and Ikhajiagbe (2021), organic crops yield 20%–40% more than those grown using conventional farming methods. This can be particularly beneficial during drought, as it helps farmers produce more. However, it is important to note that a reduction in the proportion of nutrients in the soil can result in higher levels of pesticides remaining in the treated area after spraying (Kellogg et al., 2002).

Both sorption and degradation play a role in how long pesticides remain in the soil. Based on the chemical makeup of the pesticide, these mechanisms directly control how it moves from the soil to water, which in turn controls how it moves to the air and our food. The breakdown of organic materials involves interactions between soil microbes. Sorbed compounds are less accessible to microbes; weak organic acids, which depend on the soil's organic matter to bioaccumulate, are only poorly absorbed by the soil due to their low pH and mostly acidic structure. Unknown aging mechanisms cause pesticide residues to become more resistant to extraction and degradation (Manuel et al., 2008).

12.4.2 Integrated Pest Management

Integrated pest management (IPM) is a long-term method for controlling pests in a way that is environmentally safe, socially acceptable, and commercially successful. To detect, regulate, and reduce risk from pest control methods and tactics in a way that minimizes total economic, health, and environmental concerns, this process includes biological, cultural, physical, and chemical instruments. Implementing a multitude of alternative strategies with an emphasis on the safety of people and the environment also focuses on minimizing or eliminating the reliance on chemical control choices and reducing the need for excessive chemical inputs that could lead to soil contamination. IPM is a technique that can direct specialists to utilize it as suited for their circumstances rather than a theory that rigidly and uniformly applies to every circumstance. For instance, some crops with pest and/or tolerant types can efficiently employ host plant resistance, but not other crops. For some lepidopteran and coleopteran pests, pheromones are frequently utilized for mating disruption, mass capture, or monitoring, but not for some hemipteran pests. For pests in greenhouses, biological management is frequently employed, but less frequently in the field. High-value crops like strawberries employ mechanical instruments like insect vacuums, but because they consume fossil fuels, they are not a cost-effective option for non-specialty crops and do not reduce greenhouse gas emissions. Chemical pesticides are the last resort, but in certain cases, they are the best option for shielding seeds and transplants from common and persistent pest problems or for stopping the spread of specific invasive or endemic pests and diseases; for example, they have gained popularity as a preventative approach in several crops (Dara, 2019).

The common pest management techniques listed in the next sections can be used at different stages of crop production to prevent, mitigate, or treat pest infestations and keep agricultural soils clean. Although each of them may offer some degree of control, their combined effect may be quite effective in reducing the loss of yield.

12.4.2.1 Cultural Control

Cultural control is the application of effective agricultural techniques to prevent or minimize pest infestations and damage. Using clean seed or plant material is crucial to lessen the possibility of introducing pests at the very beginning of crop production. You can avoid insect infestations or their most vulnerable times by altering planting dates. In Uganda (Karungi et al., 2000) and Nigeria (Asante et al., 2001), early cowpea planting prevented infestations of legume pod borer (*Maruca vitrata*), legume flower thrips (*Megalurothrips sjostedti*), and pod sucking insect (*Clavigralla tomentosicollis*). Plant density and row spacing will also affect pest infestations. Aphid infestations in cowpeas in Uganda (Karungi et al., 2000) and root maggot (*Delia* spp.) infestations in canola in Canada (Dosdall et al., 1996) were both reduced by high plant density. The environment may become less favorable to the pest by altering irrigation techniques, fertilizer schedules, and other agronomic practices. Micro-sprinklers are placed on strawberry beds in California, particularly for organic strawberries, as a spider mite control measure (firsthand observation). Micro-sprinkler–equipped strawberry plots also seemed to have less severe infections of powdery mildew (*Podosphaera aphanis*) and botrytis fruit rot (*Botrytis cinerea*) than strawberry plots with overhead aluminum sprinklers (Dara et al., 2016). Plants with low potassium levels produce more jasmonic acid, which makes them more resistant to some diseases and insect pests (Davis et al., 2018). Hodson and Lampinen (2018) state that higher plant N levels can aggravate arthropod infestations. Plants with high (Davis et al., 2018) or low (Snoeijers et al., 2000) N levels can also aggravate a variety of disease problems. Proper cultivation and sanitation protocols are necessary to remove plant debris, clean tools, and prevent contamination. In Iran, the number of pistachio psyllids (*Agonoscena pistaciae*) hibernating in leaf litter and weeds was decreased by winter plowing of the orchard floors. Plows are an essential control tool because they remove crop residue and expose the soil-dwelling stages of pests (Kunjwal and Srivastava, 2018). Sanitation measures including bagging unsaleable berries or even modifying the harvest cycle from every 3 days to every 2 days helped to reduce infestations of spotted-wing drosophila (*Drosophila suzukii*; Leach et al., 2017). Crop rotation that incorporates tolerant or non-host crops will interfere with pest life cycles and reduce annual pest accumulation. Crop rotation has been used in a variety of cropping systems to control diseases, weeds, and insects (Curl, 1963; Wright, 1984; Liebman and Dyck, 1993; Mohler and Johnson, 2009). Other cultural management strategies used in IPM are intercropping non-host or pest-repelling plants and using trap crops to keep pests away from the main crop (Pretty and Bharucha, 2015; Nielsen et al., 2016).

12.4.2.2 Biological Control

Microorganisms are used as natural agents for remediating the soil environment (Enerijiofi et al., 2020; Miranda-Esquivel et al., 2023). Natural enemies like predatory arthropods and parasitic wasps can be highly effective in lowering pest populations (Hajek and Eilenberg, 2018). To manage endemic pests, release naturally occurring enemies that are commercially available and safeguard natural enemy populations. Biological control is successfully used in specialty crops like strawberries grown outdoors (Zalom et al., 2018) and in greenhouses (van Lenteren, 1988). The conventional biological control approach for invasive pests entails bringing in and releasing natural enemies into the bug's new habitat (Kenis et al., 2017; Heimpel and Cock, 2018). The release of sterile, irradiated insects is another biological control method that has proven effective in managing a range of pests (Klassen and Curtis, 2005). Various methods, such as baits, traps, and mating disruption, can be used to monitor and control pest behavior (Heinz et al., 1992; Shorey and Gerber, 1996; Foster and Harris, 1997; Vladés et al., 2005; El-Sayed et al., 2009; Morrison et al., 2016). Deadly baits are substances sprayed on the ground or placed in traps to attract and kill pests. Certain colors, luminosities, and pheromone scents are attractive to pests. Devices that make use of one or more of these can be used to kill, trap, or attract pests.

12.4.2.3 Physical or Mechanical Control

This technique involves using various physical or mechanical approaches for pest exclusion, trapping (which is sometimes similar to behavioral management), removal, or destruction (Webb and Linda, 1992; Gamliel and Katan, 2012; Gogo et al., 2014, Dara et al., 2018). Pest exclusion using netting or row covers, handpicking or vacuuming to remove pests, mechanical weed-control tools, rodent pest traps, adjusting greenhouse temperature or humidity, steam sterilization or solarization, and tangible or visual bird deterrents like reflective material or sound devices are a few examples of physical or mechanical control methods.

12.4.2.4 Chemical Control

Chemical control involves using chemical pesticides to manage pests (Dodia et al., 2010). The term *chemical control* should, however, encompass both synthetic and substances with microbiological or plant origins. Despite being classified as biological or biopesticides (Lasota and Dybas, 1991; Sarfraz et al., 2005; Dodia et al., 2010), botanical extracts such as azadirachtin and pyrethrins, and microbe-derived toxic metabolites like avermectin and spinosad, are chemical compounds, just like synthetic chemicals, and carry many of the same risks for human and environmental safety as chemical pesticides, even though they are classified as biological or biopesticides (Lasota and Dybas, 1991; Sarfraz et al., 2005; Dodia et al., 2010). Government laws prohibit the use of specific chemical pesticides both in terms of quantity and frequency while also reducing the risks involved.

12.4.2.5 Integrated Nutrient Management

Integrated nutrient management (INM) reduces the amount of chemicals used in agriculture by employing both organic and inorganic fertilizers, along with other soil amendments, to enhance soil fertility and improve crop productivity. INM also contributes significantly to increasing the productivity of the soil. The addition of organic matter to the soil through the use of organic manures improves its physical, chemical, and biological qualities, as well as its nutrient cycling. INM techniques increase yield and raise a farmer's overall income. Additionally, according to

Zhang et al. (2014), INM reduces environmental contamination. So INM is a low-cost method of enhancing soil health.

The primary cause of low soil fertility is incorrect application of chemical fertilizers (Bhattacharyya et al., 2007; Mahanta et al., 2013). When paired with the appropriate amount of fertilizers, the application of natural resources directly raises soil organic carbon I, crop growth, and productivity (Jat et al., 2013; Meena et al., 2013). Several studies have demonstrated that INM lowers denitrification losses and leaching while boosting yield, decreasing mineralization, and improving fertilizer use efficiency. In general, it enhances the productivity, soil health, and long-term sustainability of the system (Kumar et al., 2014; Rahman et al., 2013). The careful application of fertilizers increases the organic C content of the soil. 50% homemade (NPK) + 50% N through industry made had the highest levels of organic C, in contrast to the control, which had the lowest levels according to Kumar et al. (2012). Applying 100% homemade alone through fertilizers increased the availability of N, P, and K. The combination of manures and inorganic fertilizers reduces the soil's electrical conductivity.

12.4.2.6 Cropping Systems and Soil Health

Various cropping techniques such as crop rotation, intercropping, and crop diversification have a significant impact on soil quality and health from different temporal and geographical angles (Vukicevich et al., 2016). Initially, cropping systems were developed to increase agricultural yield, but in recent times, their capacity to protect the environment has become increasingly important (Fargione et al., 2018). The maintenance of soil health aims to ensure stable high productivity and environmental sustainability of cropping systems in the long run, based on five essential evaluation standards: nutrient cycling, water relations, biodiversity and habitat, filtering and buffering, and physical stability and support (Hatfield et al., 2017; Enerijiofi and Ikhajiagbe, 2021). Also, soil health can be maintained by treating wastewater before discharge to the soil environment or any body of water (Bhowmik et al., 2023)

12.5 CONCLUSION

The contamination of agricultural soils has continued to have profound effect on food production and environmental health standards. Soil receives toxic compounds on a daily basis resulting from deposition of waste products. The crops may store toxic contaminants, produce poor yield, and may not grow at all, particularly with elevated concentrations of contaminants. Also, the contaminated agricultural soils have serious negative effects on human health, the environment, and other biodiversity. The effect of human health may include hunger, exposure to diseases, and outbreaks such as cancer, reproductive and respiratory disorders, as well as other debilitating diseases arising from the consumption of contaminated foods harvested in the contaminated soils. This posits that wastes such as manure and animal wastes, compost, wastewater for irrigation, mulching films and should be properly sorted and treated before being used as soil amendments to boost crop production while others such as pesticides, herbicides, fertilizers, agricultural plastic wastes, and greenhouse film and protective netting should be applied at the recommended dosage to prevent accumulation which alters soil nutrient capability. Adherence to the above recommendations ensures nutrient-rich soils, which translates to improved plant yields with the right nutrient composition for the overall benefit of humans. It also ensures environmental sustainability and renewed public health for all forms of biodiversity.

REFERENCES

Abdel-Mallek, A.Y., Moharram, A.M., Abdel-Kader, M.I. and Omar, S.A. (1994). Effect of soil treatment with the organophosphorus insecticide Profenfos on the fungal flora and some microbial activities. *Microbiological Research*, 149(2): 167–171.

Ajuzie, C.U., Atuanya, E.I. and Enerijiofi, K.E. (2015). Biodegradation potentials of microorganisms isolated from eleme petrochemical industrial effluent. *Nigerian Society for Experimental Biology Journal*, 15(4): 128–136.

Asante, S.K., Tamo, M. and Jackai, L.E.N. (2001). Integrated management of cowpea insect pests using elite cultivars, date of planting and minimum insecticide application. *African Crop Science Journal*, 9: 655–665.

Azzi, V., El Samrani, A., Lartiges, B., Kobeissi, A., Kanso, A. and Kazpard, V. (2017). Trace metals in phosphate fertilizers used in eastern Mediterranean countries. *CLEAN – Soil, Air, Water*, 45(1).

Bhattacharyya, R., Chandra, S., Singh, R.D., Kundu, S., Srivastva, A.K. and Gupta, H.S. (2007). Long-term farmyard manure application effects on properties of a silty clay Loam soil under irrigated wheat– soybean rotation. *Soil & Tillage Research*, 94: 386–396.

Bhowmik, D., Chetri, S., Enerijiofi, K.E., Naha, A., Kanungo, T.D., Shah, M.P. and Nath, S. (2023). Multitudinous approaches, challenges and opportunities of bioelectrochemical systems in conversion of wastes to energy from wastewater treatment plants. *Journal of Cleaner and Circular Bioeconomy-*Elsevier, 4: 100040.

Cameron, K.C., Di, H.J. and Moir, J.L. (2013). Nitrogen losses from the soil/plant system: A review: Nitrogen losses. *Annals of Applied Biology*, 162(2): 145–173.

Centers for Disease Control and Prevention. (2007). *Pesticides*. Wayback Machine. cdc.gov.

Curl, E.A. (1963). Control of plant diseases by crop rotation. *The Botanical Review*, 29: 413–479.

Dara, S.K. (2019). The new integrated pest management paradigm for the modern age. *Journal of Integrated Pest Management*, 10(1): 1–9, 12.

Dara, S.K., Peck, D. and Murray, D. (2018). Chemical and non-chemical options for managing twospotted spider mite, western tarnished plant bug and other arthropod pests in strawberries. *Insects*, 9: 156.

Dara, S.K., Sandoval-Solis, S. and Peck, D. (2016). Improving strawberry irrigation with micro-sprinklers and their impact on pest management. *Agricultural Sciences*, 7: 859–868.

Davis, J.L., Armengaud, P., Larson, T.R., Graham, I.A., White, P.J., Newton, A.C. and Amtmann, A. (2018). Contrasting nutrient-disease relationships: Potassium gradients in barley leaves have opposite effects on two fungal pathogens with different sensitivities to jasmonic acid. *Plant, Cell & Environment*, 41: 2357–2372.

Demirbas, A. and Ozturk, T. (2005). Anaerobic digestion of agricultural solid residues. *International Journal of Green Energy*, 1(4): 483–494.

Devi, P.I., Manjula, M. and Bhavani, R.V. (2022). Agrochemicals, environment, and human health. *Annual Review of Environment and Resources*, 47: 399–421.

Dodia, D.A., Patel, I.S. and Patel, G.M. (2010). *Botanical Pesticides for Pest Management*. Scientific Publishers (India).

Dosdall, L.M., Herbut, M.J., Cowle, N.T. and Micklich, T.M. (1996). The effect of seeding date and plant density on infestations of root maggots, Delia spp. (Diptera: Anthomyiidae), in canola. *Canadian Journal of Plant Science*, 76: 169–177.

El-Sayed, A.M., Suckling, D.M., Byers, J.A., Jang, E.B. and Wearing, C.H. (2009). Potential of "lure and kill" in long-term pest management and eradication of invasive species. *Journal of Economic Entomology*, 102: 815–835.

Enerijiofi, K.E. and Ajuzie, C.U. (2012). Health impact of some heavy metals resistance in microbes isolated from ikhueniro refuse dumpsite, Benin City. *Nigerian Journal of Microbiology*, 26: 2593–2599.

Enerijiofi, K.E. and Ekhaise, F.O. (2019a). Physicochemical and microbiological characterisation of Edaiken and Santana markets open solid waste dumpsites soils in Benin City, Nigeria. *FULafia Journal of Science and Technology*, 5(2): 136–149.

Enerijiofi, K.E. and Ekhaise, F.O. (2022). Achieving ecofriendly environment through sustainable management of solid wastes in the ecosystem. In: T. Aftab (Ed.), *Sustainable Management of Environmental Contaminants: Eco-friendly Remediation Approaches*. Springer, pp. 451–470. https://doi.org/10.1007/978-3-031-08446-1_17

Enerijiofi, K.E. and Ikhajiagbe, B. (2021). Plant-microbe interaction in attenuation of toxic waste in ecosystem. In: V. Kumar, R. Prasad, and M. Kumar (Eds.), *Rhizobiont in Bioremediation of Hazardous Waste*. Springer, pp. 291–315. https://doi.org/10.1007/978-981-16-0602-1_13 Online ISBN: 978-981-16-0601-4 Print ISBN: 978-981-16-06021-1

Enerijiofi, K.E. and Obade, E. (2018). Assessment of the impact of herbicide contaminated top soils on its physiochemical, microbiological and enzymatic properties. *FUW Trends in Science and Technology Journal*, 3(2): 802–806

Enerijiofi, K.E., Ahonsi, C.O. and Ajao, E.K. (2020). Biodegradation potentials of waste engine oil by three bacterial isolates. *Journal of Applied Sciences and Environmental Management*, 24(3): 483–487. https://dx.doi.org/10.4314/jasem.v24i3.14.

Enerijiofi, K.E., Fakeye, O.D. and Oziegbe, O.S. (2018). Physicochemical and microbial assessment of sewage from Ogwa and ebelle communities of Esan Land, Edo State. *Bayero Journal of Pure and Applied Sciences*, 11(1): 202–207. http://dx.doi.org/10.4314/bajopas.v11i1.35

Enerijiofi, K.E., Okuguni, N.N. and Ajayi, A.V. (2022). Physicochemical composition and heavy metals tolerance of bacterial isolates in leachate from Solid waste dumpsites in Delta North Senatorial District, Delta State. *Jordan Journal of Earth and Environmental Sciences* – Elsevier, 13(1): 48–52.

FAO. (2019). *The International Code of Conduct for the Sustainable Use and Management of Fertilizers*. Food and Agriculture Organization of the United Nations.

Fargione, J.E., Bassett, S., Boucher, T., Bridgham, S.D., Conant, R.T., Cook-Patton, S.C., Ellis, P.W., Falcucci, A., Fourqurean, J.W., Gopalakrishna, T., Gu, H., Henderson, B., Hurteau, M.D., Kroeger, K.D., Kroeger, T., Lark, T.J., Leavitt, S.M., Lomax, G., McDonald, R.I., Megonigal, J.P., Miteva, D.A., Richardson, C.J., Sanderman, J., Shoch, D., Spawn, S.A., Veldman, J.W., Williams, C.A., Woodbury, P.B., Zganjar, C., Baranski, M., Elias, P., Houghton, R.A., Landis, E., McGlynn, E., Schlesinger, W.H., Siikamaki, J.V., Sutton-Grier, A.E. and Griscom, B.W. (2018). Natural climate solutions for the United States. *Science Advice*, 4.

Foster, S.P. and Harris, M.O. (1997). Behavioral manipulation methods for insect pest-management. *Annual Review of Entomology*, 42: 123–146.

Gamliel, A. and Katan, J. (2012). *Solarization: Theory and Practice*. American Phytopathological Society.

.Gionfra, S. (2018). *Plastic Pollution in Soils*. Institute for European Environmental Policy, p. 18.

Gogo, E.O., Saidi, M., Ochieng, J.M., Martin, T., Baird, V. and Ngouajio, M. (2014). Microclimate modification and insect pest exclusion using agronet improve pod yield and quality of French bean. *HortScience*, 49: 1298–1304.

Gruda, N. and Popsimonova, G. (2017). Current situation and future trends of protected cultivation in South East Europe. In: W. Baudoin, A. Nersisyan, A. Shamilov, A. Hodder, D. Gutierrez, S. de Pascale, S. Nicola, N. Gruda, L. Urban, and J. Tanny (Eds.), *Good Agricultural Practices for Greenhouse Vegetable Production in the South East European countries – Principles for Sustainable Intensification of Smallholder Farms*. Plant Production and Protection Paper Edition, pp. 17–26.

Hajek, A.E. and Eilenberg, J. (2018). *Natural Enemies: An Introduction to Biological Control*. Cambridge University Press.

Hatfield, J.L., Sauer, T.J. and Cruse, R.M. (2017). Chapter one – soil: The forgotten piece of the water, food, energy Nexus. In: D.L. Sparks (Ed.), *Advances in Agronomy*. Academic Press, p. 65.

Heimpel, G.E. and Cock, M.J.W. (2018). Shifting paradigms in the history of classical biological control. *BioControl*, 63: 27–37.

Heinz, K.M., Parrella, M.P. and Newman, J.P. (1992). Time-efficient use of yellow sticky traps in monitoring insect populations. *Journal of Economic Entomology*, 85: 2263–2269.

Hodson, A.K. and Lampinen, B.D. (2018). Effects of cultivar and leaf traits on the abundance of Pacific spider mites in almond orchards. *Arthropod Plant Interact*: 1–11.

Ikhajiagbe, B., Enerijiofi, K.E. and Umendu, P.O. (2021). Mycorestoration of an oil polluted soil. *Studia Universitatis Babeş-Bolyai Biologia*, 66(1): 73–84.

Jansen, L., Henskens, M. and Hiemstra, F. (2019). *Report on Use of Plastics in Agriculture*. Schuttelaar & Partners Edition, p. 19.

Jat, L.K., Singh, S.K., Latare, A.M., Singh, R.S. and Patel, C.B. (2013). Effect of dates of sowing and fertilizer on growth and yield of wheat (Triticumaestivum) in an inceptisol of Varanasi. *Indian Journal of Agronomy*, 58: 611–614.

Jiao, W., Chen, W., Chang, A.C. and Page, A.L. (2012). Environmental risks of trace elements associated with long-term phosphate fertilizers applications: A review. *Environmental Pollution*, 168: 44–53.

Karungi, J., Adipala, E., Ogenga-Latigo, M.W., Kyamanywa, S. and Oyobo, N. (2000). Pest management in cowpea. Part 1. Influence of planting time and plant density on cowpea field pests infestation in eastern Uganda. *Crop Protection*, 19: 231–236.

Kellogg, R.L., Nehring, R., Grube, A., Goss, D.W. and Plotkin, S. (2000). Environmental indicators of pesticide leaching and runoff from farm fields. Archived 18 June 2002 at the Wayback Machine. United States Department of Agriculture Natural Resources Conservation Service.

Kenis, M., Hurley, B.P., Hajek, A.E. and Cock, M.J.W. (2017). Classical biological control of insect pests of trees: Facts and figures. *Biological Invasions*, 19: 3401–3417.

Kim, K.-H., Kabir, E. and Jahan, S.A. (2017). Exposure to pesticides and the associated human health effects. *Science of the Total Environment*, 575: 525–535.

Klassen, W. and Curtis, C.F. (2005). History of the sterile insect technique. In: V.A. Dyck, J. Hendrichs, and A. Robinson (Eds.), *Sterile Insect Technique*. Springer, pp. 3–36.

Kumar, M., Baishaya, L.K., Ghosh, D.C., Gupta, V.K., Dubey, S.K., Das, A. and Patel, D.P. (2014). Productivity and soil health of potato (*Solanum tuberosum* L.) field as influenced by organic manures, inorganic fertilizers and biofertilizers under high altitudes of eastern Himalayas. *Journal of Agriculture Science*, 4: 223–234.

Kumar, V., Tripathi, H.C. and Mishra, S.K. (2012). Impact of integrated nutrient management on yield, economics and soil fertility in hybrid rice (*Oryza sativa*) – mustard (*Brassica juncea*) cropping system. *New Agriculturist*, 23: 21–26.

Kunjwal, N. and Srivastava, R.M. (2018). Insect pests of vegetables. In: Omkar (Ed.), *Pests and Their Management*. Springer, pp. 163–221.

Lasota, J.A. and Dybas, R.A. (1991). Avermectins, a novel class of compounds: Implications for use in arthropod pest control. *Annual Review of Entomology*, 36: 91–117.

Leach, H. Moses, J., Hanson, E., Fanning, P. and Isaacs, R. (2017). Rapid harvest schedules and fruit removal as non-chemical approaches for managing spotted wing Drosophila. *Journal of Pest Science*, 91: 219–226.

Liebman, M. and Dyck, E. (1993). Crop rotation and intercropping strategies for weed management. *Ecological Applications*, 3: 92–122.

Madrid, F., López, R. and Cabrera, F. (2007). Metal accumulation in soil after application of municipal solid waste compost under intensive farming conditions. *Agriculture, Ecosystems & Environment*, 119(3): 249–256.

Mahanta, D.K., Chandrashekara, C., Arunkumar, R., Mina, B.L., Pandey, B.M., Mishra, P.K., Bisht, J.K. Srivastva, A.K. and Bhatt, J.C. (2013). Influence farmyard manure application and mineral fertilization on yield sustainability, carbon sequestration potential and soil property of Garden pea– French bean cropping system in the Indian Himalayas. *Scientia Horticulturae*, 164: 414–427.

Manuel, A.E., López-Periago, E., Martínez-Carballo, E., Gándara, J.S., Mejuto, J.C. and García-Río, L. (2008). The mobility and degradation of pesticides in soils and the pollution of groundwater resources (PDF). *Agriculture, Ecosystems and Environment*, 123(4): 247–260.

Meena, V.S., Maurya, B.R., Bohra, J.S., Verma, R. and Meena, M.D. (2013). Effect of concentrate manure and nutrient levels on enzymatic activities and microbial population under submerged rice in alluvium soil of Varanasi. *Field Crops Research*, 245: 6–12.

Miranda – Esquivel, N., de La Cruz, J.L.I., Pacheco, R.G., Enerijiofi, K.E. and Sanchez-Yanez, J.M. (2023). Bioremediation and accelerated phytoremediation of a soil impacted by a hydrocarbon mixture. *International Journal of Plant and Soil Science*, 35(23): 58–66.

Mohler, C.L. and Johnson, S.E. (2009). *Crop Rotation on Organic Farms a Planning Manual*. Natural Resource, Agriculture, and Engineering Service.

Morrison, W.R., Lee, D.-H., Short, B.D., Khrimian, A. and Leskey, T.C. (2016). Establishing the behavioral basis for an attract-and-kill strategy to manage the invasive Halyomorpha halys in apple orchards. *Journal of Pest Science*, 89: 81–96.

Musa, S.I. and Ikhajiagbe, B. (2021). Screening of bacterial isolates for phosphate solubilizing capability in a ferruginous ultisol in Benin city, Edo State Nigeria. *Bayero Journal of Pure and Applied Sciences,* 13(2): 94–106.

Nielsen, A.L., Dively, G., Pote, J.M., Zinati, G. and Mathews, C. (2016). Identifying a potential trap crop for a novel insect pest, *Halyomorpha halys* (Hemiptera: Pentatomidae), in organic farms. *Environmental Entomology*, 45: 472–478.

Poulsen, H.D. (1998). Zinc and copper as feed additives, growth factors or unwanted environmental factors. *Journal of Animal and Feed Sciences*, 7(Suppl. 1): 135–142.

Pretty, J. and Bharucha, Z.P. (2015). Integrated pest management for sustainable intensification of agriculture in Asia and Africa. *Insects*, 6: 152–182.

Qadir, M., Wichelns, D., Raschid-Sally, L., McCornick, P.G., Drechsel, P., Bahri, A. and Minhas, P.S. (2010). The challenges of wastewater irrigation in developing countries. *Agricultural Water Management*, 97(4): 561–568.

Rahman, M.H., Islam, M.R., Jahiruddin, M., Rafii, M.Y., Hanafi, M.M. and Malek, M.A. (2013). Integrated nutrient management in maize-legume-rice cropping pattern and its impact on soil fertility. *Journal of Food, Agriculture and Environment*, 11: 648–652.

Reta, G., Dong, X., Li, Z., Su, B., Hu, X., Bo, H., Yu, D.S., Wan, H., Liu, J.R., Li, Y., Xu, G., Wang, K. and Xu, S. (2018). Environmental impact of phosphate mining and beneficiation: Review. *International Journal of Hydrogen*, 2(4): 424–431.

Sarfraz, M., Dosdall, L.M. and Keddie, B.A. (2005). Spinosad: A promising tool for integrated pest management. *Outlooks on Pest Management*, 16: 78–84.

Scott, C.A., Faruqui, N.I. and Raschid-Sally, L. (Eds.). (2004). *Wastewater Use in Irrigated Agriculture: Confronting the Livelihood and Environmental Realities*. CABI.

Shorey, H.H. and Gerber, R.G. (1996). Use of puffers for disruption of sex pheromone communication among navel orangeworm moths (Lepidoptera: Pyralidae) in almonds, pistachios, and walnuts. *Environmental Entomology*, 25: 1154–1157.

Snoeijers, S.S., Pérez-García, A. Joosten, M.H.A.J. and De Wit, P.J.G.M. (2000). The effect of nitrogen on disease development and gene expression in bacterial and fungal plant pathogens. *European Journal of Plant Pathology*, 106: 493–506.

Soni Kumari and Amarnath Mishra. (2021). *Heavy Metals Contamination, Open Access Peer Reviewed Chapter*. Soni Kumari and Amarnath Mishra.

Van Lenteren, J.C. (1988). Biological and integrated pest control in greenhouses. *Annual Review of Entomology*, 33: 239–269.

Vladés, E.M.A.E., Aldana, L.L.L., Figueroa, B.R., Gutiérrez, O.M., Hernández, R.M.C. and Chavelas, M.T. (2005). Trapping of *Scyphophorus acupunctatus* (Coleoptera: Curculionidae) with two natural baits in a field of *Polianthes tuberosa* (Liliales: Agavaceae) in the state of Morelos, México. *Florida Entomologist*, 88: 338–340.

Vox, G., Loisi, R.V., Blanco, I., Mugnozza, G.S. and Schettini, E. (2016). Mapping of agriculture plastic Waste. *Agriculture and Agricultural Science Procedia*, 8: 583–591.

Vukicevich, E., Lowery, T., Bowen, P., Úrbez-Torres, J.R. and Hart, M. (2016). Cover crops to increase soil microbial diversity and mitigate decline in perennial agriculture. A review. *Agronomy for Sustainable Development*, 36.

Wang, C., Wu, S., Zhou, S., Shi, Y. and Song, J. (2017). Characteristics and source identification of polycyclic aromatic hydrocarbons (PAHs) in urban soils: A review. *Pedosphere*, 27(1): 17–26.

Wang, H., Dong, Y. and Wang, H. (2014). Hazardous metals in animal manure and their changes from 1990 to 2010 in China. *Toxicological & Environmental Chemistry*, 96(9): 1346–1355.

Webb, S.E. and Linda, S.B. (1992). Evaluation of spunbonded polyethylene row covers as a method of excluding insects and viruses affecting fallgrown squash in Florida. *Journal of Economic Entomology*, 85: 2344–235.

Weithmann, N., Möller, J.N., Löder, M.G.J., Piehl, S., Laforsch, C. and Freitag, R. (2018). Organic fertilizer as a vehicle for the entry of microplastic into the environment. *Science Advances*, 4(4): eaap8060.

Wright, R.J. (1984). Evaluation of crop rotation for control of Colorado potato beetles (Coleoptera: Chrysomelidae) in commercial potato fields on Long Island. *Journal of Economic Entomology*, 77: 1254–1259.

Xiong, X., Yanxia, L., Wei, L., Chunye, L., Wei, H. and Ming, Y. (2010). Copper content in animal manures and potential risk of soil copper pollution with animal manure use in agriculture. *Resources, Conservation and Recycling*, 54(11): 985–990.

Zalom, F.G., Bolda, M.P. Dara, S.K. and Joseph, S. (2018). *UC IPM Pest Management Guidelines: Strawberry (Insects and Mites)*. University of California Statewide IPM Program, Publication Number 3468.

Zhang, H.L., Lal, R., Zhao, X., Xue, J.F. and Chen, F. (2014). Opportunities and challenges soil carbon sequestration by conservation agriculture in China. *Advances in Agronomy*, 124: 1–36.

Section 3

Bringing Safe Foods Home through Phytoremediation

Section 3

Bringing Safe Foods Home Through Play or Mediation

13 Phytoexclusion and Phytostabilization

Protective Strategy of the Food Chain from Contamination before Phytoremediation

J. S. Ayedun, B. C. Kotun, O. A. Adewara, R. Adeoye, C. O. Iyiola, E. A. Onakoya, A. I. Domavo, and A. A. Soyemi

13.1 CONTAMINATION IN FOOD CHAINS

Contamination in food chains is a critical issue that affects the safety and quality of the food we consume. It involves the introduction of harmful substances into the various components of the food production system, from soil and water to crops and livestock. Understanding the sources and types of contaminants in food chains is essential for safeguarding human health and ensuring a secure food supply Smith and Bidstrup, 2009; Food and Agriculture Organization, 2019; Codex Alimentarius Commission, 2020; World Health Organization, 2022).

Food is a significant factor in human health and well-being and a key cause of anxiety, pleasure, and stress (Wilcock et al., 2004). The illnesses brought on by contaminated food are one of the causes of stress and worry. Food contamination can occur for several reasons (Sadiku et al., 2020).

Typical food contaminants include those that arise during food processing, those that come from the environment, and those that are not permitted as adulterants or food additives, and migrants from packing materials (Rather et al., 2017).

13.1.1 Sources of Contaminants in Food Chains

Contaminants in food chains can originate from a variety of sources, both natural and anthropogenic. These sources include the following:

Natural Sources

1. *Microbes*: Bacteria, viruses, and parasites can contaminate food chains through improper handling, inadequate sanitation, or contaminated water sources. Pathogens like *Salmonella* and *Escherichia coli* are common examples (Smith and Bidstrup, 2009; World Health Organization, 2022).
2. *Natural toxins*: Some foods naturally contain toxins. Examples include certain types of mushrooms, fish (e.g., pufferfish), and plants (e.g., poisonous berries). Certain plants and fungi produce natural toxins that can contaminate food when ingested. Examples include aflatoxins in peanuts and mushrooms containing poisonous compounds (Smith and Bidstrup, 2009; Food and Agriculture Organization, 2019). Ingesting these toxins can result in food poisoning or adverse health effects.

DOI: 10.1201/9781032683751-16

3. *Radiation*: Exposure to ionizing radiation, whether from nuclear accidents or medical applications, can lead to radioactive contamination of foodstuffs (Smith and Bidstrup, 2009).

Anthropogenic Sources

1. *Chemicals*: Chemical contaminants may come from various human-made sources, including agricultural chemicals (pesticides, herbicides), industrial pollutants (heavy metals, dioxins, polychlorinated biphenyls [PCBs]), food additives (preservatives, colorants), and veterinary drugs (antibiotics, growth promoters). They can leave residues in crops and animal products if not used correctly and can find their way into the food chain (Smith and Bidstrup, 2009; Food and Agriculture Organization, 2019; Codex Alimentarius Commission, 2020).
2. *Processing and packaging*: Contaminants can be introduced during food processing and packaging. Materials used in food containers, utensils, and processing equipment may leach harmful substances into the food. Bisphenol A plastic containers is a well-known example (Smith and Bidstrup, 2009).
3. *Cross-contamination*: Mishandling of food, such as improper storage and preparation, can lead to cross-contamination, where contaminants are transferred from one food item to another.
4. *Genetically modified organisms*: Genetically modified crops may contain unintended allergens or toxins, posing risks if not properly assessed and regulated by humans.

By categorizing these sources, it becomes evident that natural sources are inherent in the environment, while anthropogenic sources result from human activities and interventions in the food production and distribution processes.

13.1.2 Types of Contaminants in Food Chains

Contaminants in food chains encompass a wide range of substances, each with unique characteristics and potential health effects. These contaminants can be broadly categorized into the following.

1. *Chemical contaminants*: This category includes heavy metals (e.g., lead [Pb], cadmium [Cd]), pesticides, herbicides, industrial chemicals (e.g., PCBs), and food additives. Chemical contaminants can have acute or chronic toxic effects, depending on the exposure levels and duration (Smith and Bidstrup, 2009; Codex Alimentarius Commission, 2020).
2. *Microbial contaminants*: Microbes such as bacteria, viruses, and parasites can cause foodborne illnesses. Contaminated water, improper food handling, and inadequate sanitation are common pathways for microbial contamination (Smith and Bidstrup, 2009; World Health Organization, 2022).
3. *Allergens*: Allergenic substances like peanuts, tree nuts, shellfish, and gluten can trigger allergic reactions in susceptible individuals if not properly labeled and managed in food production (Smith and Bidstrup, 2009; Food and Agriculture Organization, 2019).
4. *Radiological contaminants*: Radioactive substances, such as cesium-137 and strontium-90, can contaminate food following nuclear accidents or as a result of nuclear testing. Prolonged radiation exposure can increase the risk of cancer and other health problems (Smith and Bidstrup, 2009).

13.1.3 Importance of Safeguarding Food Chains from Contamination

1. *Public health*: Contaminated food can lead to foodborne illnesses, hospitalizations, and even fatalities. Preventing contamination is crucial for protecting consumers' health.
2. *Economic impact*: Foodborne illnesses and product recalls due to contamination can result in significant economic losses for food producers, processors, and distributors. Illnesses

and recalls can damage a company's reputation and result in financial setbacks (Smith and Bidstrup, 2009; Food and Agriculture Organization, 2019).

3. *Global trade*: International trade in food products requires adherence to stringent food safety standards. Failure to meet these standards can lead to trade barriers and restrictions, affecting economies at a national level (Smith and Bidstrup, 2009).
4. *Consumer confidence*: Contamination scares can erode consumer confidence in the safety and reliability of the food supply chain. Restoring trust can be a long and challenging process (Smith and Bidstrup, 2009).
5. *Legal and regulatory compliance*: Food safety regulations exist to protect consumers and ensure the integrity of the food supply. Non-compliance with these regulations can lead to legal consequences for individuals and organizations involved in the food industry (Smith and Bidstrup, 2009; World Health Organization, 2022).

Contamination in food chains poses significant risks to public health, the economy, and global trade. Understanding the sources and types of contaminants is the first step in developing effective strategies for prevention and mitigation. Robust food safety measures, regulatory compliance, and public awareness are essential components of safeguarding our food supply from contamination (Smith and Bidstrup, 2009; Food and Agriculture Organization, 2019; World Health Organization, 2022).

13.1.4 Bioaccumulation of Contaminants in Food Webs

Bioaccumulation is the accumulation of potentially harmful chemicals in body tissues at one level of the food chain. Bioaccumulation occurs in two main ways:

1. *Bioconcentration*: This process causes a chemical to accumulate in aquatic organisms as a result of simultaneous intake from water. Once an organism absorbs a chemical from its surroundings through its skin, respiratory system, or food, the process of bioconcentration starts; as a result of successive uptake, the substance's concentration rises. The bioconcentration surpasses environmental concentration, and can only be determined in well-regulated environments without taking into account the chemical's dietary food intake (Arnot and Gobas, 2004).
2. *Biomagnification*: This occurs when a chemical moves higher up the food chain, thereby increasing the concentration of that chemical. This explains why animals at the top of the food chain accumulate the greatest amount of chemicals through biomagnification (Szynkowska et al., 2018). When harmful substances that are still present in the environment are ingested indirectly by organisms through food, biomagnification takes place. These substances can accumulate when an organism higher up the food chain eats a lower organism that contains them. In other words, it damages every rung of the food chain as it moves up. As one moves up the trophic levels of the food chain, the concentration of these poisons or substances rises. When plants come in contact with chemicals and absorb them from water or soil, they accumulate them and they cannot be decomposed. When an animal starts eating this plant, it also consumes chemicals. The more plants they eat; the more chemicals they consume.

13.1.5 Health Risks Associated with Contaminated Food

Food contamination can occur at any stage of the food production process, from processing and transportation to consumption. According to the World Health Organization (WHO), unsafe food contains bacteria, viruses, parasites, or chemicals; these agents can cause more than 200 diseases. The most common illness is diarrhea, which is mainly caused by eating contaminated food.

Foodborne illnesses are infectious and toxic. They are caused by bacteria, viruses, parasites, or chemicals that enter the body through contaminated food or water. Foodborne pathogens can be animal products, fresh vegetables, or contaminated drinking water (World Health Organization, 2022).

Some of the symptoms associated with contaminated food include fever, nausea, vomiting, and abdominal pain.

13.1.6 Effects of Contaminants on Food Quality and Safety

Food safety refers to the conditions and procedures that maintain food quality to avoid contamination and foodborne illnesses. The US Department of Agriculture's (USDA) Food Safety and Inspection Service informs consumers about the value of handling food safely and offers advice on how to lower the likelihood of contracting foodborne illness (United State Department of Agriculture, 2022).

Food safety is becoming more closely related to food quality control. It is used to prevent chemical and biological risks that could lead to contamination. All food quality control is carried out using acceptable and well-established methodologies to maintain product characteristics. It is important to understand which ingredients in food make it unsafe to consume because food quality and safety are two areas that are becoming more and more intertwined. These substances, also known as contaminants, are subject to a growing amount of regulation and control due to the potential harm they may cause to consumers' health if consumed improperly (Ahmad et al., 2023).

13.1.7 Impacts of Contaminants on Crop Yield and Nutritional Value

Crop yields and human health are negatively impacted by the contamination of soil with toxic metals like cadmium and high concentrations of aluminum (Khan et al., 2022). Metals cause oxidative stress in plants, which lowers the accumulation of biomass.

Problems can also arise if plants absorb large amounts of some toxic ions present in irrigation water, such as chloride (Cl), sodium (Na), and boron (B), and these concentrations build up to levels that can harm crops and lower yields (Shrivastava and Kumar, 2015). An important factor in toxin concentrations in water is the release of chemical wastewater into watersheds used for agriculture and/or the pumping up of irrigation water, as well as the disposal of agrochemicals on farms. However, most vegetable crops have a relatively low tolerance to B, and even very low B concentrations can harm crops (Mishra et al., 2023). Most irrigation water sources contain concentrations of elements below toxicity thresholds. Each crop experiences different levels of damage, with permanent perennial crops reportedly suffering the most.

Crop quantity and quality are significantly impacted by contaminated irrigation water. More than 10% of the world's population consumes food grown with untreated sewage or surface water tainted with feces, and most of them reside in developing nations with arid and semi-arid climates (Khalid et al., 2018). The main factors driving the rise in the use of contaminated water for irrigation are growing populations, a shortage of freshwater, and awareness of the value of wastewater as fertilizer. Foodborne disease outbreaks have been associated with the irrigation and post-harvest processes that use pathogen-contaminated urban wastewater (e.g., *Salmonella* spp., norovirus, and *E. coli*), especially with fruits and vegetables that are frequently consumed raw (Gurtler and Gibson, 2022).

13.2 PHYTOEXCLUSION

Phytoexclusion is a vital protective strategy within the realms of environmental science and agriculture. This strategy relies on the remarkable capabilities of plants to exclude or limit the uptake of contaminants from the soil or surrounding environment, thus preventing these harmful substances from entering the food chain and posing risks to human health (Khan et al., 2019).

13.2.1 Principles of Phytoexclusion

Phytoexclusion is a phytoremediation technique that focuses on using plants to immobilize, sequester, or exclude contaminants from the environment (Khan et al., 2019). It operates on the principles of selectively growing specific plant species that can absorb and accumulate contaminants, thereby preventing these substances from moving up the food chain. Phytoexclusion works by utilizing the natural mechanisms of plants to regulate the uptake of contaminants. Some plants have specific mechanisms in their roots that prevent the uptake of certain contaminants, or they may have the ability to selectively accumulate less toxic forms of contaminants. By strategically selecting plant species that are efficient in excluding contaminants, phytoexclusion can effectively reduce the concentration of pollutants in the soil and protect the surrounding ecosystem (Pandey et al., 2023).

Phytoexclusion takes advantage of the differential uptake of contaminants by various plant species. Certain plants, known as hyperaccumulators, have an innate ability to absorb and concentrate high levels of specific contaminants within their tissues without displaying adverse effects (Pandey et al., 2023). By strategically selecting hyperaccumulator plants and promoting their growth in contaminated areas, phytoexclusion can effectively limit the migration of contaminants through the food chain (Pandey et al., 2023).

13.2.2 Plant Mechanisms for Excluding Uptake of Contaminants

Plants employ a range of mechanisms to exclude or limit the uptake of contaminants, ensuring their survival while safeguarding the food chain. These mechanisms include the following.

1. *Root barriers*: Plants can develop specialized barrier tissues in their roots, such as Casparian strips and suberin layers that restrict the entry of contaminants into the plant's vascular system (Ryan et al., 2001).
2. *Selective uptake*: Different plant species exhibit varying affinities for specific contaminants. Phytoexclusion involves choosing plant species that naturally exhibit low uptake of harmful substances (Salt et al., 1995; Aransiola et al., 2023).
3. *Chemical alterations*: Some plants can release chemicals (e.g., organic acids and exudates) that modify the chemical form of contaminants in the rhizosphere, rendering them less available for uptake (Glick, 2003).

13.2.3 Case Studies Demonstrating the Effectiveness of Phytoexclusion

Numerous case studies have demonstrated the effectiveness of phytoexclusion as a protective strategy. One notable example is the use of Indian mustard (*Brassica juncea*) in remediating soil contaminated with heavy metals like Pb and Cd. Indian mustard is a well-known hyperaccumulator of these metals and can accumulate them in its above-ground tissues while limiting their translocation to edible parts (Gupta et al., 2013).

In another case, the phytoremediation of arsenic (As)-contaminated sites in Bangladesh utilized fern species like *Pteris vittata*. These ferns have shown a remarkable capacity to exclude as from their edible fronds, thus preventing the entry of this toxic element into the food chain (Meharg et al., 2008).

These case studies underscore the potential of phytoexclusion in protecting the food chain from contamination by leveraging the unique abilities of certain plant species.

13.3 PHYTOSTABILIZATION

Phytostabilization is a vital protective strategy in the field of environmental science and remediation. This approach leverages the unique capabilities of plants to stabilize contaminants within the soil, preventing their migration into the wider environment and protecting ecosystems and human health. Phytostabilization is the process by which plants absorb volatile substances through their roots. The volatile substances are either directly transpired by the plants through their leaves or transformed into metabolites that can be transpired by the plants. The pollutants are then discharged into the atmosphere (Shackira and Puthur, 2019).

This method involves using plants to stabilize and immobilize contaminants in the soil, preventing their movement and reducing their potential for leaching into groundwater or being released into the air (Lone et al., 2008). It works by reducing the mobility and bioavailability of contaminants, such as heavy metals, metalloids, and other pollutants. The plants' root systems play a crucial role in this process. Phytostabilization is effective because certain plant species have the ability to tolerate and accumulate high levels of contaminants without being significantly affected. These plants, known as hyperaccumulators, can absorb and store contaminants in their roots, stems, and leaves, effectively reducing their concentration in the soil. By establishing plant cover, phytostabilization also helps to prevent soil erosion, improve soil structure, and enhance overall soil health (Lone et al., 2008; Aransiola et al., 2019).

13.3.1 Principles of Phytostabilization

Phytostabilization is a phytoremediation technique that focuses on using plants to immobilize and contain contaminants within the soil matrix. Unlike phytoextraction, which aims to remove contaminants from the soil, phytostabilization aims to confine and limit their mobility, reducing the risk of their dispersal into the environment (Kumar et al., 2016).

The principles of phytostabilization are rooted in the ability of certain plants, often referred to as stabilizers or soil-binders, to develop extensive root systems and secrete substances that bind contaminants, preventing their leaching or uptake by other organisms (Shackira and Puthur, 2019).

13.3.2 Techniques Employed by Plants in Stabilizing Contaminants in the Soil

Phytostabilization employs a range of plant-based techniques to effectively immobilize contaminants in soil, as follows.

1. *Root encapsulation*: Some plants release organic substances from their roots, forming a protective barrier around contaminants, which prevents their movement in the soil and uptake by other organisms (Mench et al., 2009).
2. *Absorption and accumulation*: Certain plant species, known as accumulator plants, have the capacity to absorb contaminants from the soil and accumulate them within their tissues, particularly in the roots. This process reduces the concentration of contaminants in the surrounding soil (Tordoff et al., 2000).
3. *Enhancing soil structure*: The extensive root systems of stabilizer plants can improve soil structure and reduce erosion, preventing the dispersion of contaminated particles (Huang et al., 2019).
4. *Nutrient competition*: Some plants can outcompete contaminants for essential nutrients, limiting the availability of these nutrients for contaminant mobility (Berti and Cunningham, 2000).

13.3.3 Environmental Benefits of Phytostabilization

Phytostabilization offers a multitude of environmental benefits, making it a crucial protective strategy:

1. *Contaminant containment*: By immobilizing contaminants within the soil, phytostabilization prevents their spread into water bodies, air, and neighboring ecosystems, thus safeguarding surrounding environments (Kumar et al., 2016).

2. *Soil improvement*: Phytostabilization can enhance soil quality by reducing erosion and increasing soil organic matter content, which promotes overall ecosystem health (Batty et al., 2005).
3. *Biodiversity preservation*: By reducing contaminant exposure and maintaining a stable soil environment, phytostabilization helps preserve biodiversity and protects the habitats of various plant and animal species (Garbisu et al., 2019).
4. *Cost-effective remediation*: Phytostabilization is often more cost-effective than other remediation methods, making it a sustainable option for contaminated site restoration (Yadav et al., 2022).

13.4 SUCCESSFUL PHYTOEXCLUSION AND PHYTOSTABILIZATION PROJECTS

Phytoextraction and phytostabilization potential of plants grown in the vicinity of heavy metal-contaminated soils: a case study at an industrial town site (Lorestani et al., 2013).

The environment is increasingly at risk due to toxic heavy metals in the air, soil, and water. All metals are toxic at higher concentrations even though many of them are necessary because they lead to oxidative stress through the production of free radicals. Metals thus eliminate biodiversity and make the land unfit for plant growth.

The current methods of removing heavy metals from contaminated soil and water are costly, time-consuming, and environmentally damaging. Because metals cannot degrade like organic compounds can, effective cleanup calls for immunization to lessen or eliminate toxicity (Rajendran et al., 2022). The removal of pollutants from the environment may be accomplished through phytoremediation, which takes advantage of plants' extraordinary capacity to concentrate substances and elements from the environment and to metabolize different molecules in their tissues (Garbisu et al., 2002).

This case study teaches us that phytoremediation, an emerging cost-effective, non-invasive, and aesthetically pleasing technology that takes advantage of plants' remarkable capacity to concentrate elements, has the potential to be used to clean up sites that have been contaminated with metal (Abioye et al., 2017).

Both phytostabilization and phytoexclusion are sustainable and cost-effective methods for managing contaminated sites. They offer long-term solutions that promote environmental health, improve soil quality, and protect human and ecosystem well-being (Berti and Cunningham, 2000).

13.5 STRATEGIES FOR EARLY DETECTION AND PREVENTION OF CONTAMINATION

Policies are in place to control the number of various substances in food. Adulterants and unhealthy additives are not permitted by law. However, efficient monitoring and response systems are necessary to stop chemical threats from getting into the food supply and endangering the public. The US Food and Drug Administration specifies the lowest chemical concentrations that are permitted in food, such as the requirement that pesticide concentrations should not exceed the designated limit (Bajwa and Sandhu, 2011). Despite this, mistakes might still happen even according to the specified concentration and instructions. The administration of the concentration of dangerous pollutants in food remains poor, particularly in the case of underdeveloped and developing countries.

As a result of their heavy reliance on agriculture, the groundwater in some countries is contaminated with pesticides, which in turn contaminate both food and water. According to Villanueva et al. (2013), unregulated chemicals are extremely concerning, and further study is required on pollutants that are imperceptible to humans. In addition, each consumer's worries are important since they can play an important role in how they manage their health (Liang and Scammon, 2016).

Several strategies to ensure early detection and prevention of contamination in food are implemented:

1. *Hazard analysis and critical control points (HACCP)*: Starting with the definition of HACCP, there are three main sources of hazard: first, the unacceptable presence of a

biological, chemical, or physical contaminant; second, the unacceptable potential for microbial growth or survival; third, the unacceptable potential for chemical generation; and finally, the unacceptable contamination or recontamination with microbial agents, chemicals, or foreign substances. This systematic approach identifies potential hazards in food production and establishes critical control points to prevent, eliminate, or reduce contamination risks. A zero-risk specification is likely to be unrealistic to achieve at this time (Unnevehr and Jensen, 1999), without resolving prohibitive expenses. Hazards may produce foodborne illnesses with different related implications for public health. Additionally, adjustments being made to the methods and inputs used along the food chain as well as the expanding global market integration may result in unexpected dangers (Swanenburg et al., 2001).
2. *Good agricultural practices (GAPs)*: Implementing practices that promote safe and hygienic conditions during crop cultivation, including proper use of fertilizers, pesticides, and irrigation water, as well as effective waste management (Sabir et al., 2010).
3. *Good manufacturing practices (GMPs)*: Following strict guidelines in food processing facilities to maintain cleanliness, hygiene, and proper handling of ingredients, equipment, and packaging materials (Meghwal et al., 2017).
4. *Quality assurance programs*: Implementing quality control measures throughout the supply chain, including regular testing of raw materials, finished products, and environmental samples to detect any contamination (Swanenburg et al., 2001).
5. *Supplier verification*: Ensuring that suppliers meet specific quality and safety standards through audits, certifications, and regular monitoring of their production processes (Swanenburg et al., 2001).
6. *Training and education*: Providing comprehensive training to food handlers, farmers, and industry professionals on proper hygiene practices, contamination prevention, and food safety regulations (Insfran-Rivarola et al., 2020).
7. *Traceability systems*: Implementing systems that enable tracking and tracing of food products throughout the supply chain, facilitating rapid identification and recall of contaminated products if necessary (Dabbene et al., 2014).

13.5.1 Economic Benefits and Cost-Effectiveness of Implementing Protective Strategies

Implementing protective strategies such as phytostabilization and phytoexclusion can bring several economic benefits and cost-effectiveness, as follows.

1. *Cost savings*: These strategies can be more cost-effective compared to traditional remediation methods like excavation and disposal. Phytostabilization and phytoexclusion require less equipment, labor, and maintenance, resulting in lower overall costs (Mench et al., 2010).
2. *Long-term solutions*: Once established, plant-based protective strategies can provide long-term benefits. They require minimal ongoing maintenance, reducing the need for continuous monitoring and intervention (Berti and Cunningham, 2000).
3. *Environmental benefits*: By utilizing natural processes and promoting ecosystem health, protective strategies help preserve and restore the environment. This can lead to positive impacts on biodiversity, air quality, water resources, and overall ecosystem functioning (Berti and Cunningham, 2000).
4. *Land reclamation and land value*: This can help in transforming contaminated sites into usable and valuable land. The remediated land can be repurposed for various purposes like agriculture, recreation, or development, leading to economic opportunities and increased land value (Song et al., 2019).

While the specific economic benefits and cost-effectiveness may vary depending on the site and circumstances, implementing protective strategies can offer significant advantages in terms of cost savings, long-term sustainability, environmental benefits, and regulatory compliance.

13.5.2 Sustainable Practices in Food Chain Protection

Sustainable practices play a crucial role in ensuring the long-term viability of our food systems by reducing the negative impact on the environment, protecting natural resources, and protecting economic well-being. Food product creation, processing, packaging, distribution, and consumption are all included. This indicates that attention is being paid to both the processes involved in bringing the food to the consumer as well as the final product (Santhoshkumar et al., 2017). They include the following:

1. *Organic farming*: Choosing organic farming methods reduces the use of synthetic pesticides and fertilizers, minimizing environmental impacts and promoting soil health (Santhoshkumar et al., 2017).
2. *Crop rotation*: Rotating crops helps prevent the buildup of pests and diseases, reduces the need for chemical interventions, and enhances soil fertility (Santhoshkumar et al., 2017).
3. *Integrated pest management (IPM)*: IPM focuses on using a combination of biological controls, cultural practices, and targeted pesticide use to manage pests while minimizing harm to beneficial organisms and the environment (Santhoshkumar et al., 2017).
4. *Efficient water management*: Implementing water-saving techniques like drip irrigation, precision farming, and water recycling reduces water waste and conserves this precious resource (Santhoshkumar et al., 2017).
5. *Biodiversity conservation*: Preserving and enhancing biodiversity on farms helps maintain a balanced ecosystem, promotes natural pest control, and improves resilience to climate change (Santhoshkumar et al., 2017).
6. *Sustainable packaging*: Opting for eco-friendly packaging materials and reducing food waste through proper storage and distribution practices contribute to a more sustainable food chain (Villanueva et al., 2013).

13.6 FUTURE DIRECTIONS AND RESEARCH AREAS

The research world is evolving, and maintaining trends is crucial in advancing science inquiry. In this section, we discuss the emerging technologies and research trends in contamination prevention, analyze innovation in plant-based strategies that protect the food chain, and survey collaborative efforts and international initiatives for food safety.

13.6.1 Emerging Technologies and Research Trends in Contamination Prevention

Because of the hazards caused by traditional methods of decontaminating food, researchers have developed non-thermal technologies like ultrasound for the removal of food contaminants. In comparison to heat treatment which degrades food quality, non-thermal procedures ensure microbiological food safety while preserving the original aroma and flavor of the food (Ucar et al., 2021; Yuan et al., 2021).

Ultrasound is a mild, highly effective, economical, and ecologically friendly technology that enhances product quality and processing. The removal of pesticide residues and allergies from fruits, vegetables, and milk via ultrasound processing is effective despite the growth of this technique (Yuan et al., 2021)

Minimal heating is required during ultra-high-pressure hydrostatic processing to kill the vegetative cells of spoilage organisms and pathogens. For fruit goods like jams and jellies, this method was initially made commercially available in Japan. It can also be used for heat-sensitive fruits and vegetables. Applying electric current to conductive food results in quick heating and the eradication of microorganisms (Conway, 2000).

13.6.2 Innovations in Plant-Based Strategies for Food Chain Protection

1. Revegetation of mine tailings is a preventive practice of contaminant dispersal. Metal-tolerant grasses have been used to make mine tailings stable. Metal tolerance in plants is enhanced by symbiotic, root-colonizing, arbuscular mycorrhizal fungi (AMF) through metal sequestration in the AMF hyphae (Greipsson, 2011).
2. Plants can also reduce the bioavailability of pollutants or transform them into less harmful forms. Natural chelates, such as siderophores, organic acids, and phenolics, generated by certain plants' roots can form complexes with metals in the rhizosphere (Greipsson, 2011).
3. Poplar trees (*Populus* spp.) have been used to degrade toxic and recalcitrant organic compounds. Trichloroethylene has been volatilized using hybrid poplar trees, which convert it to chlorinated acetates and carbon dioxide. Plants can volatilize metals like selenium by converting them into dimethyl selenide (Greipsson, 2011).
4. Soil microorganisms can reduce the harmful effects of pollutants in the soil. Exudates (peptides) from the bacteria *Pseudomonas putida*, for example, can reduce Cd toxicity in plants (Greipsson, 2011).
5. The use of low heavy metal accumulating cultivars is an economical technique which has been employed for restricting root uptake of heavy metals by limiting the secretion of organic acid (Wang et al., 2021).

13.6.3 Collaborative Efforts and International Initiatives for Food Safety

The WHO collaborates with the Food and Agriculture Organization of the United Nations (FAO), the World Organisation for Animal Health, and the UN Environment Programme to corroborate the safety of food across the food supply chain. The WHO uses the IFOSAN network to manage threats concerning food safety and to acknowledge its crisis (World Health Organization, 2022).

FAO is part of the Standards and Trade Development Facility, a world partnership held by the World Trade Organization (WTO) to build the implementation of international sanitation and phytosanitary standards in developing countries, to profit and sustain market access, agricultural productivity, and food safety. The FAO and the International Atomic Energy Agency help in the development and adoption of nuclear and related techniques that provide regulation of food safety through science-based techniques (Food and Agriculture Organization, 2023).

The Codex Alimentarius Commission is a body that was established in 1963 by WHO and FAO with more than 170 members of the Joint Food Standard Program. It aims to protect consumers' health and enable fair practices in the food trade (Food and Drug Administration, 2023).

13.7 CONCLUSION AND RECOMMENDATIONS

Contaminants imperil human health and are detrimental to the environment. Phytoexclusion and phytostabilization have been used as a protective strategy to prevent contamination in foods. In this section, we summarize the significance of phytoexclusion and phytostabilization, review recommendations for the protection of food chain from contamination, and encourage responsible practices for a safe and secure environment.

Reducing contaminants from the soil and the environment is essential. Plants prevent these harmful substances from entering the food supply chain and endangering human health. Phytoexclusion has been proven to be eco-friendly and cost-effective in restoring the environment, so it is preferable when compared to engineering procedures that are destructive to the soil. Some plants also help to maintain contaminants in the soil, preventing their dispersal in the environment and protecting human health. Phytostabilization uses the potential of plants to decrease the action of toxic substances in the soil, also making contaminants less toxic or decreasing their bioavailability.

In the food chain, food safety should include the three core elements of risk analysis: assessment, management, and communication. There should be an institutional separation of science-based data in this analysis process. Risk management, or the regulation and control of risk, is distinct from risk assessment. Additionally, a sensible strategy to risk management and evaluation should be used. Traceability methods are necessary to improve treatment, food processing, and consumer distribution. Harmonization of food safety regulations, which denotes more expansion and further development of equivalence in food safety systems, is necessary particularly in light of the Agreement on the Application of Sanitary and Phytosanitary Measures (SPS) of the WTO. Equivalence refers to the achievement of identical levels of protection against foodborne risks regardless of the means of control utilized.

Ex ante risk avoidance or prevention at source along the entire food chain, from farm or sea to plate, needs to be given more attention in order to complement the traditional ex post approach to food safety management, which is based on regulation and control.

There are biological, chemical, and physical agents that contaminate food along the food chain. As a result, prevention of contaminants must be used at all points along the continuum. The food safety policy should be enhanced to promote a safety network. Inspection and monitoring programs must be significant for enforcement of food safety regulations. The National Codex Committee should be strengthened to enable the development of better national positions and to solve national standardization issues efficiently. Risk analysis, principles, and procedures should be prepared in order to be on alert in cases of food safety emergencies.

REFERENCES

Abioye, O. P., Ijah, U. J. J., & Aransiola, S. A. (2017). Phytoremediation of Soil Contaminants by Biodiesel Plant Jatropha curcas, Chapter 4 of Bauddh, K., et al. (eds.), *Phytoremediation Potential of Bioenergy Plants*. Springer Nature Singapore Pte Ltd. [https://link.springer.com/book/10.1007%2F978-981-10-3084-0].

Ahmad, R. S., Munawar, H., Saima, H., & Siddique, F. (2023). Introductory Chapter: Food Safety. In *Food Safety-New Insights*. IntechOpen. London, United Kingdom.

Aransiola, S. A., Ijah, U. J. J., Abioye, O. P., & Bala, J. D. (2019). Microbial-Aided Phytoremediation of Heavy Metals Contaminated Soil: A Review. *European Journal of Biological Research*, 9(2), 104–125. [http://www.journals.tmkarpinski.com/index.php/ejbr/article/view/15].

Aransiola, S. A., Ikhumetse, A. A., Babaniyi, B. R., Abioye, O. P., Oyedele, O. J., & Falade, N. O. (2023). Phytoaccumulation of Micro and Nanoplastics: Root Uptake. In Maddela, N. R., Reddy, K. V., & Ranjit, P. (eds.), *Micro and Nanoplastics in Soil*. Springer, Cham. https://doi.org/10.10007/978-3-031-21195-9_8.

Arnot, J. A., & Gobas, F. A. (2004). A Food Web Bioaccumulation Model for Organic Chemicals in Aquatic Ecosystems. *Environmental Toxicology and Chemistry: An International Journal*, 23(10), 2343–2355.

Bajwa, U., & Sandhu, K. (2011). Effect of Handling and Processing on Pesticide Residues in Food – a Review. *Journal of Food Science and Technology*, 51, 201–220.

Batty, L. C., Baker, A. J., & Wheeler, B. D. (2005). Juncus Subnodulosus: Can Plant Age Affect Metal Tolerance and Uptake? *New Phytologist*, 165(3), 801–810.

Berti, W. R., & Cunningham, S. D. (2000). *Phytostabilization of Metals. Phytoremediation of Toxic Metals: Using Plants to Clean Up the Environment* (pp. 71–88). Wiley, New York.

Codex Alimentarius Commission. (2020). Codex Alimentarius: International Food Standards. [http://www.fao.org/fao-who-codexalimentarius/en/].

Conway, G. (2000). Food for All in the [2]1st Century. *Environment: Science and Policy for Sustainable Development*, 42(1), 8–18.

Dabbene, F., Gay, P., & Tortia, C. (2014). Traceability Issues in Food Supply Chain Management: A Review. *Biosystems Engineering*, 120, 65–80.

Food and Agriculture Organization. (2019). The State of Food and Agriculture: Moving Forward on Food Loss and Waste Reduction. [http://www.fao.org/state-of-food-agriculture/en/].

Food and Agriculture Organization. (2023). Safe Food for Everyone: Working Together for Food Safety. [https://www.fao.org/3/cc4347en/online/safe-food-for-everyone-2023/international-cooperation-community-awareness.html].

Food and Drug Administration. (2023). International Cooperation on Food Safety. [https://www.fda.gov/food/international-interagency-coordination/international-cooperation-food-safety], last accessed January 20, 2024.

Garbisu, C., Allica, J. H., Barrutia, O., Alkorta, I., & Becerril, J. M. (2002). Phytoremediation: A Technology Using Green Plants to Remove Contaminants From Polluted Areas. *Reviews on Environmental Health*, 17(3), 173–188.

Garbisu, C., Urra, J., Mench, M., & Kidd, P. (2019). *Guide To Best Phytomanagement Practices for the Recovery of Biodiversity in Degraded and Contaminated Sites*. [https:www.phytosudoe.eu/wp-content/uploads/2020/05/PRODUCT-WP-2-GUIDE-TO-BEST-PHYTOMANAGEMENT-PRACTICES.pdf].

Glick, B. R. (2003). Phytoremediation: Synergistic Use of Plants and Bacteria to Clean Up the Environment. *Biotechnology Advances*, 21(5), 383–393.

Greipsson, S. (2011). Phytoremediation. *Nature Education Knowledge*, 3(10), 7.

Gupta, D. K., Nicoloso, F. T., & Schetinger, M. R. C. (2013). Lead Tolerance and Phytoremediation Potential of Brazilian Leguminous Tree Species. *Environmental Science and Pollution Research*, 20(8), 5689–5701.

Gurtler, J. B., & Gibson, K. E. (2022). Irrigation Water and Contamination of Fresh Produce With Bacterial Foodborne Pathogens. *Current Opinion in Food Science*, 47, 100889.

Huang, J. W., Chen, J., & Berti, W. R. (2019). Phytoextraction of Lead from Contaminated Soils. *Environmental Science and Technology*, 33(18), 3828–3833.

Insfran-Rivarola, A., Tlapa, D., Limon-Romero, J., Baez-Lopez, Y., Miranda-Ackerman, M., Arredondo-Soto, K., & Ontiveros, S. (2020). A Systematic Review and Meta-Analysis of the Effects of Food Safety and Hygiene Training on Food Handlers. *Foods*, 9(9), 1169.

Khalid, S., Shahid, M., Natasha, Bibi, I., Sarwar, T., Shah, A. H., & Niazi, N. K. (2018). A Review of Environmental Contamination and Health Risk Assessment of Wastewater Use for Crop Irrigation With a Focus on Low and High-Income Countries. *International Journal of Environmental Research and Public Health*, 15(5), 895.

Khan, M. A., Nizami, A. S., & Tariq, R. (2019). Phytoremediation: An Environmentally Sustainable Green Technology for Management of Heavy Metal Contaminated Soils. *Ecological Engineering*, 127, 497–507.

Khan, Z., Fan, X., Khan, M. N., Khan, M. A., Zhang, K., Fu, Y., & Shen, H. (2022). The Toxicity of Heavy Metals and Plant Signaling Facilitated By Biochar Application: Implications for Stress Mitigation and Crop Production. *Chemosphere*, 308, 136466.

Kumar, P. B. A. N., Dushenkov, V., Motto, H., & Raskin, I. (2016). Phytoextraction: The Use of Plants to Remove Heavy Metals from Soils. *Environmental Science and Technology*, 29(5), 1232–1238.

Liang, B., & Scammon, D. (2016). Food Contamination Incidents: What Do Consumers Seek Online? Who Cares? *International Journal on Nonprofit and Voluntary Sector Marketing*, 21, 227–241.

Lone, M. I., He, Z., Stoffella, Peter, J., & Yang, X. (2008). Phytoremediation of Heavy Metal Polluted Soils and Water: Progress and Perspectives. *Journal of Zhejiang University Science B.*, 9(3), 210–220.

Lorestani, B., Yousefi, N., Cheraghi, M., & Farmany, A. (2013). Phytoextraction and Phytostabilization Potential of Plants Grown in the Vicinity of Heavy Metal-Contaminated Soils: A Case Study at an Industrial Town Site. *Environmental Monitoring and Assessment*, 185, 10217–10223.

Meghwal, M., Heddurshetti, U., & Biradar, R. (2017). Good Manufacturing Practices for Food Processing Industries: Principles and Practical Applications. In *Food Technology*. Applied Research and Production Techniques, 1st Edition. Apple Academic Press, New York, pp 1–26.

Meharg, A. A., Rahman, M. M., & Hossain, M. (2008). Phytoremediation of Arsenic-Contaminated Paddy Soils with the Wetland Fern Species *Pteris Vittata L. Environmental Science and Technology*, 42(21), 8011–8016.

Mench, M., Lepp, N., Bert, V., Schwitzguébel, J. P., Gawronski, S. W., Schröder, P., & Vangronsveld, J. (2010). Successes and Limitations of Phytotechnologies at Field Scale: Outcomes, Assessment and Outlook from COST Action 859. *Journal of Soils and Sediments*, 10, 1039–1070.

Mench, M., Schwitzguébel, J. P., Schroeder, P., Bert, V., Gawronski, S. W., Gupta, S., . . . Vangronsveld, J. (2009). Assessment of Successful Experiments and Limitations of Phytotechnologies: Contaminant Uptake, Detoxification and Sequestration, and Consequences for Food Safety. *Environmental Science and Pollution Research*, 16(7), 876–900.

Mishra, S., Kumar, R., & Kumar, M. (2023). Use of Treated Sewage or Wastewater as an Irrigation Water for Agricultural Purposes-Environmental, Health, and Economic Impacts. *Total Environment Research Themes*, 100051.

Pandey, V., Mishra, D., Yadav, R., Siddiqui, A., Hiremath, C., Kumar, B., & Khare, P. (2023). Phyto-Exclusion of Pb and Cd By Different Genotypes of Andrographis Paniculata (Burm. F.) Nees: A Novel Approach for Safe Cultivation. *Industrial Crops and Products*, 191, 115977.

Rajendran, S., Priya, T. A. K., Khoo, K. S., Hoang, T. K., Ng, H. S., Munawaroh, H. S. H., Karaman, C., Orooji, Y., & Show, P. L. (2022). A Critical Review on Various Remediation Approaches for Heavy Metal Contaminants Removal From Contaminated Soils. *Chemosphere*, 287, 132369.

Rather, I. A., Koh, W. Y., Paek, W. K., & Lim, J. (2017). The Sources of Chemical Contaminants in Food and Their Health Implications. *Frontiers in Pharmacology*, 8, 308465.

Ryan, P. R., Raman, H., Gupta, S., & Horst, W. J. (2001). Deliberating Suberization. *Plant Science*, 160(3), 405–413.

Sabir, N., Singh, B., Hasan, M., Sumitha, R., Deka, S., Tanwar, R., Ahuja, D. B., Tomar, B. S., Bambawale, O. M., & Khah, E. M. (2010). Good Agricultural Practices (GAP) for IPM in Protected Cultivation. *Technical Bulletin*, 23, 1–14.

Sadiku, M. N., Ashaolu, T. J., & Musa, S. M. (2020). Food Contamination: A Primer. *International Journal of Advanced Science and Research*, 6(03), 01–07.

Salt, D. E., Smith, R. D., & Raskin, I. (1995). Phytoremediation. *Annual Review of Plant Biology*, 46(1), 15–36.

Santhoshkumar, M., Reddy, G. C., & Sangwan, P. S. (2017). A Review on Organic Farming-Sustainable Agriculture Development. *International Journal of Pure and Applied Bioscience*, 5(4), 1277–1282.

Shackira, A. M., & Puthur, J. T. (2019). Phytostabilization of Heavy Metals: Understanding of Principles and Practices. *Plant-Metal Interactions*, 263–282.

Shrivastava, P., & Kumar, R. (2015). Soil Salinity: A Serious Environmental Issue and Plant Growth

Promoting Bacteria as One of the Tools for Its Alleviation. *Saudi Journal of Biological Sciences*, 22(2), 123–131.

Smith, R. A., & Bidstrup, P. L. (2009). Food Contamination: Sources and Surveillance. *Food Security in Australia*, 67–85.

Song, Y., Kirkwood, N., Maksimovic, C., Zheng, X., O'Connor, D., Jin, Y., & Hou, D. (2019). Nature Based Solutions for Contaminated Land Remediation and Brownfield Redevelopment in Cities: A Review. *Science of the Total Environment*, 663, 568–579.

Swanenburg, M., Rijsman, V. M. C., Teeuw, J., Mengelers, M. J. B., Noordam, M. Y., & Schwarz-Bovee, A. (2001). *A Framework for Defining Hazard Indicators in the Food Supply Chain*. Agricultural Economics Research Institute, Wageningen, The Netherlands.

Szynkowska, M. I., Pawlaczyk, A., & Mackiewicz, E. (2018). Bioaccumulation and Biomagnification of Trace Elements in the Environment. *Recent Advances in Trace Elements*, 251–276.

Tordoff, G. M., Baker, A. J., Willis, A. J., & Massacci, A. (2000). Heavy Metal Tolerance in Plants: Strategies for Phytoremediation. *Environmental Science and Technology*, 34(18), 4505–4511.

Ucar, Y., Ceylan, Z., Durmus, M., Tomar, O., & Cetinkaya, T. (2021). Application of Cold Plasma Technology in the Food Industry and Its Combination with Other Emerging Technologies. *Trends in Food Science & Technology*, 114, 355–371.

United States Department of Agriculture. (2022). A Citizen's Guide to Phytoremediation. [https://www.da.gov/remedytech/citizens-guide-phytoremediation].

Unnevehr, L. J., & Jensen, H. H. (1999). The Economic Implications of Using HACCP as a Food Safety Regulatory Standard. *Food Policy*, 24, 625–635.

Villanueva, C., Kogevinas, M., Cordier, S., Templeton, M., Vermeulen, R., & Nuckols, J. (2013). Assessing Exposure and Health Consequences of Chemicals in Drinking Water: Current State of Knowledge and Research Needs. *Environmental Health Perspectives*, 122, 213–221.

Wang, L., Zhang, Q., Liao, X., Li, X., Zheng, S., & Zhao, F. (2021). Phytoexclusion of Heavy Metals Using Low Heavy Metal Accumulating Cultivars: A Green Technology. *Journal of Hazardous Materials*, 413, 125427.

Wilcock, A., Pun, M., Khanona, J., & Aung, M. (2004). Consumer Attitudes, Knowledge and Behaviour: A Review of Food Safety Issues. *Trends in Food Science & Technology*, *15*(2), 56–66.

World Health Organization. (2022). Food Safety. [https://www.who.int/news-room/fact-sheets/detail/food-safety].

Yadav, R., Singh, S., Kumar, A., & Singh, A. N. (2022). Phytoremediation: A Wonderful Cost-Effective Tool. In *Cost Effective Technologies for Solid Waste and Wastewater Treatment* (pp. 179–208). Elsevier, Amsterdam, The Netherlands.

Yuan, S., Li, C., Zhang, Y., Yu, H., Xie, Y., Guo, Y., & Yao, W. (2021). Ultrasound as an Emerging Technology for the Elimination of Chemical Contaminants in Food: A review. *Trends in Food Science & Technology*, 109, 374–385.

14 Phytoremediation
Strategies for Protecting the Food Chain from Contamination

Haya Fathima, Harshitha Shettigar, Srivasta Udupa, Sachin Ashok Thorat, Nisha Govender, and Annamalai Muthusamy

14.1 INTRODUCTION

Globally over 20 million hectares of land are considered as soil-polluted sites, more than 50% of which are contaminated by heavy metals and metalloids (He et al. 2017). As per the reports from the Central Ground Water Boards (CGWB) India, heavy metal toxicity has affected more than 718 districts. The emission of heavy metals has increased over the decades reaching 22,000 t (metric tonnes) for cadmium (Cd), 939,000 t for copper (Cu), 1,350,000 t for zinc (Zn), and 738,000 t for lead (Pb) per year globally (Ahmadpour et al. 2012). Contaminated soil and water can significantly disrupt the natural ecosystem and subsequently human health through the food chain (Kumar et al. 2019). The primary causes of soil and water pollution are industrialization, improper disposal of the industrial waste directly into freshwater bodies or land areas, and extensive use of herbicides, pesticides, and fertilizers, which ultimately leads to the accumulation of toxic metals in soil and water bodies (Mishra et al. 2019). Heavy metals like Cd, Cu, mercury (Hg), and Zn are known to pose serious health issues when transferred through the food chain. Thus these accumulated heavy metals are considered a serious threat to humans and other animals (Aransiola et al. 2013; Sonone et al. 2020).

In response to these concerning issues, there is a need for effective methods for the decontamination of the affected location. Although remediation can be accomplished by physical, chemical, or biological techniques (Abioye et al. 2017; Ossai et al. 2020; Song et al. 2022) phytoremediation has emerged as a promising, sustainable, and effective solution for mitigating contamination and food chain protection. Phytoremediation utilizes the natural ability of plants to absorb, accumulate, degrade, or immobilize the contaminants present in soil, water, and air by various biochemical and physiological mechanisms (Ali, Khan, and Sajad 2013; Awa and Hadibarata 2020; Babaniyi et al. 2023). With this background, this chapter aims to explore the various strategies and applications of phytoremediation in protecting the food chain from contamination by hazardous pollutants. Through a comprehensive review of current research and advancements in phytoremediation techniques, this paper aims to investigate the effectiveness, challenges, and prospects of utilizing plant remediation agents (Aransiola et al. 2022). Understanding the various mechanisms underlying phytoremediation can play a significant role in the protection of the food chain from contamination.

14.2 PHYTOREMEDIATION STRATEGIES FOR TREATMENT OF CONTAMINATION

Plants can naturally remove various pollutants (organic or inorganic) from soil and water utilizing various mechanisms depending on environmental condition and type of plant species (Antoniadis et al. 2017; Aransiola et al. 2019). The selection of plant species for phytoremediation plays a significant role in successful phytoremediation (Figure 14.1). Along with plant selection, understanding

DOI: 10.1201/9781032683751-17

the mechanism of plants' responses to different metals, concentrations, and toxicity levels is also crucial for the effective implementation of phytoremediation technology. Furthermore, it is important to understand the mechanism of plant accumulation and uptake of metals. Depending on the contaminant types, forms, and medium, different plants use various techniques to remediate contaminants from soil and water. The mechanism of phytoremediation is described in Figure 14.2. Rhizofiltration, phytodegradation, phytovolatilization, and phytodegradation are some of the methods that can be used to treat contaminated groundwater. Rhizofiltration, phytodegradation, and rhizodegradation can be used to treat contaminants from surfaces and wastewater. The processes of phytoextraction, phytodegradation, phytostabilization, rhizodegradation, or phytovolatilization are used to remediate contaminations in soil, sediments, or sludges (Kafle et al. 2022). Various plants are used to eliminate the pollutants from the surrounding environment using different mechanisms of phytoremediation, which are listed in Table 14.1.

14.2.1 Phytoextraction

Phytoextraction is a commonly employed phytoremediation method to eliminate heavy metals and metalloids from polluted soil matrix through their uptake by a plant (Suman et al. 2018). Plants concentrate the absorbed pollutants and precipitate hazardous metals from contaminated soils into shoots and leaves to eliminate the pollutants from soil. Phytoextraction efficiency is affected by various factors, including soil characteristics, metal availability to plants, metal speciation, and the properties of the relevant plant (Sarwar et al. 2017). Phytoextraction is dependent on the yield of plants and effective contamination transmission from roots to aerial parts (Evangelou, Ebel, and Schaeffer 2007). The type of plants used and the volume of water (as well as any dissolved materials, such as heavy metals) that pass through them in a time determine the effectiveness of this process. Further plants including *Cannabis sativa, Helianthus annuus, Nicotiana tabacum*, and *Zea mays* have been effectively used in this procedure to remove trace elements from the soil

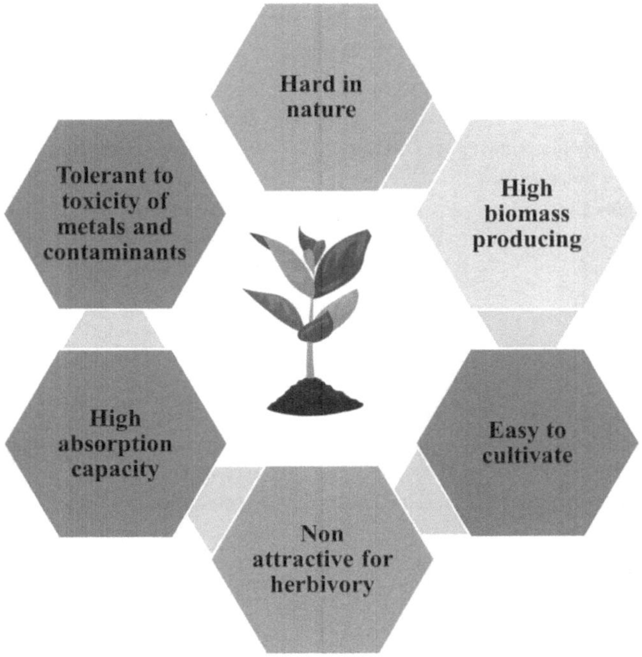

FIGURE 14.1 Features of an ideal plant for phytoremediation.

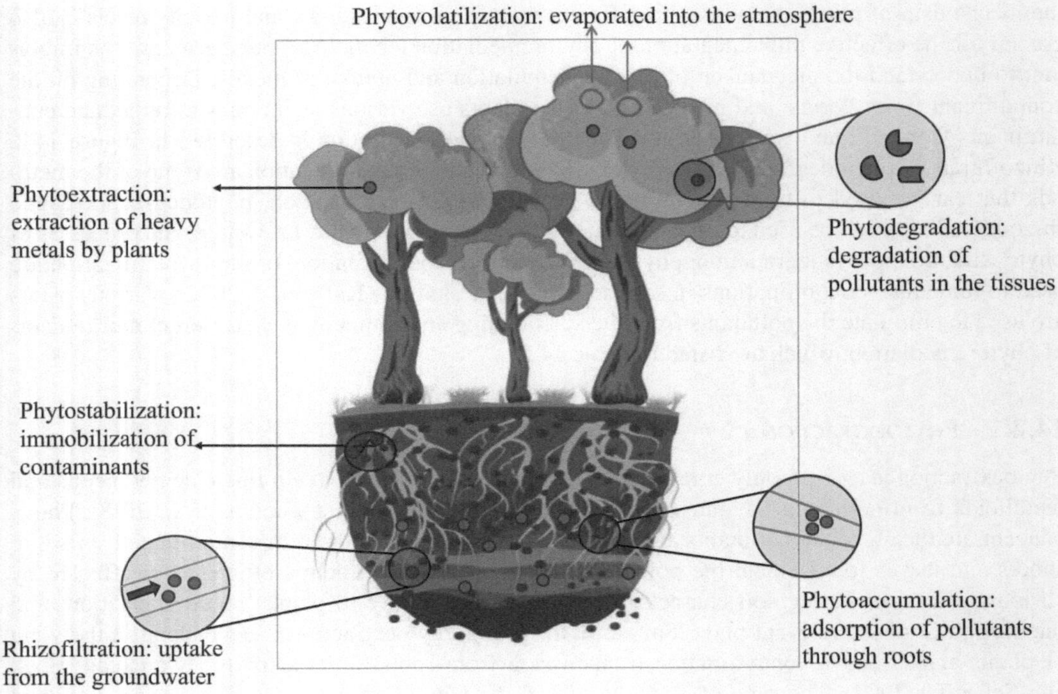

FIGURE 14.2 Various mechanism of phytoremediation.

TABLE 14.1
List of Plants Used for Phytoremediation of Various Pollutants

Scientific Name	Common Name	Contaminants	Remarks	References
Avena sativa	Common oat	Cr (Chromium)	Tolerant to higher Cr accumulation	(Amin et al. 2019)
Azolla caroliniana	Carolina mosquito fern/water velvet	As (Arsenic)	Serves as an As indicator	(Favas, Pratas, and Prasad 2012)
Blumea malcolmii	Blumea	Malachite green	Phytodegradation of industrial waste	(Kagalkar et al. 2011)
Brassica juncea	Mustard	Se (Selenium)	Phytovolatilization of Se	(Banuelos et al. 1997)
Brassica oleracea var. *italica*	Broccoli	Se	Volatilization of methyl selenate favored by the addition of chelating agents	(Raskin, Smith, and Salt 1997)
Callitriche lusitanica	Water starwort	As	A high concentration of As accumulation and acts as indicator	(Favas, Pratas, and Prasad 2012)
Callitriche stagnalis	Pond water-starwort	U (Uranium)	Accumulation of U	(Pratas et al. 2012)
Eichhornia crassipes	Water hyacinth	Fe (Iron), Cr, Cu, Cd, Zn, Ni (Nickel), As	Translocation factor: Fe: 0.69, Cr: 0.56, Cu: 0.79, Cd: 0.78, Zn: 0.67, Ni: 0.55, As: 0.45	(Rai 2019)
Erythrina crista-galli	Cockspur coral tree	Petroleum waste	Petroleum degradation is higher in vegetated contaminated soil	(de Farias et al. 2009)

TABLE 14.1 *(Continued)*
List of Plants Used for Phytoremediation of Various Pollutants

Scientific Name	Common Name	Contaminants	Remarks	References
Eupatorium cannabinum	Holy rope/ hemp-agrimony	As	Citric acid 20 mg/L addition favored phytostabilization	(Gonzalez et al. 2019)
Fontinalis antipyretica	Water moss	U	Accumulation of U	(Pratas et al. 2012)
Helianthus annuus	Sunflower	Cu, Zn, Pb, Hg (Mercury), As, Cd, Ni	Vermicompost as a supplement, usually for low levels of metal contaminants	(Chhotu and Fulekar 2008)
Lactuca sativa	Lettuce	Ni, Co (Cobalt), Fe	Greater survival time	(Hernández et al. 2019)
Lemna minor	Duckweed	U	Accumulation of U	(Pratas et al. 2012)
Myriophyllum aquaticum	Parrot's feather watermilfoil	Co, Cu, Fe, Mn (Manganese), Ni, Pb, Zn	Accumulation causes oxidative stress and degradation of chlorophyll	(Harguinteguy, Schreiber, and Pignata 2013)
Myriophyllum brasiliensis	Parrot's feather	Se	Wetland plants	(Pilon-Smits et al. 1999)
Nicotiana tabacum	Tobacco	Cd	Cd higher in stems and leaves	(Yang et al. 2019)
Pelargonium × *hortorum*	Geranium	Pb	Bacteria tolerant to Pb used	(Manzoor et al. 2019)
Phragmites australis	Perennial reed grass	Ibuprofen (IBP)	IBP degradation	(He et al. 2017)
Polypogon monspeliensis	Rabbitsfoot grass	As	$AsCl(CH_3)_2$ and $As(CH_3)_5$ is volatilized	(Ruppert et al. 2013)
Pontederia crassipes	Water hyacinth	Sodium dodecyl sulfate (SDS)	Degradation of SDS was enhanced by ascorbate peroxidase (APX) activity in the presence of *Chromolaena odorata* L. extract	(Gong et al. 2019)
Solanum nigrum	Black nightshade	Cu, Zn, Cd	10% biochar/attapulgite addition increased uptake of metals	(Li et al. 2019)
Spirodela polyrhiza	Duckweed	ofloxacin (OFX)	93.73%–98.36% reduction in OFX	(Singh, Pandey, and Suthar 2019)
Vigna unguiculata	Cowpea	Pb, Zn	Phytostabilization of mine waste	(Kshirsagar and Aery 2007)
Xanthium strumarium	Common Cocklebur	Cd, Pb, Ni, Zn	Highest bioconcentration factor for Ni, 1.651; Cd, 1.574; and Pb, 1.048	(Khalid et al. 2018)
Zea mays	Corn	Pb, Ti (Titanium)	Chelators induced the phytoextraction of Pb and Ti	(Huang et al. 2019)

(Herzig et al. 2014; Tlustoš et al. 2006; Vangronsveld et al. 2009). Compared to other phytoremediation approaches, phytoextraction is more suited for commercial use. It is cost-effective and does not adversely alter the soil matrix (Sun 2011). Studies have shown that chelates play a significant role in enhancing the uptake of heavy metals from soil and its transport to shoots (Evangelou, Ebel, and Schaeffer 2007). A recent study revealed that ornamental plants have promising phytoextraction capacity for cleaning up surroundings polluted by heavy metals. This process seems to be more environmentally friendly than conventional cleaning methods (Bhat et al. 2022). Numerous studies have demonstrated that the use of synthetic chelators increased the translocation of heavy metals from roots to shoots, hence promoting the phytoextraction of the metals (Hsiao, Kao, and Hseu 2007; Luo et al. 2006; Turgut, Pepe, and Cutright 2005). According to a study (Sun 2011), phytoextraction of Cd-hyperaccumulator from contaminated soil by treatment of mature *Rorippa globosa* with 1.0 g/kg ethylene diamine tetraacetic acid was found to be the promising method. A study by Hernández et al. (2019) was conducted on the removal of galvanoplastic waste (containing nickel [Ni], cobalt [Co], and iron [Fe]) from polluted soil using *Lactuca sativa* L. and *Lolium perenne*. The study revealed that *L. sativa* L. has a strong absorption capacity, and *L. perenne* survived for a longer time. Further study validates *N. tabacum* as one of the best candidates for phytoextraction of contaminants from soil (Cd) along with recovery of nutrients and nicotine from biomass (Yang et al. 2019). Furthermore, studies conducted on chelators (diethylene triamine pentaacetic acid and oxalic acid) induced phytoextraction of Pb and titanium (Ti) from contaminated soil using *Z. mays* as a tolerant plant. The results demonstrate a significant increase in uptake of Pb (86%) and Ti (43%) when compared to that of the control (Huang et al. 2019). Results of a study of phytoremediation ability of *Xanthium strumarium* on Cd, Pb, Ni, Zn metals demonstrated a significant increase in the concentration of metals in the plant, when compared to that of the control (Khalid et al. 2018). Studies also demonstrate that the association of Pb-tolerant bacteria enhances the phytoextraction of Pb by ornamental species like *Pelargonium × hortorum* and *Mesembryanthemum criniflorum* (Manzoor et al. 2019).

14.2.2 Phytoaccumulation

Phytoaccumulation mechanisms employ the absorption of water, other nutrients, and pollutants, usually metals through roots. Instead of being eliminated, the pollution is embedded in the plant shoots and leaves (Susarla 2002). It has been identified that *Avena sativa*, *Abelmoschus esculentus*, and *Cyamopsis tetragonoloba* could be highly potential plants for decontamination of chromium (Cr)-contaminated soils due to their higher tolerance and higher Cr accumulation (Amin et al. 2019). Further, the use of polluted water for irrigation led to buildup of heavy metals in crops and soil, which can cause serious adverse effects in humans (Irshad, Ruqia, and Hussain 2015). Elevated levels of heavy metals have been observed in municipal water in comparison to that of groundwater. Further, it was concluded that the treatment of municipal water led to phytoaccumulation of heavy metals in two leafy cultivars (Waheed et al. 2019). Furthermore, the phytoremediation potential of water mint (*Mentha aquatica*), creeping primrose (*Ludwigia palustris*), and parrot's feather (*Myriophyllum aquaticum*) in the accumulation of heavy metals from contaminated water was studied. All three plants effectively removed Fe, Zn, Cu, and Hg from the contaminated water. Among them, parrot's feather exhibited the highest efficiency in metal accumulation, followed by water mint and creeping primrose (Kamal et al. 2004). When compared to other plant diets, many leafy and root crops can accumulate higher levels of Cd. Cd levels in grain crops tend to accumulate rather high. It was discovered that several aquatic plant groups, including those present in the Barak Valley of South Assam and northeastern India, including the Nelumbonaceae, Araceae, Lemnaceae, Scrophulariaceae, Convolvulaceae, and Hydrocharitaceae, can hyperaccumulate Cu (Muthusaravanan et al. 2018).

14.2.3 Phytostabilization

This remediation can be accomplished by immobilizing or inactivating contaminants or toxicants in the rhizosphere or inside the roots (Kafle et al. 2022). Reduced toxic effects are the result of pollutant mobility and bioavailability being limited by the stabilizing action of plants. The process of phytostabilization can take place through various mechanisms like precipitation, metal valence reduction and complexation (Etim 2012). Plants that are used for phytostabilization should have reasonable rate of translocation of metals to the shoots and high rates of bioconcentration. Furthermore, they must possess a strong resistance to soil pollution and generate a substantial amount of root biomass (Li et al. 2019). Further, the association of microorganisms like bacteria and mycorrhiza facilitates penetration and increases the surface area covered in the soil. In addition, these associations play a beneficial role in protection of plants from the translocation of contaminants/trace elements from roots to shoots (Göhre and Paszkowski 2006).

This is useful for the treatment of Pb, arsenic (As), Cr, Cu, Zn, and Cd. Stabilizing these wastes can be accomplished economically and sustainably with phytomanagement schemes like phytostabilization. Completing the promising plant list requires field experiments (Zine et al. 2020). Many plants can withstand exposure to lead and successfully phytostabilize the toxicant. Several plant species are suitable for Pb phytostabilization since it has been demonstrated that they can hold large amounts of Pb in their roots without moving it to shoots. Limiting Pb movement using plant roots safeguards surrounding ecosystems. This technique can efficiently replace contaminated soil layers with new soil, making it a time- and cost-efficient approach for a variety of locations (Egendorf et al. 2020). Another major global issue is the copper contamination in soil, which is reducing agricultural production and having negative health impacts on people through the food chain. Further, using *Festuca rubra* L. (Radziemska, Vaverková, and Baryla 2017), demonstrated a glass house pot experiment to examine the potential application of halloysite as an immobilizing ingredient in the assisted phytostabilization of Cu contaminated soil. In general, halloysite raises the pH of the soil and promotes the precipitation of metal carbonate and oxides, which reduce the solubility of metals. Hence this study provided evidence that applying halloysite and *F. rubra* together could be an effective way to lower environmental risk (Radziemska, Vaverková, and Baryla 2017).

The study conducted by (Kshirsagar and Aery 2007) demonstrated the phytostabilization potential of *Vigna unguiculata* (L.) Walp. (cowpea) from mine waste under greenhouse conditions. Further, heavy metals at lower concentrations (5, 10, and 20 ppm) were found to stimulate the growth of *Helianthus annuus* (sunflower) in combination with vermicompost soil. The findings reveal that sunflower could be an effective crop in remediating heavy metals from contaminated soil. Vermicompost could be used to remediate contaminants and along with providing nutrients for plant growth (Chhotu and Fulekar 2008). Furthermore, *Solanum nigrum* L. was able to remediate multiple metals from contaminated soil under greenhouse conditions. The results showed that the addition of 10% attapulgite (MA_2) and 10% biochar (MB_2) increased the absorption of metals (Cu, Zn, Cd, Hg, Pb, and manganese [Mn]) from contaminated soil (Li et al. 2019). Additionally, *Eupatorium cannabinum* was able to accumulate a high amount of As in the roots from contaminated soil; thus it is considered a promising species for the remediation of contaminated soil. Further, there was an increase in the biomass with the addition of citric acid 20 mg L^{-1} (Gonzalez et al. 2019).

14.2.4 Phytovolatilization

Phytovolatilization is a process in which plants take up pollutants from the contaminated sites and convert them to volatile forms, which are then released into the atmosphere. The pollutants absorbed by the during phytovolatilization are subsequently transpired from the leaves along with water vapor as a gaseous form of the contaminant or volatile degradation product (Etim 2012). Volatile organic chemicals can be released into the air through either direct phytovolatilization from stems or leaves or indirect phytovolatilization from plant root activities through the soil. A higher likelihood of phytovolatilization for compounds with poor octanol-air partitioning coefficients (log KOA < 5) is observed (Limmer and Burken 2016). A study reported the phytovolatilization of 2,4-dibromoanisole (2,4-DBA) and 2,4-dibromophenol (2,4-DBP) by conducting experiments on rice plants using hydroponic exposure and cultivation solution (Zhang et al. 2020). In addition to being a potent neurotoxin, Hg is another toxicant whose dangerous characteristics endanger crop productivity. Mercury is mobile, which makes it more volatile in soil and causes it to settle on sediments (Gworek, Dmuchowski, and Baczewska-Dąbrowska 2020). It therefore seeps into the groundwater table. Contaminated sites have been cleaned using various techniques, such as phytovolatilization and the employment of microorganisms. It will be necessary to grow plant species on contaminated sites to complete this process. *Brassica juncea, Brassica oleracea* var. *italica*, and *Myriophyllum brasiliensis* have been found to have potential in phytovolatization of selenium (Se) from contaminated soil (Banuelos et al. 1997; Pilon-Smits et al. 1999; Raskin, Smith, and Salt 1997). Furthermore, *Polypogon monspeliensis* volatilized dimethylchloroarsine [$AsCl(CH_3)_2$] and pentamethylarsine [$As(CH_3)_5$] from Arsenic polluted sites (Ruppert et al. 2013).

14.2.5 Phytodegradation and Rhizodegradation

Plants can break down and eliminate pollutants from their tissues by a process called phytodegradation. Plant constituents may significantly affect phytodegradation either directly or indirectly. Under the direct process, pollutants are taken up by the plant, which then metabolizes them or breaks them down into smaller, less toxic chemicals to be distributed throughout the plant tissue. Certain organic contaminants, including herbicides, explosives, and chlorinated solvents, are remedied by phytodegradation. It can also deal with chemicals found in soil, sediment, or groundwater. Additionally, a wide range of organic chemicals, including petroleum hydrocarbons, polycyclic aromatic hydrocarbons (PAHs), benzene, toluene, ethylbenzene, chlorinated solvents, xylenes, and pesticides were treated effectively. This process is also referred to as plant-assisted bioremediation, which encourages fungal and bacterial breakdown by releasing enzymes and exudates into the rhizosphere (Etim 2012). Ibuprofen's metabolism-based detoxification and certain intermediate production in Australia were disclosed by a phytodegradation investigation (He et al. 2017). Phytodegradation of malachite green in industrial waste is observed in *Blumea malcolmii* (Kagalkar et al. 2011). Duckweed (*Spirodela polyrhiza*) reduces ofloxacin (93.73%–98.36%) from contaminated soil (Singh, Pandey, and Suthar 2019). Further, water hyacinth (*Pontederia crassipes*) degrades sodium dodecyl sulfate in the presence of *Chromolaena odorata* L. extract by enhancing the activity of ascorbate peroxidase (Gong et al. 2019). *Phragmites australis* demonstrated its potential use in the remediation of ibuprofen from wetlands (He et al. 2017). The difference in the anatomic structure of the roots of *Erythrina crista-galli* helps it to degrade petroleum. The study was performed in greenhouse conditions, and it revealed that the survival rate and degradation of petroleum waste were higher in vegetated contaminated soil (de Farias et al. 2009).

14.2.6 Rhizofiltration

Rhizofiltration is a widely employed mechanism in phytoremediation that cleans up contaminated groundwater resources. Plants precipitate contaminants within roots by absorbing them from the surroundings (rhizosphere). Plants used for this process often have long root hairs and a high potential for the accumulation of heavy metals (Dhanwal et al. 2017). For rhizofiltration, plants are usually grown in freshwater for the development of extensive root systems; later the clean water is replaced by contaminated water to acclimatize the plant. Then once the plants are successfully acclimatized, they are transferred to a contaminated site (Wuana and Okieimen 2011). Moreover, rhizofiltration has been implemented in a variety of technical ways, for example by using mats floating on the surface of water and keeping the plant roots submerged (*Helianthus* sp.), as with aquatic plants like *P. australis* (Cav.) Trin. ex Steud, *Eichhornia crassipes* (Mart.) Raf, *Typha laifolia* (L.), and *Lemna minor* (L.) (Lee and Yang 2010). Plant roots release exudates, or secondary metabolites, which cause acidification of the rhizosphere. This helps in the effective absorption of trace elements or heavy metals by the roots. Further biogeochemical processes lead to the precipitation of pollutants onto the roots or into the water body. After that, adsorption depends on the kind of plant, contaminant species, and concentration; there may be an accumulation of contaminants on the roots, within the roots, or translocation to the phyllosphere (plant organs above ground) (Kristanti et al. 2021). Rhizofiltration tests were carried out with *Raphanus sativus* L., *L. sativa*, *Brassica campestris* L. and *Oenanthe javanica* contaminated with uranium (U). The tests conducted, using an electron-dispersive x-ray spectrophotometer and scanning electron microscopy, reveal that U is adsorbed on the root surface of polluted waters. These findings show that *R. sativus* L. exhibits a remarkable accumulation of U in addition to a strong tolerance to high U concentrations and low pH circumstances (Han et al. 2020). Another study stated that *Pistia stratiotes* are an excellent candidate for rhizofiltration of heavy metals, especially Ni, from polluted waters. The plant can accumulate several metals. The plant's increased uptake of Ni serves as a defense against infections and grazing. *P. stratiotes* collects metals at a concentration that is more than 300 times higher than that of typical plants. *P. stratiotes* is, therefore, a prime contender for rhizofiltration of contaminated water to remove heavy metals (Abubakar et al. 2014). *Azolla caroliniana* and *Callitriche lusitanica* act on bioconcentration factors of 0.000397 and 0.002346, respectively (Favas, Pratas, and Prasad 2012). Rhizofiltration is seen in *Fontinalis antipyretic, Callitriche stagnalis*, and *Lemna minor*, in U-contaminated soil. The results revealed that the plants were able to accumulate U at high concentrations (Pratas et al. 2012). The natural wetland plant *E. crassipes* demonstrated remediation of Fe, Cr, Cu, Cd, Zn, Ni, and As from wastewater (Rai 2019).

14.3 STRATEGIES FOR PROTECTION OF FOOD CHAINS FROM CONTAMINATION

Globally, climate change is turning multilevel severe impacts on the agriculture, livestock, and food systems (Change 2014). In coming years, changes in climatic conditions will increase, which will spread several foodborne diseases through bacteria, viruses, pathogens, and several other biotic factors. In addition, using chemical fertilizers after the green revolution for agricultural practices will add hazardous chemicals into the food chain that further increase contamination in food supply (Change 2020). Therefore, protecting the food chain from contamination is important to avoid their severe impacts globally. To address this, there must be a safety standard implemented in food chain management through various safety standard programs. To implement this, it is important to adopt the three Rs (risk management, risk assessment, and risk communication); this can be helpful in mitigating food chain contamination in larger perspective (WHO and Food and Agriculture Organization of the United Nations 2006). Moreover, these strategies can lead significantly to implications of safety standards in all food and agricultural industries if maintained and assessed regularly. However, safety guidelines and follow-up are very important for mitigating contamination in the food chain.

14.4 IMPROVING FOOD SAFETY THROUGH RISK ASSESSMENT

Analysis of risk is a structured approach that ensures risk assessment and safety management decisions have been granted based on the scientific evidence such as health risk assessment (WHO and Food and Agriculture Organization of the United Nations 2006). Risk assessment as a tool can be implemented in four ways: hazard characterization, hazard identification, risk characterization, and exposure assessment in food products (Codex Alimentarius Commission 1999). Further, the microbial and chemical risk assessment strategies are important to implicate for accessing contaminants from microbes, food additives, flavoring agents, and chemical residues. To address this, several US Food and Drug Administration guidelines, tools, software, and quality assessment council have been formed in all the food processing industries, to ensure the proper follow-up of guidelines in the in-house International Standards Organization Committee has been established (Feliciano et al. 2022).

Food safety has been vulnerable to the several factors of changing climatic conditions. Recently, several analyses figured out that the cause of the vulnerability is due to the increase of average temperatures globally, which has created a fruitful environment for the proliferation of airborne and vector-borne diseases (Tirado et al. 2010). To safeguard the food chain from changing climatic conditions, there should be an early detection warning system through weather reports and environmental data analysis systems (Feliciano et al. 2022).

14.5 IMPROVE FOOD CHAIN VALUE THROUGH THE EVALUATION OF ENVIRONMENTAL CHANGES

The direct and indirect effects of environmental changes on natural resources include keeping an assessment of consumption and productivity to ensure the reduction of waste and food loss. Moreover, the greenhouse gas emissions from the animal-based food chain are comparatively higher than the vegetable-based food chain; this concludes that livestock management has higher risk to the environment and climate change (Willett et al. 2019). However, the field grown vegetables and their pre- and post-processing result in an increase accumulation of greenhouse gases in the environment; therefore, to mitigate this there should be the development of certain tools and software to manage sustainable production to reduce the climatic impact (Ridoutt, Hendrie, and Noakes 2017). The life cycle assessment of vegetable or food products can help mitigate these factors; it includes the four steps of inventory analysis, scope and goal, interpretation, and impact on life cycle. These assessments ensure troubleshooting impacts such as eutrophication, global warming, particulate matter, human toxicity, and ionizing radiation (ISO 14040 2006). Through several risk assessments indicators, the midpoint and endpoint analysis for sustainable emissions of these toxic contaminants in food chain can lower the damaging effects. However, calculating these damaging factors through comparison-based studies can significantly improve the impact of food chains on Earth.

14.6 CONCLUSION

In summary, the soil and water pollution induced by heavy metals and metalloids can be mitigated by the application of suitable phytoremediation methods, which is a viable, long-term, and efficient approach. Techniques like phytoextraction, phytoaccumulation, phytostabilization, phytovolatilization, phytodegradation, and rhizofiltration can be used for remediation of contaminated sites depending on the nature of contaminants and environmental conditions. It is essential to comprehend the mechanisms underlying plant reactions to various metal concentrations and toxicity levels to successfully apply phytoremediation technologies. It has a diverse range of applications in the remediation of heavy metals, petroleum waste, pesticides, mine waste, and other organic pollutants. Supplementation of other constituents can enhance the biomass and uptake of pollutants during the process of phytoremediation. In addition, the microbial association also plays a crucial role in the uptake of pollutants from the contaminated sites by increasing the surface area for adsorption. Proper selection of plants for phytoremediation can overcome some of the limitations like seasonal variations, plant growth, and developmental time. Overall, phytoremediation can be considered a sustainable, economical, and effective technique for remediation of soil, water, and air contamination.

14.7 FUTURE PROSPECTS

Further exploration into the combined effects of various phytoremediation methods is essential to enhance overall remediation efficacy. By combining techniques, the efficiency of remediation processes can be significantly increased. Additionally, the implementation of plant genetic engineering to strengthen stress tolerance and metal-accumulating capabilities holds promise for developing hyperaccumulator plants depending on specific pollutants and environmental conditions. A comprehensive understanding and utilization of plant-microorganism interactions, including beneficial bacteria and mycorrhizal association, are essential to maximize the effectiveness of phytoremediation. To validate the practical efficacy of phytoremediation, a transition from laboratory experiments to extensive field testing is necessary, accounting for diverse pollutant types and environmental factors. Moreover, assessing the economic feasibility of phytoremediation compared to conventional techniques is crucial for its widespread adoption, necessitating exploration into effective strategies and potential revenue sources such as biomass utilization.

14.8 ACKNOWLEDGMENTS

We thank Manipal Academy of Higher Education (MAHE), Manipal, Karnataka, India, TIFAC-CORE and FIST, DST New Delhi, DBT BUILDER grant (BT/INF/22/SP43065/2021), DBT, New Delhi, and K-FIST, VGST, Govt. of Karnataka for the facilities. The authors (S. U. and S.A.T.) are thankful to MAHE for the T.M.A. Pai PhD scholarship, and S.A.T. is thankful to DST-SERB (CRB/2022/003498 dated 20.06.2023) for JRF. We are grateful to Dr. B. S. Satish Rao, Professor and Director, Manipal School of Life Sciences, MAHE for encouragement and support.

REFERENCES

Abioye, O. P., U. J. J. Ijah, and S. A. Aransiola. 2017. "Phytoremediation of soil contaminants by biodiesel plant jatropha curcas." Chapter 4 of K. Bauddh, et al. (eds.), *Phytoremediation Potential of Bioenergy Plants*. Springer Nature Singapore Pte Ltd. https://link.springer.com/book/10.1007%2F978-981-10-3084-0.

Abubakar, M. M., M. M. Ahmad, and B. U. Getso. 2014. "Rhizofiltration of heavy metals from eutrophic water using Pistia stratiotes in a controlled environment." *IOSR J Environ Sci Toxicol Food Technol* 8 (6):3.

Ahmadpour, P., F. Mahmud, T. M. M., Abdu, A., Soleimani, M, and Tayefeh, F. H. 2012. "Phytoremediation of heavy metals: A green technology." *Afr J Biotechnol* 11 (76):14036–14043.

Ali, H., E. Khan, and M. A. Sajad. 2013. "Phytoremediation of heavy metals – concepts and applications." *Chemosphere* 91 (7):869–881.

Amin, H., B. Ahmed Arain, M. S. Abbasi, F. Amin, T. M. Jahangir, and N. U. Soomro. 2019. "Evaluation of chromium phyto-toxicity, phyto-tolerance, and phyto-accumulation using biofuel plants for effective phytoremediation." *Int J Phytorem* 21 (4):352–363.

Antoniadis, V., E. Levizou, S. M. Shaheen, Y. S. Ok, A. Sebastian, C. Baum, M. N. V. Prasad, W. W. Wenzel, and J. Rinklebe. 2017. "Trace elements in the soil-plant interface: Phytoavailability, translocation, and phytoremediation–A review." *Earth-Sci Rev* 171:621–645.

Aransiola, S. A., U. J. J. Ijah, and O. P. Abioye. 2013. "Phytoremediation of lead polluted soil by glycine max L." *Appl Environ Soil Sci* 2013. Article ID 631619. doi: 10.1155/2013/631619. https://www.hindawi.com/journals/aess/2013/631619/abs/.

Aransiola, S. A., U. J. J. Ijah, O. P. Abioye, and J. D. Bala. 2019. "Microbial-aided phytoremediation of heavy metals contaminated soil: A review." *Eur J Biol Res* 9 (2):104–125. http://www.journals.tmkarpinski.com/index.php/ejbr/article/view/157.

Aransiola, S. A, U. J. J. Ijah, O. P. Abioye, and J. D. Bala. 2022. "Vermicompost-assisted phytoremediation of toxic trace element-contaminated soil in Madaka, Nigeria using Melissa officinalis L and Sida acuta." *Int J Environ Sci Technol*. http://dx.doi.org/10.1007/s13762-022-04105-y.

Awa, S. H., and T. Hadibarata. 2020. "Removal of heavy metals in contaminated soil by phytoremediation mechanism: A review." *Wat, Air, Soil Poll* 231 (2).

Babaniyi, B. R., O. D. Ogundele, A. Bisi-Omotosho, E. E. Babaniyi, and S. A. Aransiola. 2023. "Remediation approaches in environmental sustainability." In Naga Raju Maddela, Lizziane Kretli Winkelstroter Eller, and Ram Prasad (eds.), *Microbiology for Cleaner Production and Environmental Sustainability*. CRC Press. ISBN: 9781032496061. https://www.routledge.com/Microbiology-for-Cleaner-Production-and-Environmental-Sustainability/Maddela-Eller-Prasad/p/book/9781032496061.

Banuelos, G. S., H. A. Ajwa, N. Terry, and A. Zayed. 1997. "Phytoremediation of selenium laden soils: A new technology." *J Soil and Wat Conserv* 52 (6):426–430.

Bhat, S. A., O. Bashir, S. A. Ul Haq, T. Amin, A. Rafiq, M. Ali, J. H. P. Americo-Pinheiro, and F. Sher. 2022. "Phytoremediation of heavy metals in soil and water: An eco-friendly, sustainable and multidisciplinary approach." *Chemosphere* 303 (Pt 1):134788.

Change, F. C. 2014. *Synthesis Report of the IPCC Fifth Assessment Report*. (AR5)/Eds. R. K. Pachauri, and L. A. Meyer. IPCC: Geneva, Switzerland. 151 p.

Change, F. C. 2020. *Unpacking the Burden on Food Safety*. FAO – Food and Agriculture Organization of the United Nations: Rome, Italy.

Chhotu, D. J., and M. H. Fulekar. 2008. "Phytoremediation: The application of vermicompost to remove zinc, cadmium, copper, nickel and lead by sunflower plant." *Environ Eng Manag J* 7 (5): 547–558.

Codex Alimentarius Commission. 1999. "Principles and guidelines for the conduct of microbiological risk assessment." https://www.fao.org/fao-who-codexalimentar ius/sh-proxy/en/?lnk=1&url=https%253A%252F%252Fworkspace.fao.org%252Fs ites%252Fcodex%252FStandards%252FCXG%2B30–1999%252FCXG_030e_2014. pdf (Accessed 7 February 2024).

de Farias, V., L. T. Maranho, E. C. de Vasconcelos, M. A. da Silva Carvalho Filho, L. G. Lacerda, J. A. Azevedo, A. Pandey, and C. R. Soccol. 2009. "Phytodegradation potential of Erythrina crista-galli L., Fabaceae, in petroleum-contaminated soil." *Appl Biochem Biotechnol* 157 (1):10–22.

Dhanwal, P., A. Kumar, S. Dudeja, V. Chhokar, and V. Beniwal. 2017. "Recent advances in phytoremediation technology." *Adv Environ Biotechnol*:227–241.

Egendorf, S. P., P. Groffman, G. Moore, and Z. Cheng. 2020. "The limits of lead (Pb) phytoextraction and possibilities of phytostabilization in contaminated soil: A critical review." *Int J Phytorem* 22 (9):916–930.

Etim, E. E. 2012. "Phytoremediation and its mechanisms: A review." *Int J Environ Bioener* 2 (3):120–126.

Evangelou, M. W., M. Ebel, and A. Schaeffer. 2007. "Chelate assisted phytoextraction of heavy metals from soil. Effect, mechanism, toxicity, and fate of chelating agents." *Chemosphere* 68 (6):989–1003.

Favas, P. J., J. Pratas, and M. N. Prasad. 2012. "Accumulation of arsenic by aquatic plants in large-scale field conditions: Opportunities for phytoremediation and bioindication." *Sci Total Environ* 433:390–397.

Feliciano, R. J., P. Guzmán-Luna, G. Boué, M. Mauricio-Iglesias, A. Hospido, and J. M. Membré. 2022. "Strategies to mitigate food safety risk while minimizing environmental impacts in the era of climate change." *Trends Food Sci Technol* 126:180–191.

Göhre, V., and U. Paszkowski. 2006. "Contribution of the arbuscular mycorrhizal symbiosis to heavy metal phytoremediation." *Planta* 223 (6):1115–1122.

Gong, Y., J. Chen, and R. Pu. 2019. "The enhanced removal and phytodegradation of sodium dodecyl sulfate (SDS) in wastewater using controllable water hyacinth." *Int J Phytorem* 21 (11):1080–1089.

Gonzalez, H., D. Fernandez-Fuego, A. Bertrand, and A. Gonzalez. 2019. "Effect of pH and citric acid on the growth, arsenic accumulation, and phytochelatin synthesis in Eupatorium cannabinum L., a promising plant for phytostabilization." *Environ Sci Pollut Res Int* 26 (25):26242–26253.

Gworek, B., W. Dmuchowski, and A. H. Baczewska-Dąbrowska. 2020. "Mercury in the terrestrial environment: A review." *Environ Sci Eur* 32 (1).

Han, Y., J. Lee, C. Kim, J. Park, M. Lee, and M. Yang. 2020. "Uranium rhizofiltration by Lactuca sativa, brassica campestris L., raphanus sativus L., Oenanthe javanica under different hydroponic conditions." *Minerals* 11 (1).

Harguinteguy, C. A., R. Schreiber, and M. L. Pignata. 2013. "Myriophyllum aquaticum as a biomonitor of water heavy metal input related to agricultural activities in the Xanaes River (Córdoba, Argentina)." *Ecol Indic* 27:8–16.

He, Y., A. A. M. Langenhoff, N. B. Sutton, H. H. M. Rijnaarts, M. H. Blokland, F. Chen, C. Huber, and P. Schroder. 2017. "Metabolism of ibuprofen by phragmites australis: Uptake and phytodegradation." *Environ Sci Technol* 51 (8):4576–4584.

Hernández, A., N. Loera, M. Contreras, L. Fischer, and D. Sánchez. 2019. "Comparison between Lactuca sativa L. and Lolium perenne: Phytoextraction capacity of Ni, Fe, and Co from galvanoplastic industry." *Ener Technol 2019*:137–147.

Herzig, R., E. Nehnevajova, C. Pfistner, J. P. Schwitzguebel, A. Ricci, and C. Keller. 2014. "Feasibility of labile Zn phytoextraction using enhanced tobacco and sunflower: Results of five- and one-year field-scale experiments in Switzerland." *Int J Phytorem* 16 (7–12):735–754.

Hsiao, K. H., P. H. Kao, and Z. Y. Hseu. 2007. "Effects of chelators on chromium and Nickel Uptake by Brassica Juncea on serpentine-mine tailings for phytoextraction." *J Hazard Mater* 148 (1–2):366–376.

Huang, X., D. Luo, X. Chen, L. Wei, Y. Liu, Q. Wu, T. Xiao, X. Mai, G. Liu, and L. Liu. 2019. "Insights into heavy metals leakage in chelator-induced phytoextraction of Pb- and Tl-contaminated soil." *Int J Environ Res Public Health* 16 (8).

Irshad, M., B. Ruqia, and Z. Hussain. 2015. "Phytoaccumulation of heavy metals in natural vegetation at the municipal wastewater site in Abbottabad, Pakistan." *Int J Phytorem* 17 (12):1269–1273.

ISO 14040. 2006. "Environmental management – life cycle assessment – principles and framework." ISO 14040. https://www.iso.org/obp/ui#iso:std: Iso:14040:ed-2:v1:en (Accessed 7 February 2024).

Kafle, A., A. Timilsina, A. Gautam, K. Adhikari, A. Bhattarai, and N. Aryal. 2022. "Phytoremediation: Mechanisms, plant selection and enhancement by natural and synthetic agents." *Environ Adv* 8.

Kagalkar, A. N., M. U. Jadhav, V. A. Bapat, and S. P. Govindwar. 2011. "Phytodegradation of the triphenylmethane dye Malachite Green mediated by cell suspension cultures of Blumea malcolmii Hook." *Bioresour Technol* 102 (22):10312–10318.

Kamal, M., A. E. Ghaly, N. Mahmoud, and R. Cote. 2004. "Phytoaccumulation of heavy metals by aquatic plants." *Environ Int* 29 (8):1029–1039.

Khalid, N., A. Noman, M. Aqeel, A. Masood, and A. Tufail. 2018. "Phytoremediation potential of Xanthium strumarium for heavy metals contaminated soils at roadsides." *Int J Environ Sci Technol* 16 (4):2091–2100.

Kristanti, R. Ayu, W. J. Ngu, A. Yuniarto, and T. Hadibarata. 2021. "Rhizofiltration for removal of inorganic and organic pollutants in groundwater: A review." *Biointerface Res Appl Chem* 11 (4):12326–12347.

Kshirsagar, S., and N. C. Aery. 2007. "Phytostabilization of mine waste: Growth and physiological responses of Vigna unguiculata (L.) Walp." *J Environ Biol* 28 (3):651–654.

Kumar, S., S. Prasad, K. K. Yadav, M. Shrivastava, N. Gupta, S. Nagar, Q. V. Bach, H. Kamyab, S. A. Khan, S. Yadav, and L. C. Malav. 2019. "Hazardous heavy metals contamination of vegetables and food chain: Role of sustainable remediation approaches – A review." *Environ Res* 179 (Pt A):108792.

Lee, M., and M. Yang. 2010. "Rhizofiltration using sunflower (Helianthus annuus L.) and bean (Phaseolus vulgaris L. var. vulgaris) to remediate uranium contaminated groundwater." *J Hazard Mater* 173 (1–3):589–596.

Li, X., X. Zhang, X. Wang, and Z. Cui. 2019. "Phytoremediation of multi-metal contaminated mine tailings with Solanum nigrum L. and biochar/attapulgite amendments." *Ecotoxicol Environ Saf* 180:517–525. doi: 10.1016/j.ecoenv.2019.05.033.

Limmer, M., and J. Burken. 2016. "Phytovolatilization of organic contaminants." *Environ Sci Technol* 50 (13):6632–6643.

Luo, C., Z. Shen, L. Lou, and X. Li. 2006. "EDDS and EDTA-enhanced phytoextraction of metals from artificially contaminated soil and residual effects of chelant compounds." *Environ Pollut* 144 (3):862–871.

Manzoor, M., I. Gul, I. Ahmed, M. Zeeshan, I. Hashmi, B. A. Z. Amin, J. Kallerhoff, and M. Arshad. 2019. "Metal tolerant bacteria enhanced phytoextraction of lead by two accumulator ornamental species." *Chemosphere* 227:561–569.

Mishra, S., Bharagava, R. Nandkishor More, Ashutosh Yadav, Surabhi Zainith, Sujata Mani, and Pankaj Chowdhary. 2019. "Heavy metal contamination: An alarming threat to environment and human health." *Environ Biotechnol Sustain Future*:103–125.

Muthusaravanan, S., N. Sivarajasekar, J. S. Vivek, T. Paramasivan, M. Naushad, J. Prakashmaran, V. Gayathri, and Omar K. Al-Duaij. 2018. "Phytoremediation of heavy metals: Mechanisms, methods and enhancements." *Environ Chem Lett* 16 (4):1339–1359.

Ossai, I. C., A. Ahmed, A. Hassan, and F. S. Hamid. 2020. "Remediation of soil and water contaminated with petroleum hydrocarbon: A review." *Environ Technol Innov* 17. doi: 10.1016/j.eti.2019.100526.

Pilon-Smits, E. A. H., M. P. de Souza, G. Hong, A. Amini, R. C. Bravo, S. T. Payabyab, and N. Terry. 1999. "Selenium volatilization and accumulation by twenty aquatic plant species." *J Environ Qual* 28 (3):1011–1018.

Pratas, J., P. J. Favas, C. Paulo, N. Rodrigues, and M. N. Prasad. 2012. "Uranium accumulation by aquatic plants from uranium-contaminated water in Central Portugal." *Int J Phytorem* 14 (3):221–234.

Radziemska, M., M. D. Vaverková, and A. Baryla. 2017. "Phytostabilization-management strategy for stabilizing trace elements in contaminated soils." *Int J Environ Res Public Health* 14 (9).

Rai, P. K. 2019. "Heavy metals/metalloids remediation from wastewater using free floating macrophytes of a natural wetland." *Environ Technol Innov* 15.

Raskin, I. I., R. D. Smith, and D. E. Salt. 1997. "Phytoremediation of metals: Using plants to remove pollutants from the environment." *Curr Opin Biotechnol* 8 (2):221–226.

Ridoutt, B. G., G. A. Hendrie, and M. Noakes. 2017. "Dietary strategies to reduce environmental impact: A critical review of the evidence base." *Adv Nutr* 8 (6):933–946. doi: 10.3945/an.117.016691.

Ruppert, L., Z. Q. Lin, R. P. Dixon, and K. A. Johnson. 2013. "Assessment of solid phase microfiber extraction fibers for the monitoring of volatile organoarsinicals emitted from a plant-soil system." *J Hazard Mater* 262:1230–1236.

Sarwar, N., M. Imran, M. R. Shaheen, W. Ishaque, M. A. Kamran, A. Matloob, A. Rehim, and S. Hussain. 2017. "Phytoremediation strategies for soils contaminated with heavy metals: Modifications and future perspectives." *Chemosphere* 171:710–721.

Singh, V., B. Pandey, and S. Suthar. 2019. "Phytotoxicity and degradation of antibiotic ofloxacin in duckweed (Spirodela polyrhiza) system." *Ecotoxicol Environ Saf* 179:88–95.

Song, P., D. Xu, J. Yue, Y. Ma, S. Dong, and J. Feng. 2022. "Recent advances in soil remediation technology for heavy metal contaminated sites: A critical review." *Sci Total Environ* 838 (Pt 3):156417.

Sonone, S. S., S. Jadhav, M. S. Sankhla, and R. Kumar. 2020. "Water contamination by heavy metals and their toxic effect on aquaculture and human health through food chain." *Lett Appl NanoBioSci* 10 (2):2148–2166.

Suman, J., O. Uhlik, J. Viktorova, and T. Macek. 2018. "Phytoextraction of heavy metals: A promising tool for clean-up of polluted environment?" *Front Plant Sci* 9:1476.

Sun, Y. 2011. "The role of EDTA on cadmium phytoextraction in a cadmium-hyperaccumulator Rorippa globosa." *J Environ Chem Ecotoxicol* 3 (3):45–51.

Susarla, S. 2002. "Phytoremediation: An ecological solution to organic chemical contamination." *Ecol Eng* 18 (5):647–658.

Tirado, M. C., R. Clarke, L. A. Jaykus, A. McQuatters-Gollop, and J. M. Frank. 2010. "Climate change and food safety: A review." *Food Res Int* 43 (7):1745–1765.

Tlustoš, P., J. Száková, J. Hrubý, I. Hartman, J. Najmanová, J. Nedělník, D. Pavlíková, and M. Batysta. 2006. "Removal of As, Cd, Pb, and Zn from contaminated soil by high biomass producing plants." *Plant, Soil and Environ* 52 (9):413–423.

Turgut, C., M. K. Pepe, and T. J. Cutright. 2005. "The effect of EDTA on Helianthus annuus uptake, selectivity, and translocation of heavy metals when grown in Ohio, New Mexico and Colombia soils." *Chemosphere* 58 (8):1087–1095.

Vangronsveld, J., R. Herzig, N. Weyens, J. Boulet, K. Adriaensen, A. Ruttens, T. Thewys, A. Vassilev, E. Meers, E. Nehnevajova, D. van der Lelie, and M. Mench. 2009. "Phytoremediation of contaminated soils and groundwater: Lessons from the field." *Environ Sci Pollut Res Int* 16 (7):765–794.

Waheed, H., N. Ilyas, N. Iqbal Raja, T. Mahmood, and Z. Ali. 2019. "Heavy metal phyto-accumulation in leafy vegetables irrigated with municipal wastewater and human health risk repercussions." *Int J Phytorem* 21 (2):170–179.

Willett, W., J. Rockström, B. Loken, M. Springmann, T. Lang, S. Vermeulen, T. Garnett, D. Tilman, F. DeClerck, A. Wood, M. Jonell, M. Clark, L. J. Gordon, J. Fanzo, C. Hawkes, R. Zurayk, J. A. Rivera, W. De Vries, L. Majele Sibanda, A. Afshin, A. Chaudhary, M. Herrero, R. Agustina, F. Branca, A. Lartey, S. Fan, B. Crona, E. Fox, V. Bignet, M. Troell, T. Lindahl, S. Singh, S. E. Cornell, K. Srinath Reddy, S. Narain, S. Nishtar, and C. J. L. Murray. 2019. "Food in the anthropocene: The EAT-lancet commission on healthy diets from sustainable food systems." *Lancet* 393 (10170):447–492.

Wuana, R. A., and F. E. Okieimen. 2011. "Heavy metals in contaminated soils: A review of sources, chemistry, risks and best available strategies for remediation." *ISRN Ecol* 2011:1–20.

Yang, Y., Y. Ge, P. Tu, H. Zeng, X. Zhou, D. Zou, K. Wang, and Q. Zeng. 2019. "Phytoextraction of Cd from a contaminated soil by tobacco and safe use of its metal-enriched biomass." *J Hazard Mater* 363:385–393.

Zhang, Q., W. Kong, L. Wei, Y. Wang, Y. Luo, P. Wang, J. Liu, J. L. Schnoor, and G. Jiang. 2020. "Uptake, phytovolatilization, and interconversion of 2,4-dibromophenol and 2,4-dibromoanisole in rice plants." *Environ Int* 142:105888.

Zine, H., L. Midhat, R. Hakkou, M. El Adnani, and A. Ouhammou. 2020. "Guidelines for a phytomanagement plan by the phytostabilization of mining wastes." *Sci Afr* 10.

15 Phytoremediation and Food Safety Case Studies

Srinivasan Kameswaran, Ramesh B. Kasetti, and Bellamkonda Ramesh

15.1 INTRODUCTION

Natural resources including forests, wildlife, land, soil, air, water, plants, and animals abound on the planet Earth. Humanity began as a race when people began to lead settled lives as opposed to wandering ones. Ever since the beginning of civilization, there has been exploitation and misuse of natural resources, which has resulted in their depletion. Consequently, according to Gosavi et al. (2004), 50% of our natural resources are now either depleted or in danger of being depleted. But the current situation of our resources demonstrates a lack of caution and diligence in their use. There are several instances that show how natural resources are being exploited, such as when chemical fertilizers are used in agriculture, industrial waste is released, and human activity occurs. Though crop yields and productivity have increased due to the use of chemical fertilizers, insecticides, and herbicides in agriculture, this has resulted in the soil and terrestrial environment absorbing harmful levels of phosphorus (P) and nitrogen (N). Likewise, the discharge of harmful substances from diverse artificial origins led to the pollution of natural assets, resulting in a deficiency of uncontaminated water and a reduction in soil fertility (Kumari et al., 2013; Chen et al., 2013).

Agricultural fields have become contaminated due to industrial and human activity, which has resulted in a decline in biodiversity. The diversity of plant and animal species contributes significantly to the growth of a robust and productive ecosystem, which in turn benefits both the environment and humans economically. Regrettably, several species are in jeopardy due to the swiftly expanding human population and heightened human activities. Natural processes including the production of many natural activities, including the production of crude oil, the soil fixation of nitrogen, biological insect control, manufacturing of medicines, fruit dispersion, and pollination, are carried out by the enormous range of species that can be found all over the world (Yerima et al. 2012). Because of the expected large increase in polluted sites, which is becoming a global crisis, these issues are extremely concerning. There were over 40,000 polluted sites in the United States as of May 2004, according to a study by the Environmental Protection Agency (EPA). Some industrialized Western European nations have significantly more polluted sites in a relatively smaller region (Prokop et al., 2000). Natural processes including the production of many over the past few decades, several significant events have shown how important it is to prevent effluents from escaping into the environment. These events include the Exxon Valdez oil spill, Minamata disease in Japan, the Union Carbide Bhopal disaster, extensive pollution of the Rhine River, radioactive debris released during the Chernobyl accident, and the gradual degradation of conifer forests and aquatic ecosystems in the northeastern United States, Canada, and parts of Europe (Kumar et al., 2011). A much better approach to overcoming these disadvantages would be to either totally eliminate the pollutants or convert them into biodegradable materials. Bioremediation is a technology that may be used to accomplish this strategy. The use of natural biological activities is a way to eliminate different toxins from the environment and protect its resources. It is regarded as more environmentally friendly, affordable, safer, and clean technology that is widely accepted by the public and may frequently be used at any location. The term *bioremediation* refers to the process of breaking down environmental

DOI: 10.1201/9781032683751-18

pollutants into less harmful forms using a variety of biological agents, mostly microbes, according to Van Dillewijn et al. (2007). Using a strain of *Pseudomonas putida* (Prescott et al., 2002) to break down petroleum, the first biological remediation agent patent was filed in 1974. In 1991, it was revealed that around 70 different microbial species could break down petroleum compounds (US Congress Office of technology assessment, 1991). "Over the next two decades", almost the same number of new genera has been added to this list (Kumar et al., 2011). Bioremediation agents are classified by the EPA as nutritional additions, enzymes, and microbiological cultures that dramatically speed up the pace of biodegradation to lessen the impact of different contaminants. The primary benefits of bioremediation are its cheap cost, high efficiency, ability to recover metals, selectivity towards certain metals, minimal chemical and biological sludge, regeneration of biosorbent, and lack of extra nutritional demand (Kratochvil and Volesky, 1998). Naturally occurring bioremediation, also known as intrinsic bioremediation or natural attenuation, can also be accelerated by adding fertilizers to increase the medium's bioavailability (biostimulation). Bioremediation techniques include land farming, phytoremediation, rhizofiltration, bioaugmentation, bioleaching, bioreactors, composting, biostimulation, and bioventing (Li and Li, 2011). Both in situ and ex situ bioremediation technologies may be categorized based on the removal as well as the transport of wastes. In situ bioremediation treats polluted substances at the same place as ex situ bioremediation, which involves eradicating polluted substances from one area and transferring them to a different location where biological agents have been utilized for treatment. In situ and ex situ approaches differ in the consistency of the process output and the rate of biodegradation, according to a comparison of the two approaches. Ex situ bioremediation is more expensive due to the requirement for the excavation of contaminated samples for treatment. Both in situ and ex situ bioremediation techniques rely primarily on microbial metabolism; however, for the ecological restoration of polluted soil, water, and environment, in situ techniques are now recommended above ex situ techniques (Vidali, 2001). The uses of different plants to break down, remove, contain, or immobilize pollutants from soil and water is known as phytoremediation. As an inventive and affordable replacement for the more traditional treatment techniques utilized at hazardous waste sites, this technology has recently drawn notice. Protecting the ecosystem and the public's health from the risks associated with hazardous waste sites is the mission of the EPA. To achieve this goal, the agency supports the advancement of cutting-edge technologies like phytoremediation. Technologies for the treatment of organic and inorganic contaminants that use phytoremediation have attracted a lot of interest since they are both economical and environmentally benign (Pilon-Smits, 2005). Scientists in academia and industry generally agree that plant-based environmental remediation is a sustainable cleanup method that can be used in developing and developed countries (Raskin and Ensley, 2000; Robinson et al., 2003).

15.2 PHYTOREMEDIATION MECHANISMS

A rapidly developing discipline of research and technology called phytoremediation (Salt et al., 1998) employs plants to purify wastewater, groundwater, and contaminated soil. Phytoremediation is the process of using plants – grasses as well as woody plants – to remove, incorporate, or transform radioactive materials, metals, metalloids, trace elements, organic compounds, and other environmental contaminants in water and soil into harmless forms. In this description, plant-influenced biological (Zouboulis and Katsoyiannis, 2005; Aransiola et al., 2013; Abioye et al., 2017), chemical, and physical processes are mechanisms that help plants, soil microbes, or interactions between plants and microbes absorb, sequester, break down, and metabolize pollutants. Utilizing the distinct and targeted absorption capacities of plant roots in addition to the translocation, bioaccumulation, and pollutant storage/degradation capacities of the complete plant, phytoremediation makes use of these advantages. The root system, which is a vast and self-extending absorption network that improves the subsurface ecosystem for later productive use, is what makes plant-based soil remediation systems similar to biological treatment systems. In addition to having the ability to remediate locations contaminated with several types of pollutants, phytoremediation reduces the danger of contamination spreading by avoiding the excavation and transportation of contaminated medium. The use of phytoremediation has several limitations, including the need for the plant to grow in specific conditions (temperature, altitude, geology, etc.); large-scale operations need access to agricultural knowledge and equipment; the plant's tolerance to the pollutant affects the remediation's success; pollutants obtained in senescing organs can be released returned to the environment during specific seasons; the remediation process takes far longer than it does for other technologies; and the solubility of the contaminants may increase, increasing the likelihood of leaching and causing more environmental harm (Mudhoo et al., 2010).

The mechanisms and efficacy of phytoremediation are dependent on the kind of pollution, bioavailability, and soil properties (Cunningham and Ow, 1996). In many ways, plants may restore or clean up contaminated habitats. The main defense mechanisms against toxicity are situated in the root system, which is also where plants absorb pollutants. The vast surface area of the root system allows it to collect and retain non-essential pollutants as well as water and nutrients needed for development (Raskin and Ensley, 2000). Plants can influence the mass of contaminants in soil, sediments, and water through several different ways. Even if the terminology for some of these systems varies, there are numerous commonalities between them. As the use of phytoremediation is intended to achieve, each of these processes will have an impact on the amount, mobility, or toxicity of pollutants (USEPA, 2000; Aransiola et al., 2019).

Many researchers have studied plant-based remediation, a type of biological remediation, in the management of AMD, and found it proved more cost-effective than the conventional therapeutic approach (Pilon-Smits, 2005). The term refers to any physical or chemical technique in which plants are used to break down or immobilize environmental pollutants, particularly those found in groundwater and soil. The six main categories of phytoremediation are phytodegradation, phytohydraulics, phytoextraction, phytovolatization, and phytostabilization. Phytoextraction and phytostabilization are the main topics of this chapter.

The term *phytoaccumulation* also refers to phytoextraction. The technique by which pollutants are eliminated in this botanical remediation approach involves the hyperaccumulation of toxins via root systems and their subsequent storage in tissues of stems or leaves. When plants are harvested, the pollutants also escape out of the surrounding area. Metals removal from soils represent an application of this botanical remediation technique that is extremely helpful. Additionally, it offers a way to recover the metals by burning the plants – a procedure known as phytomining. When plants absorb volatile substances via root systems, this process is known as phytovolatilization. Utilizing their leaves, plants may exchange the volatile substances directly or transform them into metabolites that they can transpire. The pollutants are discharged into the atmosphere.

Phytosequestration is another term for phytostabilization. This process entails the pollutants being absorbed by roots and then adsorbed on the root surface. In addition, the plant produces biochemicals that are subsequently discharged into the groundwater or soil around the roots. The adjacent pollutants can be immobilized by the biochemicals by sequestration, precipitation, or other means. The underground water or soil around the roots of a plant is where rhizodegradation takes place. Initiating the biodegradation of pollutants in the soil, plant exudates encourage the rhizosphere bacteria that surround plant roots. The breakdown and accumulation of contaminants after coming into contact with a tree's root system is another aspect of phytohydraulics. Poplar trees were employed for holding the methyl-tert-butyl-ether plume in groundwater in research by Hong et al. (2001). When pollutants are absorbed into plant tissues, their metabolism or biotransformation is applied through a process called phytodegradation. Under the kind of plant engaged in plant degradation, the pollutants may change in the roots, leaves, or stem of the plant.

Among the plants grown in the island nation of Sri Lanka, *Brassica juncea* (Brassicaceae), sometimes known as Indian mustard, is a particularly attractive and well-studied extensive species for phytoextraction. According to Prasad and Freitas (2003), this weedy mustard is effectively grown or naturalized in Central Asia, Western Egypt, India, and Europe. Significant amounts of chromium (Cr), copper (Cu), cadmium (Cd), nickel (Ni), zinc (Zn), and lead (Pb) can be simultaneously accumulated and translocated to the plant system, yielding 18 tonnes of biomass per hectare/crop (Kumar et al., 1995; Blaylock et al., 1997; Begonia et al., 1998; Zhu et al., 1999). This plant has been shown to be capable of accumulating heavy metals by numerous researchers, and a field trial conducted in the United States has demonstrated its ability to be used for heavy metal and radionuclide remediation (Huang et al., 1998; Lasat et al., 1998; Tucker and Shaw, 2000; Dushenkov, 2003; Qadir et al., 2004). Biochar and technosol combined may improve the effectiveness of Indian mustard as a bioremediator (Forján et al., 2017) by increasing its biomass (Forján et al., 2017). Vetiver grass (*Chrysopogon zizanioides*) can be extensively cultivated and applied to gold mine tailings phytoremediation in South Africa (Melato et al., 2016). An excellent account of the bioremediating properties of vetiver grass can be found in Banerjee et al. (2018). In temperate and dry settings, the study examined both phytoextraction and phytostabilization of mine tailings, as well as the impact of acid mine drainage (AMD) on macrophytes living in water in a phytoremediation system.

15.3 *EICHHORNIA CRASSIPES, PISTIA STRATIOTES,* AND *SPIRODELA POLYRHIZA* EXPOSED TO ACID MINE DRAINAGE: A CASE STUDY

Around the world, mining operations contaminate the surrounding environment. Sulfur-containing minerals like arsenopyrite may be present in trash piles and mining tailings. Perhaps the most common pollutant in mining is arsenopyrite. This oxidizes when exposed to water and oxygen in the atmosphere, which releases potentially hazardous compounds into the environment utilizing the AMD system. According to Duker et al. (2005), sulfur (S), iron (Fe), manganese (Mn), and arsenic (As) are frequently found in AMD as well and are thought to be extremely dangerous even in small amounts.

By removing species from the food chain, AMD is thought to have an impact on aquatic systems and change the structure of biological communities. Accordingly, using flora to remediate polluted surroundings at a cheap cost is a technique known as botanical remediation, or in situ remediation (Pilon-Smits, 2005). Phytoextraction and phytostabilization are the most widely applied types of phytoremediation technique among all of them. Because they can grow quickly, accumulate large amounts of heavy metals and metalloids, withstand extreme conditions, and withstand significant amounts of contaminants, macrophytes – plant species that float freely in water – are the most commonly used for phytoremediation (Mishra et al., 2008a, 2008b).

Studies on phytoremediation have a significant drawback in that they are restricted to a small number of metals and a few specific plant species (Hasan et al., 2007; Rahman et al., 2007). The concurrent absorption of many pollutants, particularly in AMD, is so poorly understood. Species such as *E. crassipes, P. stratiotes*, and *S. polyrhiza* have the capacity to partially absorb several elements while concurrently accumulating heavy metals and hazardous metalloids from AMD. The main characteristic of AMD is the low pH present in the water, which can harm *S. polyrhiza* and *P. stratiotes*, according to a preliminary study done by Michelle et al. (2010). Additionally, they discovered that *E. crassipes* had a higher tolerance for acidic conditions. Another consequence of AMD on plants was reported to be a reduction in chlorophyll content, as seen by the necrosis on the leaves of *P. stratiotes* and *S. polyrhiza*.

Additionally, it was shown that *E. crassipes* plants accumulated some soluble As from AMD in their various tissues, with the roots acquiring a greater amount of the heavy metal. Additionally, Mokhtar et al. (2011) found that *E. crassipes* showed 97.3% efficacy for Cu removal in an aqueous solution containing varying quantities of Cu. AMD kills *S. polyrhiza* and *P. stratiotes*. Michelle et al. (2010) provided information about the capacity of certain aquatic macrophytes to collect heavy metals. Further research is required to determine the simultaneous uptake of various metals through aquatic macrophytes and their interactions, to improve the invention of more effective phytoremediation technology for AMD while maintaining the aquatic macrophytes capable of accumulating heavy metals.

15.4 CONSTRAINTS ON UTILIZING AQUATIC VEGETATION FOR PHYTOREMEDIATION

According to several studies (Sood et al., 2012; Sharma et al., 2014; Newete and Byrne, 2016), phytoremediation is a more economical and ecologically friendly approach than traditional approaches for treating wastewater and contaminated settings. The capability of vegetation to develop quickly provided further benefits to this remediation strategy; nevertheless, its incapacity to withstand elevated levels of metals and the impact of seasonal variations on its efficacy are drawbacks (Rai, 2008; Mannino et al., 2008).

According to Sharma et al. (2014), due to the high concentration and toxicity of heavy metals, phytoremediation is utilized as a secondary or tertiary treatment technique for the treatment of mine or industrial wastewater. To prevent the dispersion of the absorbed nutrients back to the environment during the decomposition of the plant, Rai (2008) promoted the regular harvest and safe disposal of the biomass for an efficient phytoremediation treatment.

15.5 A CASE STUDY OF SRI LANKA AND SOUTH AFRICA CONCERNING PLANTS ASSOCIATED WITH PHYTOEXTRACTION AND PHYTOSTABILIZATION OF HEAVY METALS

The plant species that are accessible determine the effectiveness of phytoextraction and phytostabilization. Additionally, unique to the area of enthusiasm, plants that can withstand and accumulate high concentrations of toxic metals are the ones that can perform both of those phytoremediation procedures while avoiding AMD, according to Baker and Whiting (2002). Furthermore, Baker et al. (2000) noted that some heavy metal hyperaccumulators, including Cu, cobalt (Co), aluminum (Al), Zn, Pb, Mn, selenium (Se), Cd, and Ni, are unique in that they are invariably native to substrates that are rich in metals and can tolerate large concentrations of these heavy metals. Approximately 45 plant families contain around 420 species of toxic metal hyperaccumulators (Cobbett, 2003). Furthermore, according to Krämer (2003), new hyperaccumulators are being found in field collections, and a small percentage of these have been subjected to laboratory testing to determine their behavior and associated processes. Since the invention of hyperaccumulators, some intriguing trends have been noticed (Ernst, 2000; Baker and Whiting, 2002). First, it was discovered that numerous plant families, including the Asteraceae, Brassicaceae, Euphorbiaceae, Fabaceae, Flacourtiaceae, and Violaceae, included a significant amount of hyperaccumulators. It was proposed some genera within the families would do better than others in the presence of elevated heavy metal concentrations. A substantial proportion of these hyperaccumulators were found in tropical parts of the planet, which was the second trend identified. Of the approximately 320 species of Ni hyperaccumulators identified, two-thirds were found in tropical locations worldwide (Reeves, 2003; Proctor, 2003). A third conclusion indicates that over 80% of the discovered hyperaccumulators absorb Ni at a higher rate than the other metals. The following components working together might be the cause of the reported pattern of hyperaccumulators occurring:

- Time is available in the tropical areas for the emergence of these hyperaccumulator species.
- Tropical locations see high levels of weathering of naturally existing mineral deposits, which puts more selection pressure on native plants.
- Due to the hyperaccumulators' heavy metal accumulation features, they offer a selection advantage in coping with additional pressures associated with tropical climates.
- Nickel is more readily available than other metals because Ni-enriched serpentine outcrops are found along continental borders in large tracts.
- The enthusiasm scientists have in identifying hyperaccumulators in tropical climates.

More than 50 species of Ni hyperaccumulators were identified as indigenous to the metal-rich serpentine outcrops of the island of New Caledonia, according to Reeves (2003). These include species from genera such as *Cleidion* (Euphorbiaceae), *Argophyllum* (Grossulariaceae), *Geissois* (Cunoniaceae), *Casearia* (Flacourtiaceae), *Hybanthus* (Violaceae), *Homalium* (Flacourtiaceae), *Pancheria* (Cunoniaceae), *Oncotheca* (Oncothecaceae), *Phyllanthus* (Euphorbiaceae), and *Xylosma* (Flacourtiaceae; Jaffré et al., 1979; Reeves, 2003). More than 20% Ni by dry weight was discovered on the island in the latex of a tiny endemic tree named *Sebertia acuminata* (Sapotaceae), according to Jaffré et al. (1976). Few hyperaccumulators have been found in Asia, according to Proctor (2003), and the majority of them are closely linked to those found in Sri Lanka. In Indonesia, the genera of Ni hyperaccumulators are *Planchonella* (Sapotaceae), *Myristica* (Myristicaceae), and *Trichospermum* (Tiliaceae). Two Ni-hyperaccumulating species are found in the Asian genus *Rinorea* (Violaceae) (Brooks and Wither, 1977a, 1977b; Wither and Brooks, 1977). *Rinorea bengalensis*, which was taken from the herbarium of Sri Lanka, was found by Wither and Brooks (1977) to have 1% Ni hyperaccumulation by dry weight. Rajakaruna and Baker (2004) suggest that such species may have been introduced within the serpentinized region of Katupotha, which is located in the south-central part of the island. Sabah was also home to the same species, which was discovered to hyperaccumulate Ni (Jopony and Tongkul, 1999). The genera such as *Brackenridgea* (Ochnaceae), *Dichapetalum* (Dichapetalaceae), and *Walsura* (Meliaceae), as well as *Phyllanthus* (Euphorbiaceae) and *Shorea* (Dipterocarpaceae), widely represented in the Sri Lankan flora, were found to hyperaccumulate Ni in Sabah and the Philippines (Proctor et al., 1989; Baker et al., 1992). Jayasekera and Rossbach (1994, 1996) hypothesized that there may be numerous species in Sri Lanka with the ability to collect significant quantities of heavy metals.

High levels of whole-plant Ni hyperaccumulation were observed in Sri Lankan endemic plants, including *Crotalaria biflora* (Fabaceae), *Evolvulus alsinoides* (Convolvulaceae), and *Hybanthus enneaspermus* (Violaceae); similarly, high levels of whole-plant Cu hyperaccumulation were observed in *Croton bonplandianus* (Euphorbiaceae), *Clerodendrum infortunatum* (Verbenaceae), *Tephrosia villosa* (Fabaceae), *Geniosporum tenuiflorum* (Lamiaceae), and *Waltheria indica* (Sterculiaceae; Rajakaruna and Bohm, 2002). Notably, Cu hyperaccumulation is uncommon, and more research is necessary to validate these initial results for the species that have been demonstrated to accumulate Cu. In the Ni- and Cr-enriched serpentine outcrop of Ussangoda, *Cassia kleinii* (Fabaceae), a woody shrub that is well adapted, has demonstrated Ni hyperaccumulation (Rajakaruna and Bohm, 2002; Iqbal et al., 2006).

River bushwillow, or *Combretum erythrophyllum*, is a native tree species of South Africa that is being utilized as a phytoextraction and phytostabilization option on several tailings dams in Johannesburg. It can grow to a height of about 5 m in 5 years and up to 15 m in 50 years (Stoffberg et al., 2008). The species produces seeds and may thrive in highly salinized environments. McGaw et al. (2001) observed that the leaves and bark of this tree are utilized in traditional medicine in southern Africa. Because of this, while using this tree for phytoremediation, caution must be utilized when harvesting it for traditional medicine.

Australia is the natural habitat of *Eucalyptus camaldulensis*, also known as river redgum, an invasive species. It can be found in South Africa as well. This tree can withstand droughts, high salinity, acidic environments, and other challenging circumstances. In an experiment conducted to evaluate the mineral uptake of *Acacia nilotica* L. (babool), *Dalbergia sissoo* L. (sissoo), and *E. camaldulensis* seedlings, Singh et al. (2010) discovered that *E. camaldulensis* showed high productivity and great reduction in most soil mineral concentrations. It is a hyperaccumulator of Mn and Pb. The greatest alternative for soil rehabilitation was found to be *E. camaldulensis*, since the results of this investigation showed that it could withstand high concentrations of salt in both soil and water. In order to limit the danger of environmental invasion, it is crucial to establish sterile plants for phytoextraction.

A native of South Africa, *Tamarix usneoides*, or wild tamarix, does well in conditions that are extremely salty. In dry settings, the plant's well-adapted salt glands may secrete surplus salts, and its root system can reach deep water tables. In South Africa, the species is utilized for phytoremediation with success around tailings dams. Because it is a perennial grass, *Sporobolus spicatus*, often known as salt grass, may grow along river, marsh, and salt pan borders. The salt collection by the shoots of this grass rises with increasing soil salinity (Ramadan, 2001). Because this plant can withstand harsh and damp environments, it is suggested that it be employed in phytoremediation.

15.6 MINING TAILINGS PHYTOEXTRACTION OF METALS: A CASE INVESTIGATION OF TEMPERATE AND ARID ENVIRONMENTS

Utilizing metal-hyperaccumulating plants that store metals in their roots, leaves, or stems is known as phytoextraction of metals, particularly from mine tailings. This buildup of contaminants is followed by the harvesting of the plants. The gathered heavy metals can be recycled and utilized again, or the collected plants can be destroyed. Hyperaccumulators are plants that concentrate metals at extremely high concentrations into their above-ground biomass. Metal accumulation by these hyperaccumulators typically occurs at concentrations between 0.01% and 1% of total biomass generated by plants. The particular metal that these plants are absorbing will determine this (Brooks, 1998; Baker et al., 2000). Maximal metal accumulation is enhanced by these hyperaccumulators' quick growth rates and high biomass output. Several metrics are used to identify and characterize hyperaccumulators, such as the accumulation factor (the ratio of metals in the shoot tissue to the metal in the contaminated environment) and the translocation factor (the ratio of the metals in the shoot to the metals in the plant root). Over 400 species of hyperaccumulators belonging to 45 families have been identified as effective hyperaccumulators of metals from contaminated soils based on accumulation factors, translocation factors, high biomass, and fast-growing rate (Baker et al., 2000; Deng et al., 2007; Turnau et al., 2007). Species that are specifically capable of phytoextraction in mine tailings have also been identified. The concentration of metal in plant shoot tissues is a common way to detect and characterize hyperaccumulators. Moreover, accumulation factor is also required to determine whether phytoextraction is feasible. This feasibility of phytoextraction denotes the number of cropping cycles necessary for the metal removals to the level that is allowed (Zhao et al., 2003).

To find viable plants for phytoextraction in AMD, several investigations involving plants that colonize mine tailings and hyperaccumulators that are established there have been conducted. Significant obstacles must be overcome for the phytoextraction of metals from mine tailings. Due to the low metal bioavailability in tailings treated with compost to promote plant development, phytoextraction of metals from mine tailings is known to be typically poor (Knight et al., 1997; Meerts and Van Isacker, 1997; Salt et al., 1997; Bennett et al., 2003). Additionally, certain plant species may live and colonize mine tailings, but they cannot be employed as hyperaccumulators due to insufficient metal accumulation, according to Krzaklewski and Pietrzykowski (2002). Metal chelators have been investigated and utilized to increase metal hyperaccumulation by hyperaccumulators. But the plants' phytotoxicity was a concern when they were used. According to Berti and Cunningham (2000), chelators are often used following the achievement of excellent plant biomass output. Another difficulty addressed when chelators are applied is the leaching of metals into the groundwater especially in temperate areas owing to enhanced solubility of metals in the mine tailings (Wu et al., 1999; Schmidt, 2003). Pollution from metals is caused by their seeping into groundwater.

Many nations across the world have investigated and studied the use of transgenic hyperaccumulator plants for mining tailings cleanup. The negative effects of these transgenic hyperaccumulators on species diversity, soil ecosystems, and the food chain, however, are their main disadvantage (Wolfenbarger and Phifer, 2000; Wolfe and Bjornstad, 2002). Moreover, a significant worry with the adoption of transgenic hyperaccumulators is their potential for invasion, their capacity to settle disturbed regions in desert settings, and their propensity to produce late successional species (Castellanos et al., 2005). To improve the appropriate handling of these transgenic hyperaccumulators, it has been suggested to utilize sterile clones, harvest plants before blooming, and use species that won't outcompete native plants (Pilon-Smits and Pilon, 2002; Snow et al., 2005). In 2002, Wolfe and Bjornstad made the last observation that an issue with transgenic hyperaccumulators' application may arise from low public acceptability.

The low pH, lack of soil structure, metal toxicity, and poor nutrient content that plants grow in temperate areas with mine tailings face are obstacles to their growth. Low biomass output is a common issue for plants that may thrive in temperate climates and overcome their obstacles. In temperate conditions, phytoextraction is hampered by this poor biomass output since plant accumulation of metals depends on biomass production. *Thlaspi caerulescens* species have been seen to prosper in metal-contaminated locations and mine tailing sites in Europe, where they can collect metals. This region serves as a typical illustration of the phenomenon. *T. caerulescens* has been discovered to flourish in mining spoils and tailings in the United Kingdom. The plant was reported to collect 50–160 mg kg^{-1} Cd and 13,000–19,000 mg kg^{-1} Zn with accumulation factors of Zn > 1 (Baker et al., 1994). The accumulation factors of the plant was shown to rise up to 73 and 40 for Cd and Zn, respectively, when the plant was subjected to a lower ratio of Cd to Zn (Zhao et al., 2003).

Research carried out in a greenhouse employing *T. caerulescens* exposed to a mining ruin containing up to 3300 mg kg^{-1} Zn and 58 mg kg^{-1} Cd to test its phytoextraction capability indicated that the plant absorbed 250 mg Cd kg plant $biomass^{-1}$ and 8,000 mg Zn kg plant $biomass^{-1}$ (Knight et al., 1997). Due to the plant's poor biomass output in the mine tailings, the investigated mine tailings would require around 100–1200 cropping cycles to remove Cd and 200–600 cropping cycles to remove Zn. The full cleanup of the mining tailings took decades as a result of several cropping cycles. The acquired results suggested that while *T. caerulescens* would not be appropriate for the more severely polluted mining waste that was tested, it might be helpful for cleaning up environments contaminated with low levels of Zn and Cd.

In the effort to increase phytoextraction efficiency in mine tailings, two transgenic strains of *B. juncea* (L.) Czern were identified and reported to accumulate up to 1.5- to 3-fold more metals (Cd, Cr, Cu, Pb, Zn) than the wild type strain in a greenhouse study (Bennett et al., 2003). The transgenic strains overproduced the enzymes glutathione synthetase and glutamylcysteine synthetase, which influence the synthesis of the metal-binding thiol peptides phytochelatin and glutathione, respectively. The strains of this transgenic *B. juncea* decreased 25% and 5% of Cd and Zn, respectively, from tailings supplemented with compost; nonetheless, leaching was shown to be the main cause of metal reduction in this study (Ernst, 2005). According to Bennett et al. (2003), the transgenic plants' total phytoextraction efficiency was reduced by the carbon and lime added to the mine tailings, which was necessary for *B. juncea* to flourish.

Along with low pH, metal toxicity, poor nutrients, lack of structure in the soil, and an absence of water, salty conditions are major obstacles to plant development in dry environments. Many natural hyperaccumulators that can colonize mining waste are being identified, despite the difficulties the plants face in this environment. Few of these species have, however, been studied in a greenhouse to determine their potential for hyperaccumulation (Leita et al., 1989; Bech et al., 2002; Gonzalez and Gonzalez-Chavez, 2006). The natural colonizers of mine tailings in Spain, *Cichorium intybus* L. and *Cynodon dactylon* (L.) Pers., were assessed for Pb accumulation. Potting soils that had been intentionally polluted with Pb were used for this investigation. According to observations, *C. intybus* and *C. dactylon* had shoot Pb accumulations of 800–1500 and 400–1200 mg kg^{-1} Pb, respectively. It was also noted that the buildup of metal in the shoots did not limit the development or biomass output of the plants. The work was limited, nonetheless, by the authors' inability to demonstrate how these plants' phytoextraction capabilities might apply to a mining tailings setting (Del Rio-Celestino et al., 2006).

To assess *Lygeum spartum* L.'s phytoextraction potential, Conesa, Robinson et al. (2007) conducted both field and greenhouse research on the grass, which is known to colonize mine tailings sites in Mediterranean areas. Except for zinc, which was found to be exceptionally high (between 3000 and 6000 mg kg^{-1}) when plants were watered and fertilized in the greenhouse, it was noted that the metal accumulation was modest in both the field and the greenhouse. In contrast to the common perception, Frérot et al. (2006) reported that they detected Pb hyperaccumulation at 1,000–2,000 mg kg^{-1} in the field in *Armeria arenaria* (Pers.) Schult., *Festuca arvernensis* Auquier, and *Koeleria vallesiana* (Honck.) Gaudin.

In dry areas, the most prevalent feature of mine tailings is elevated salinity. Salinity is thought to affect the electrical conductivity of the environment; in dry areas with high salinity (22 dS m^{-1}), electrical conductivity was found to exceed the amount that plants could tolerate (Munshower, 1994; Marschner, 1995). According to Osmond et al. (1980) and Mendez et al. (2007), *Atriplex* spp. are called halophytes because they absorb salt and potassium (P) in high quantities in mine waste and unpolluted soil. *Atriplex halimus* L. was used by Lutts et al. (2004) in a hydroponic greenhouse research. The results revealed that the plant was able to absorb 830 and 440 mg kg plant $tissue^{-1}$ of Cd and Zn, respectively, in shoot biomass, and that it did not exhibit any hazardous symptoms while growing. The low quantities of cations (Na^+ and K^+) in the hydroponic growing medium may have contributed to the high concentration of accumulated metals. It is unknown, therefore, if the field investigation produced the same outcome.

15.7 THE DRAWBACKS OF PHYTOEXTRACTION

Significant disadvantages exist for the extensive application of phytoextraction as a remediation technique for mine tailings. Complete phytoextraction may take a very long period, particularly when dealing with severely metal-polluted mine tailings. While compost amendments are necessary for plant development, the distribution of various metals in mine tailings may result in a reduction in plant metal bioavailability and metallic extraction efficiency. A further obstacle to the effectiveness of phytoextraction, according to Ernst (2005), is the polymetallic character of mine tailings. This is because several plant species and ecotypes are needed, as hyperaccumulators can only collect one or two metals. In dry areas where mine tailings have a high saline content and many halophytes prefer Na^+ to K^+, this disadvantage is exacerbated (Osmond et al., 1980; Munshower, 1994; Marschner, 1995). Plant traits may also reduce a plant's ability to absorb metals from mining tailings. Hyperaccumulators often have shallow root zones, which limit the amount of metal that may be removed by plants by rooting deep (Keller et al., 2003). Tailings with high quantities of metals may have decreased metal extraction effectiveness. Generally speaking, the unforgiving conditions found in mine tailings affect plant biomass production and, in turn, phytoextraction efficiency (Ernst, 2005; Audet and Charest, 2007).

Metals are accumulating in the shoot system of the hyperaccumulators, which wildlife animals feed, putting them at risk of exposure. The worst case occurs in dry areas with few food sources, as noted by Wood et al. (1995). The implementation of phytoextraction is thought to be more expensive as a result of all these disadvantages, and more expenses may be incurred for process optimization. Moreover, the hyperaccumulator's harvesting, disposal, and reseeding costs add to the technology's overall cost as they may be necessary for several cropping cycles (Cunningham and Berti, 2000). Ultimately, it is imperative to take wildlife safety into account. This does not eliminate the use of phytoextraction for cleanup in mine tailings. But it is important to weigh the expenses of this tactic against the land's prospective worth when it is recovered.

15.8 A CASE STUDY OF TEMPERATE AND ARID ENVIRONMENTS FOR THE PHYTOSTABILIZATION OF MINE TAILINGS

Establishing plants with a deep root system is necessary for the phytostabilization of mine tailings because it allows the plants to form a stable vegetative cover that prevents metals from building up in above-ground tissues. Established plants in mine tailings operate by solubilizing metals onto their root surfaces, accumulating them in their root tissue, and stabilizing them by precipitation (Cunningham et al., 1995; Wong, 2003). In order to improve the soil structure, established plants in the mine tailings for phytostabilization should also be able to promote diverse heterotrophic microbial community interaction, which will stimulate plant growth and facilitate the transformation of the materials in the tailings (Mummey et al., 2002a; Glick, 2003; Wong, 2003; Mendez et al., 2007). A diverse and self-sustaining vegetative cover that lowers water erosion, leaching processes, and eolian processes is left in the environment by the established plants, which also supply the initial plant cover that promotes the development and establishment of other plants in the area.

Depending on the kind of metal contamination in the mine tailings, different plants are utilized for phytostabilization in different areas. In addition, physicochemical variables affect the variety and stability of naturally occurring plant communities in the same areas (Conesa, Garcia et al., 2007). The physicochemical properties of specific mine tailings sites must thus be taken into account prior to the growth of plants necessary for phytostabilization. In order to eliminate any harmful effects and reduce the bioavailability of metals, it is also necessary to amend the mine tailings with compost manure or lime, which is necessary for plant development. Metals do not need to be translocated to shoot tissues by the plants chosen for phytostabilization. This prevents surface pollution and safeguards the lives of wildlife species that could rely on them for survival. Shoot metal concentrations should thus have a ratio of less than 1 to either root metal concentration or soil metal concentration. According to Wood et al. (1995) and the NRC (2005), the plants must also not expose animals that forage for food to the US domestic animal metal toxicity limits.

Because of their large root systems and rapid development, grasses are mostly utilized for phytostabilization in temperate climates. The two natural grass species utilized most frequently for phytostabilization in Europe are *Festuca* spp. and *Agrostis* spp. In the United States and China, they are also employed for phytostabilization (Smith and Bradshaw, 1979; Ye et al., 1999; Pierzynski et al., 2002; Von Willert and Stehouwer, 2003). Pichtel and Salt's (1998) greenhouse experiment conducted in Scotland shown that NPK (nitrogen, phosphorus, and potassium) fertilizer administration only stimulates the development of *Festuca ovina* L. and *Festuca rubra* in a moderately acidic Pb/Zn mining waste sample.

In Pb/Zn and Cu mine tailings, a mixture of fertilizers and lime was found to successfully grow *F. rubra* and *Agrostis tenuis* Sibthorp in many field studies conducted in the United Kingdom (Smith and Bradshaw, 1979). Numerous research have also looked into using native plants in conjunction with soil cover to achieve phytostabilization. This was observed in the work conducted in Washington State by Kramer et al. (2000) using two plants, *Anaphalis margaritacea* (L.) and *Alnus viridis* subsp. sinuata (Regel). The plants were grown in Cu mine tailings amended with a mixture of biosolids and woodchips prior to planting, and the researchers saw that the two plants demonstrated adequate plant biomass and shoot tissue exclusion of high concentrations of Al and Zn. An analogous investigation was conducted in China by Shu et al. (2001), who used 41 native seed species from a soil seed bank that was sourced from adjacent uncontaminated regions. Three of the nine plants that they found to have sustained both good plant coverage and minimal metal buildup were those that had been grown in a Pb/Zn mine tailings after an 8 cm topsoil layer had been applied over the tailings material.

In arid and semi-arid conditions, successful phytostabilization of mine tailings requires plants that are resistant of drought, salt, and metals. Arid environments benefit from irrigation and the addition of an organic amendment to improve re-vegetation. According to Osmond et al. (1980) and Jefferson (2004), *Atriplex* spp. are recognized pioneer plants on mine tailings in semi-arid Western Australia. The development of *Atriplex lentiformis* (Torr.) in moderately acidic and severely acidic Pb/Zn mine tailings needed at least 10% compost input, according to a research by Mendez et al. (2007) done in an irrigated greenhouse. In addition, they noted that the shoot absorption of metals did not exceed the domestic animal metal toxicity limits in the United States.

In a greenhouse research utilizing biosolids, Sabey et al. (1990) also found that *Atriplex canescens* (Pursh) grows best in copper mine tailings when sludge is applied at a rate of 60 mg ha^{-1}. Reduced Cu concentrations in the leaf tissues that were five times higher than the US domestic animal metal toxicity limits were one drawback of the heavy sludge application they saw. In an alkaline mine tailing site in southern Arizona, Rosario et al. (2007) carried out a field experiment to investigate the effect of compost on the development of *A. canescens*. The individual plants received 44 L of water during the first 8 months of development. Regarding composted and uncomposted tailings, similar plant growth outcomes were observed, with 80% of the transplants surviving. In addition, Pb was found to be up to three times more toxic than the US domestic animal metal toxicity limits in several plants in this study. The data gathered from these investigations suggested that the physicochemical characteristics of the mine tailings and plant development needs dictate the kind and quantity of amendments that plants need. As such, needs for amendments vary depending on the place and type of plant. Results from field tests conducted in dry environments showed promise for re-vegetation using direct seeding. Piha et al. (1995) conducted a study in Zimbabwe using the direct seeding method without irrigation in a tin mine tailings site. The seeds were initially subjected to 7 weeks of drought, and after the first growing season, the seeds produced 40% plant cover with a mixture of grasses and legumes using only an initial addition of NPK fertilizer. Van Rensburg and Morgenthal's (2004) study in South Africa found that an average of 2% plant cover was established in platinum (Pt) tailings using seeds from a variety of grasses and fertilizer amendments. The scientists hypothesized that these subpar outcomes are the result of inadequate irrigation and bad timing that missed the opportunity to benefit from the seasonal rains. The results of Windsor and Clements' (2001) seeding study at a gold mine tailing site under irrigation revealed that the density of seedlings in irrigated plots was three times higher than in non-irrigated plots. According to this, irrigation greatly boosts plant establishment in dry and semi-arid conditions. In direct seeding research, exact timing is necessary to capitalize on the regular rainy season if irrigation is not practical.

15.9 THE LIMITS OF MINE TAILINGS PHYTOSTABILIZATION

It is challenging to assess the overall effectiveness of phytostabilization because it is a long-term field study. Mummey et al. (2002b) conducted a 20-year research which revealed that the plant and microbial diversity in a re-vegetated mining site was lower than that of an adjacent undisturbed site in a semi-arid grassland. But since the amount of plant cover was greater than that of the undisturbed site, the adoption of this technique showed promise in that regard. The environment affects plant coverage in its entirety. Comparing semi-arid and arid shrublands to temperate conditions, the former are unlikely to yield a full covering of plants. This becomes a barrier to the effective remediation of acid mine tailings by phytostabilization. The supply of N, the efficiency of water consumption, and light all affect how much more or less plant cover there is in dry versus temperate environments. While light and N availability affect plant dynamics in temperate environments, N availability and water usage efficiency affect plant dynamics in semi-arid and arid environments (Tilman, 1988; Gebauer and Ehleringer, 2000). For this reason, in semi-arid and dry environments, occasional rainfall may lead to a higher death rate of secondary successional plant communities (Watson et al., 1997; Castellanos et al., 2005). In order to address some of the constraints, it is advised that a thorough strategy utilizing a range of tactics be implemented. In order to biotransform polluted soils, Emenike et al. (2018) investigate the employment of bacteria and plants in tandem. Das (2018) suggests that comparable techniques, such as nutrient addition and nanoremediation, may be sustainable.

15.10 INDIA CASE STUDIES FOR PHYTOREMEDIATION

Water hyacinth (*E. crassipes*) and water lettuce (*P. stratiotes*) are alien plants that have the potential to be used as phytoremediation aquatic macrophytes for the removal of nutrients (N and P) and algal interception from open ponds contaminated with domestic sewage. Qin et al. conducted experiments to investigate this potential. For the extensive purification of home sewage with high N concentrations, water hyacinth was a better choice than water lettuce because it shown a higher accumulating capacity for N (58.64% of total reductions). This outcome might be explained by the fact that water hyacinth has more root activity, root biomass, and net photosynthetic rate than water lettuce, as well as a bigger total root surface, active absorption area, and leaf area (Qin et al., 2016).

The efficacy of water lettuce (*P. stratiotes*) and water hyacinth (*E. crassipes*) in treating aquaculture effluent in Malaysia was examined by Akinbile and Yusoff. After sampling the wastewater from a fish farm in Semanggol Perak, Malaysia, several parameters were measured, such as pH, turbidity, dissolved oxygen (DO), chemical oxygen demand (COD), biochemical oxygen demand (BOD), phosphate (PO_4^{3-}), nitrate (NO_3^-), nitrite (NO_2^-), ammonia (NH_3), and total Kjeldahl nitrogen (TKN). Additionally, throughout the 30-day experiment, a hydroponics system was built up and new plants weighing 150 ± 20 g of *E. crassipes* and 50 ± 10 g of *P. stratiotes* were introduced. In every parameter that the phytoremediators were used to treat, there was a noticeable percentage reduction. In comparison to *P. stratiotes*, which had percentage reductions of 92.70% and 93.69%, *E. crassipes* had reductions of 85.26% and 87.05% in turbidity. In COD, TKN, NO_3^-, NH_3, and PO_4^{3-}, comparable decreases were noted. According to Akinbile and Yusoff's (2012) study, these plants are capable of extracting nutrients.

Research on water hyacinth and water lettuce by Sivasankari and Ravindran (2016) demonstrated that these two macrophytes are easily broken down in anaerobic conditions to produce sizable amounts of hydrogen (H). One practical method for using the chemical energy these weeds contain is anaerobic digestion. In this study, the potential as biohydrogen generation substrates for water hyacinth and water lettuce was assessed. The outcome demonstrated the potential of aquatic plants as a biomass for the generation of biohydrogen. Future developments will be greatly impacted by biohydrogen as it can employ sustainable energy sources (Sivasankari and Ravindran, 2016).

Duckweed (*S. polyrhiza*), water lettuce (*P. stratiotes*), and algae (natural colonization) were used as treatment agents in a bench-scale continuous-flow wastewater treatment system with three parallel lines by Awuah et al. (2004) in order to assess environmental parameters, fecal coliform profiles, and overall treatment efficacy. Each line was made up of four series-connected ponds that were supplied with diluted sewage. Environmental conditions, turbidity, COD and BOD, NO_3^-, NO_2^-, NH_3, and total P were among the influent and effluent characteristics that were assessed. The duckweed system removed the most BOD, with *Pistia* and algae coming in second and third at 95%, 93%, and 25%, respectively. Duckweed and *Pistia* had COD removal rates of 65% and 59%, respectively, but COD rose by 56% in algal ponds. The percentages of NO_3^-removed from duckweed, *Pistia*, and algal ponds were 72%, 70%, and 36%, respectively. In the *Pistia* and duckweed systems, total P reductions were 33% and 9%, respectively, but a 19% increase was noted in the algal treatment system. In algal systems, NH_3 reductions were 93%, while in *Pistia* and duckweed, they were 95%. According to Awuah et al. (2004), the removal of total dissolved solids (TDS) for *Pistia* was 70%, for duckweed it was 15%, and for algae it was 9%.

Five aquatic species (*E. crassipes, Alternanthera philoxeroides, Egeria densa, Najas flexilis*, and *Potamogeton crispus*) were cultivated in plastic pools filled with well water, and Dixit et al. (2011) investigated the effects of sewage effluent on their development with and without the addition of 25% of sewage effluent. Out of the five test plants, *A. philoxeroides* and *E. crassipes* had the greatest growth response to the sewage effluent. The water hyacinth took out 6.9 g of N, 2.9 g of P, and 8.7 g of K from the sewage pools, outcompeting other species and covering 71% of the water's surface (Dixit et al., 2011).

In their research of a bench-scale continuous-flow wastewater treatment system with sewage input, Awuah et al. (2004) utilized lettuce. It was shown that lettuce reduced TDS by 70%, fecal coliform by 99%, BOD by 93%, COD by 59%, NO_3^- by 70%, total P by 33%, and NH_3 by 95%. It has been observed that water lettuce lowers ammonium ions in the water because it uses ammonium (NH_4-N) as a nitrogen source before nitrate (NO3-N) and doesn't start using NO_3-N until all of the NH_4-N has been used (Awuah et al., 2004).

The NO_3^-removal effectiveness of water lettuce was observed by Ingersoll and Baker (1998) to range from 31% to 51%. In batch culture tests, water lettuce was shown to have a comparable

NO_3^-removal ability to water hyacinth at an initial NO_3^- concentration of 5.5 mg/L by Aoi and Hayashi (1996). Metals like Zn, Ni, and Cu have been removed from the water column using it quite a bit. However, after 3 days, lettuce plants with 20 mg/L Cr revealed 100% mortality (Aoi and Hayash, 1996).

Increased temperature promotes their development and proliferation, according to Girija et al. (2011). After a month after planting, low pH values approach neutrality. The dissolved mineral matter level of contaminated water is closely correlated with its electrical conductivity (EC), and following vetiver planting, the EC dropped to an extremely low value (Girija et al., 2011).

According to Mane et al. (2011), at 200 mM NaCl concentration, vetiver grass shoot length rose by 18.6%; however, at 50 mM NaCl, root length increased by around 24.8%. In salinized circumstances, the mean leaf area also rose. Under salt stress, biomass in terms of both dry weight and fresh weight was less effective. Additionally, they saw higher polyphenol concentrations at higher salinities as a result of secondary metabolite buildup. As salinity increases, there is a linear increase in the soil's EC and TDS. Additionally, vetiver can withstand up to 100 mM of salt due to an increase in growth and photosynthetic parameter (Mane et al., 2011).

Vegetable growth was shown to be greatly enhanced by both N and P inputs, according to Wagner et al. (2003). N rates up to 10,000 kg/ha did not negatively impact vetiver development, but at rates more than 6000 kg/ha/year, very little growth response happened. The need for P by *C. zizanioides* (vetiver) was not as great as that of N, and at rates greater than 250 kg/ha/year, there was no growth response. But at P up to 1000 kg/ha/year, its development was not negatively impacted (Wagner et al., 2003).

In contaminated river water, Anon (1997) and Zheng et al. (1997) discovered that 98% of total P was removed in 4 weeks and 74% of total N was removed after 5 weeks (Anon, 1997; Zheng et al., 1997).

After 4 days of treating residential wastewater with vetiver, Truong and Hart (2001) found that 94% of total nitrogen, 90% of total P, 50% of EC, a pH shift (from 7.26 to 5.98), 44% of fecal coliform, and 91% of *Escherichia coli* were removed. According to Truong and Hart (2001), vetiver thus has a great deal of promise for application in the treatment of industrial wastewater.

15.11 CONCLUSION

Using phytoremediation techniques to remediate mining tailings is a viable tactic. Research on phytoextraction revealed that the technique will only be used on properties with high land values, and that the process of recovering a site would take a while because of the high expense and extended treatment periods. Because it is less expensive and easier to execute, phytostabilization offers a more appropriate generation of phytoremediation techniques. Nevertheless, more examination and refinement of these methods are required to improve AMD repair in its entirety. It has been determined that the aquatic macrophyte *E. crassipes, B. juncea* (Brassicaceae), *C. zizanioides*, and Indian mustard are effective in mitigating AMD. Further investigation into the processes behind heavy metal remediations, the impact of climate change, and the nutritional needs for the growth of sufficient biomass for phytoremediation are recommended by this review. More study is necessary to determine how these three plant species may be combined to remediate most regions or mine tailings polluted with heavy metals.

REFERENCES

Abioye, O. P., Ijah, U. J. J. and Aransiola, S. A. 2017. Phytoremediation of soil contaminants by biodiesel plant Jatropha curcas. In Chapter 4 of Bauddh, K., et al. (eds.), *Phytoremediation Potential of Bioenergy Plants, DOI 10.1007/978-981-10-3084-0_4*. Springer Nature Singapore Pte Ltd. https://link.springer.com/book/10.1007%2F978-981-10-3084-0.

Akinbile, C. O. and Yusoff, M. S. 2012. Assessing water hyacinth (*Eichhornia crassipes*) and lettuce (Pistia stratiotes) effectiveness in Aquaculture wastewater treatment. *International Journal of Phytoremediation* 14(1), 201–211.

Anon, A., 1997. Consideration and preliminary test of using vetiver for water eutrophication control in Taihu Lake in China. Proceedings of the International Vetiver Workshop, Fuzhou, China.

Aoi, T. and Hayashi, T. 1996 Nutrient removal by water lettuce (*Pistia stratiotes*). *Water Science Technology* 34(7–8), 407–412.

Aransiola, S. A., Ijah, U. J. J. and Abioye, O. P. 2013. Phytoremediation of Lead Polluted Soil by Glycine max L. *Applied and Environmental Soil Science* 2013, Article ID 631619. doi: 10.1155/2013/631619. https://www.hindawi.com/journals/aess/2013/631619/abs/.

Aransiola, S. A., Ijah, U. J. J., Abioye, O. P. and Bala, J. D. 2019. Microbial-aided phytoremediation of heavy metals contaminated soil: A review. *European Journal of Biological Research* 9(2), 104–125. http://www.journals.tmkarpinski.com/index.php/ejbr/article/view/157.

Audet, P. and Charest, C. 2007. Heavy metal phytoremediation from a meta-analytical perspective. *Environmental Pollution* 147, 231–237.

Awuah, E., Peprah, O. M., Lubberding, H. J. and Gijzen, H. J. 2004. Comparative performance studies of water lettuce, duck weed and algal based stabilization ponds using low strength sewage. *Journal of Toxicology and Environmental Health* 67(1), 1727–1739.

Baker, A. J. M. and Whiting, S. M. 2002. In search for the Holy Grail – another step in understanding metal hyperaccumulation? *New Phytologist* 155, 1–7.

Baker, A. J. M., McGrath, S. P., Reeves, R. D. and Smith, J. A. C. 2000 Metal hyperaccumulator plants: A review of the ecology and physiology of a biological resource for phytoremediation of metal-polluted soils. In Terry, N. and Banuelos, G. (eds.), *Phytoremediation of Contaminated Soil and Water*. Lewis Publishers, Boca Raton.

Baker, A. J. M., Proctor, J. and Reeves, R. D. 1992. The vegetation of ultramafic (serpentine) soils. Proceedings of the First International Conference on Serpentine Ecology, University of California, Intercept, Hampshire, UK.

Baker, A. J. M., Reeves, R. D. and Hajara, A. S. M. 1994. Heavy metal accumulation and tolerance in British populations of the metallophyte *Thlaspi Caerulescens* J. & C. Presl (Brassicaceae). *New Phytology* 127, 61–68.

Banerjee, R., Goswami, P. and Mukherjee, A. 2018. Chapter 22 – Stabilization of Iron ore mine spoil dump sites with vetiver system. In *Bio-Geotechnologies for Mine Site Rehabilitation*. Elsevier, Kolkata, 393–413.

Bech, J., Poschenrieder, C., Barcelo, J. and Lansac, A. 2002. Plants from mine spoils in the South American area as potential sources of germplasm for phytoremediation technologies. *Acta Biotechnology* 1–2, 5–11.

Begonia, G. B., Davis, C. D., Begonia, M. E. T. and Gray, C. N. 1998. Growth responses of Indian Mustard (*Brassica juncea* (L.) Czern.) and its phytoextraction of lead from a contaminated soil. *Bulletin of Environmental Contamination and Toxicology* 61, 38–43.

Bennett, L. E., Burkhead, J. L., Hale, K. L., Terry, N., Pilon, M. and Pilon-Smits, E. A. H. 2003. Analysis of transgenic Indian mustard plants for phytoremediation of metalcontaminated mine tailings. *Journal of Environmental Quality* 32, 432–440.

Berti, W. W. R. and Cunningham, S. D. 2000: Phytostabilization of metals. – In: Raskin, I. and Ensley, B. D. (eds.), *Phytoremediation of Toxic Metals – Using Plants to Clean Up the Environment*. John Wiley & Sons, Inc., New York, 71–88.

Blaylock, M. J., Salt, D. E., Dushenkov, S., Zakharova, O., Gussman, C., Kapulnik, Y., Ensley, B. D. and Raskin, I. 1997. Enhanced accumulation of Pb in Indian mustard by soil applied chelating agents. *Environmental Science and Technology* 31, 860–856.

Brooks, R. R. 1998. Plants that hyperaccumulate heavy metals: Their role in phytoremediation, microbiology, archaeology, mineral exploration and phytomining. – CAB International, Wallingford.

Brooks, R. R. and Wither, E. D. 1977a. Nickel hyperaccumulation by *Rinorea bengalensis* (Wall.) O.K. *Journal of Geochemical Exploration* 7, 295–300.

Brooks, R. R. and Wither, E. D. 1977b. Balt and nickel in Rinorea species. *Plant and Soil* 47, 707–712.

Castellanos, A. E., Martinez, M. J., Llano, J. M., Halvorson, W. L., Espiricueta, M. and Espejel, I. 2005. Successional trends in Sonoran Desert abandoned agricultural fields in northern Mexico. *Journal of Arid Environment* 60, 437–455.

Chen, J., Qing-Xuan Xu, Yi Su, Zhi-Qi Shi and Fengxiang, X. 2013. Phytoremediation of organic polluted soil. *Journal of Bioremediation & Biodegradation*4, 1–3.

Cobbett, C. 2003). Heavy metals and plants – model systems and hyperaccumulators. *New Phytologist* 159, 289–293.

Conesa, H. M., Garcia, G., Faz, A. and Arnaldos, R. 2007. Dynamics of metal tolerant plant communities' development in mine tailings from the Cartagena-La Union Mining District (SE Spain) and their interest for further revegetation purposes. *Chemosphere* 68, 1180–1185.

Conesa, H. M., Robinson, B. H., Schulin, R. and Nowack, B. 2007. Growth of *Lygeum spartum* in acid mine tailings: Response of plants developed from seedlings, rhizomes, and at field conditions. *Environmental Pollution* 145, 700–707.

Cunningham, S. D. and Berti, W. W. R. 2000. Phytoextraction and phytostabilization: Technical, economic, and regulatory considerations of the soil-lead issue. –. In: Terry, N. and Banuelos, G. (eds.), *Phytoremediation of Contaminated Soil and Water*. CRC Press LLC, Boca Raton.

Cunningham, S. D. and Ow, D. W. 1996. Promises and prospect of phytoremediation. *Plant Physiology* 110, 715–719.

Cunningham, S. D., Berti, W. R. and Huang, J. W. W. 1995. Phytoremediation of contaminated soils. *Trends in Biotechnology* 13, 393–397.

Das, P. K. 2018. Phytoremediation and Nanoremediation: Emerging Techniques for treatment of acid mine drainage water. *Defence Life Science Journal* 3(2), 190–196.

Del Rio-Celestino, M., Font, R., Moreno-Rojas, R. and De Haro-Bailon, A. 2006. Uptake of lead and zinc by wild plants growing on contaminated soils. *Industrial Crop and Products* 24, 230–237.

Deng, J. C., Liao, B., Ye, M., Deng, D. M., Lan, C. Y. and Shu, W. S. 2007. The effects of heavy metal pollution on genetic diversity in zinc/cadmium hyperaccumulator Sedum alfredii populations. *Plant Soil* 297: 83–92.

Dixit, A., Dixit, S. and Goswamy, D. S. 2011. Process and plants for wastewater remediation – a review. *Science Reviews and Chemical Communication* 1(1), 71–77.

Duker, A. A., Carranza, E. J. M. and Hale, M. 2005. Arsenic geochemistry and health. *Environmental International* 31, 631–641.

Dushenkov, S. 2003. Trends in phytoremediation of radionuclides. *Plant and Soil* 249, 167–175.

Emenike, C. U., Jayanthi, B., Agamuthu, P. and Fauziah, S. H. 2018. Biotransformation and removal of heavy metals: A review of phytoremediation and microbial remediation assessment on contaminated soil. *Environmental Reviews* 26(2), 156–168.

Ernst, W. H. O. 2000. Evolution of metal hyperaccumulation and the phytoremediation hype. *New Phytologist* 146, 357–358.

Ernst, W. H. O. 2005. Phytoextraction of mine wastes – options and impossibilities. *Chemie der Erde-Geochemistry* 65(1), 29–42.

Forján, R., Rodríguez-Vila, A. and Covelo, E. F. 2017. Increasing the nutrient content in a mine soil through the application of technosol and biochar and grown with *Brassica juncea* L. *Waste and Biomass Valorization*, 1–17.

Frérot, H., Lefébvre, C., Gruber, W., Collin, C., Santos, A. and Escarré, J. 2006. Specific interactions between local metallicolous plants improve the phytostabilization of mine soils. *Plant Soil* 282, 53–65.

Gebauer, R. L. E. and Ehleringer, J. R. 2000. Water and nitrogen uptake patterns following moisture pulses in a cold desert community. *Ecology* 81, 695–712.

Girija, N., Pillai, S.S and Koshy, M. 2011. Potential of vetiver for phytoremediation of waste in retting area. *The Ecoscan* 1, 267–273.

Glick, B. R. 2003. Phytoremediation: Synergistic use of plants and bacteria to clean up the environment. *Biotechnology Advances* 21, 383–393.

Gonzalez, R. C. and Gonzalez-Chavez, M. C. A. 2006. Metal accumulation in wild plants surrounding mining wastes: Soil and sediment remediation (SSR). *Environmental Pollution* 144, 84–92.

Gosavi, K., Sammut, J., Gifford, S. and Jankowski, J. 2004. Macroalgal biomonitors of trace metal contamination in acid sulfate soil aquaculture ponds. *Science of the Total Environment* 324, 25–39.

Hasan, S. H., Talat, M. and Rai, S. 2007. Sorption of cadmium and zinc from aqueous solutions by water hyacinth (*Eichhornia crassipes*). *Bioresource Technology* 98(4), 918–928.

Hong, M. S., Farmayan, W. F., Dortch, I. J., Chiang, C. Y., McMillan, S. K. and Schnoor, J. L. 2001. Phytoremediation of MTBE from groundwater plume. *Environmental Science and Technology* 35, 1231–1239.

Huang, J. W., Blaylock, M. J., Kapulnik, Y. and Ensley, B. D. 1998. Phytoremediation of uranium-contaminated soils: Role of organic acids in triggering uranium hyperaccumulation in plants. *Environmental Science and Technology* 32, 2004–2008.

Ingersoll, T. and Baker, L. A. 1998. Nitrate removal in wetland microcosms. *Water Research* 32, 677–684.

Iqbal, M. C. M., Kulasekara, L., Rajakaruna, N. and Iqbal, S. S. 2006. Plant-soil relations of a serpentine site in the southern coast of Sri Lanka. 5th International Conference on Serpentine Ecology. Siena, Italy, 09–13 May. (Abstract).

Jaffré, T., Brooks, R. R., Lee, J. and Reeves, R. D. 1976: *Sebertia acuminata*: A hyperaccumulator of nickel from New Caledonia. *Science* 193, 579–580.

Jaffré, T., Kersten, W., Brooks, R. R. and Reeves, R. D. 1979. Nickel uptake by Flacourtiaceae of New Caledonia. *Proceedings of the Royal Society of London* B205, 385–394.

Jayasekera, R. and Rossbach, M. 1994. Trace elements in plants growing in contrasting ecosystems in Sri Lanka: Analytical aspects. *Environmental Geochemistry and Health* 16, 171–179.

Jayasekera, R. and Rossbach, M. 1996. Background levels of heavy metals in plants of different taxonomic groups from a montane rain forest in Sri Lanka. *Environmental Geochemistry and Health* 18, 55–62.

Jefferson, L. V. 2004. Implications of plant density on the resulting community structure of mine site land. *Restoration Ecology* 12, 429–438.

Jopony, M. and Tongkul, F. 1999. Hyperaccumulation of nickel by *Rinorea benghalensis* in ultrabasic soils of Sabah, Malaysia. 3rd International Conference on Environmental Chemistry and Geochemistry in the Tropics. Hong Kong, 24–26 November. (Abstract).

Keller, C., Hammer, D., Kayser, A., Richner, W., Brodbeck, M. and Sennhauser, M. 2003. Root development and heavy metal phytoextraction efficiency: Comparison of different plant species in the field. *Plant Soil* 249, 67–81.

Knight, B., Zhao, F. J., McGrath, S. P. and Shen, Z. G. 1997. Zinc and cadmium uptake by the hyperaccumulator *Thlaspi caerulescens* in contaminated soils and its effects on the concentration and chemical speciation of metals in soil solution. *Plant Soil* 197, 71–78.

Kramer, P. A., Zabowski, D., Scherer, G. and Everett, R. L. 2000. Native plant restoration of copper mine tailings: I. Substrate effect on growth and nutritional status in a greenhouse study. *Journal of Environmental Quality* 29, 1762–1769.

Krämer, U. 2003. Phytoremediation to phytochelatin – plant trace metal homeostasis. *New Phytologist* 158, 4–6.

Kratochvil, D. and Volesky, B. 1998. Advances in the biosorption of heavy metals. *Trends in Biotechnology* 16, 291–300.

Krzaklewski, W. and Pietrzykowski, M. 2002. Selected physicochemical properties of zinc and lead ore tailings and their biological stabilisation. *Water Air Soil Pollution* 141, 125–142.

Kumar, A, Bisht, B. S., Joshim, V. D. and Dhewa, T. 2011. Review on bioremediation of polluted environment: A management tool. *International Journal of Environmental Science* 1, 1079–1093.

Kumar, P. B. A. N., Dushenkov, V., Motto, H. and Raskin, I. 1995. Phytoextraction: The use of plants to remove heavy metals form soils. *Environmental Science and Technology* 29, 1232–1238.

Kumari, R., Kaur, I. and Bhatnagar, A. K. 2013. Enhancing soil health and productivity of *Lycopersicon esculentum* Mill. using *Sargassum johnstonii* Setchell and Gardner as a soil conditioner and fertilizer. *Journal of Applied Phycology* 25(4), 1225–1235.

Lasat, M. M., Fuhrmann, M., Ebbs, S. D., Cornish, J. E. and Kochian, L. V. 1998. Phytoremediation of a radiocesium-contaminated soil: Evaluation of cesium-137 bioaccumulation in the shoots of three plant species. *Journal of Environmental Quality* 27, 165–169.

Leita, L., De Nobili, M., Pardini, G., Ferrari, F. and Sequi, S. 1989. Anomalous contents of heavy metals in soils and vegetation of a mine area in S.W. Sardinia, Italy. *Water Air Soil Pollution* 48, 423–433.

Li, Y, and Li, B. 2011. Study on fungi-bacteria consortium bioremediation of petroleum contaminated mangrove sediments amended with mixed biosurfactants. *Advanced Materials Research* 183, 1163–1167.

Lutts, S., Lefevre, I., Delperee, C., Kivits, S., Dechamps, C., Robledo, A., Correal, E. (2004): Heavy metal accumulation by the halophyte species Mediterranean saltbush. *Journal of Environmental Quality* 33: 1271–1279.

Mane, A. V., Saratale, G. D., Karadge, B. A. and Samant, J. S. 2011. Studies on the effects of salinity on growth, polyphenol content and photosynthetic response in *Vetiveria zizanioides* (L.) Nash. *Emirates Journal of Food and Agriculture*23(1), 59–70.

Mannino, I., Franco, D., Piccioni, E., Favero, L., Mattiuzzo, E. and Zanetto, G. 2008. A cost-effectiveness analysis of seminatural wetlands and activated sludge wastewater treatment systems. *Environmental Management* 41, 118–129.

Marschner, H. 1995. *Mineral Nutrition of Higher Plants*. Academic Press Inc., San Diego.

McGaw, L. J., Rabe, T., Sparg, S. G., Jäger, A. K., Eloff, J. N. and van Staden, J. 2001. An investigation on the biological activity of *Combretum* species. *Journal of Ethnopharmacology* 75(1), 45–50.

Meerts, P. and Van Isacker, N. (1997): Heavy metal tolerance and accumulation in metallicolous and non-metallicolous populations of *Thlaspi caerulescens* from continental Europe. *Plant Ecology* 133, 221–231.

Melato, F. A., Mokgalaka, N. S. and McCrindle, R. I. 2016. Adaptation and detoxification mechanisms of Vetiver grass (*Chrysopogon zizanioides*) growing on gold mine tailings. *International Journal of Phytoremediation* 18(5), 509–520.

Mendez, M. O., Glenn, E. P. and Maier, R. M. 2007. Phytostabilization potential of quailbush for mine tailings: Growth, metal accumulation, and microbial community changes. *Journal of Environmental Quality* 36, 245–253.

Michelle, B.da-C., Rosane, A. and Jaime, W. V.de-M. 2010. Phytoremediation of acid mine drainage by aquatic floating macrophytes. – Institute of Science and Technology for Mineral Resource, Water and Biodiversity, INCT-ACQUA – Annual Report 1–4.

Mishra, V. K., Upadhyaya, A. R., Pandey, S. K. and Tripathi, B. D. 2008a. Heavy metal pollution induced due to coal mining effluent on surrounding aquatic ecosystem and its management through naturally occurring aquatic macrophytes. *Bioresource Technology* 99, 930–936.

Mishra, V. K., Upadhyaya, A. R., Pandey, S. K. and Tripathi, B. D. 2008b. Phytoremediation of mercury and Arsenic from Tropical open cast coalmine effluent through naturally occurring aquatic macrophytes. *Water, Air and Soil Pollution* 192, 303–314.

Mokhtar, H., Morad, N. and Ahmad Fizri, F. F. 2011. Hyperaccumulation of copper by two species of aquatic plants. International Conference on Environmental Science and Engineering IPCBEE 8. IACSIT Press, Singapore.

Mudhoo, A., Sharma, S. K., Lin, Z. Q. and Dhankher, O. P. 2010. Phytoremediation of arsenic-contaminated environment an overview. In Sharma, S. K. and Mudhoo, A. (eds.), *Green Chemistry for Environmental Sustainability*. Taylor and Francis Group, Boca ratan London, New York, 127.

Mummey, D. L., Stahl, P. D. and Buyer, J. S. 2002a. Microbial biomarkers as an indicator of ecosystem recovery following surface mine reclamation. *Applied Soil Ecology* 21, 251–259.

Mummey, D. L., Stahl, P. D. and Buyer, J. S. 2002b. Soil microbiological properties 20 years after surface mine reclamation: Spatial analysis of reclaimed and undisturbed sites. *Soil Biology and Biochemistry* 34, 1717–1725.

Munshower, F. F. 1994. *Practical Handbook of Disturbed Land Revegetation*. Lewis Publishing, Boca Raton.

National Research Council. 2005. *Mineral Tolerance of Animals*. National Academies Press, Washington.

Newete, S. W. and Byrne, M. J. 2016. The capacity of aquatic macrophytes for phytoremediation and their disposal with specific reference to water hyacinth. *Environmental Science and Pollution Research* 23(11), 10630–10643.

Osmond, C. B., Bjorkman, O. and Anderson, D. J. (1980). *Physiological Processes in Plant Ecology: Toward a Synthesis with Atriplex*. Springer-Verlag, New York.

Pichtel, J., Salt, C. A. (1998): Vegetative growth and trace metal accumulation on metalliferous wastes. – *Journal of Environmental Quality* 27: 618–624.

Pierzynski, G. M., Lambert, M., Hetricck, B. A. D., Weeney, D. W. and Erickson, L. E. 2002. Phytostabilization of metal mine tailings using tall fescue. *Practical Periodical Hazard and Toxic Radioactive Waste Manage* 6, 212–217.

Piha, M. I., Vallack, H. W., Michael, N. and Reeler, B. M. 1995. A lowinput approach to vegetation establishment on mine and coal ash wastes in semiarid regions. 2. Lagooned pulverized fuel ash in Zimbabwe. *Journal of Applied Ecology* 32, 382–390.

Pilon-Smits, E. A. H. 2005. Phytoremediation. *Annual Review of Plant Biology* 56, 15–39.

Pilon-Smits, E. and Pilon, M. 2002. Phytoremediation of metals using transgenic plants. *Critical Review in Plant Science* 21, 439.

Prasad, M. N. V. and Freitas, H. M. O. 2003. Metal hyperaccumulation in plants – Biodiversity prospecting for phytoremediation technology. *Electronic Journal of Biotechnology* 6, 285–321.

Prescott, L. M., Harley, J. P. and Klein, D. A. 2002. *Microbiology*. McGraw-Hill, New York, 5, 1014.

Proctor, J. 2003. Vegetation and soil and plant chemistry on ultramafic rocks in the tropical Far East. *Perspectives in Plant Ecology, Evolution, and Systematics* 6, 105–124.

Proctor, J., Phillipps, C., Duff, G. K., Heaney, A. and Robertson, F. M. 1989. Ecological studies in Gunung Silam, a small ultrabasic mountain in Sabah, Malaysia. II. Some forest processes. *Journal of Ecology* 77, 317–331.

Prokop, G., Schamann, M. and Edelgaard, I. 2000. Management of contaminated sites in western Europe. *European Environment Agency, Copenhagen, Denmark* 171.

Qadir, S., Qureshi, M. I., Javed, S. and Abdin, M. Z. 2004. Genotypic variation in phytoremediation potential of *Brassica juncea* cultivars exposed to Cd stress. *Plant Science* 167, 1171–1181.

Qin, H., Zhang, Z., Liu, M., Liu, H., Wang, Y., Wen, X., Zhang, Y. and Yan, S. 2016. Site test of phytoremediation of an open pond contaminated with domestic sewage using water hyacinth and water lettuce. *Ecological Engineering* 95(1), 753–762.

Rahman, M. A., Hasegawa, H., Ueda, K., Maki, T., Okumura, C. and Rahman, M. M. 2007. Arsenic accumulation in duckweed (*Spirodela polyrhiza* L.): A good option for phytoremediation. *Chemosphere* 69(3), 493–499.

Rai, P. K. 2008. Heavy metal pollution in aquatic ecosystems and its phytoremediation using wetland plants: An Eco-sustainable approach. *International Journal of Phytoremediation* 10(2), 131–158.

Rajakaruna, N. and Baker, A. J. M. 2004. Serpentine: A model habitat for botanical research in Sri Lanka. *Ceylon Journal of Science (Biological Sciences)* 32, 1–19.

Rajakaruna, N. and Bohm, B. A. 2002. Serpentine and its vegetation: A preliminary study from Sri Lanka. *Journal of Applied Botany* 76, 20–28.

Ramadan, T. 2001. Dynamics of Salt Secretion by *Sporobolus spicatus* (Vahl) Kunth from Sites of Differing Salinity. *Annals of Botany* 87, 259–266.

Raskin, I. and Ensley, B. D. 2000. *Phytoremediation of Toxic Metals – Using Plants to Clean up the Environment.* J. Wiley & Sons, New York, 304.

Reeves, R. D. 2003. Tropical hyperaccumulators of metals and their potential for phytoextraction. *Plant and Soil* 249, 57–65.

Robinson, B. H., Green, S., Mills, T., Clothier, B., Velde, M, Laplane, R., Fung, L., Deurer, M., Hurst, S., Thayalakumaran, T. and Dijssel, C. 2003. Phytoremediation: Using plants as blowups to improve degraded environments. *Australian Journal of Soil Research* 41, 599–611.

Rosario, K., Iverson, S. L., Henderson, D. A., Chartrand, S., McKeon, C., Glenn, E. P. and Maier, R. M. 2007. Bacterial community changes during plant establishment at the San Pedro River mine tailings site. *Journal of Environmental Quality* 36, 1249–1259.

Sabey, B. R., Pendleton, R. L. and Webb, B. L. 1990. Effect of municipal sewage-sludge application on growth of two reclamation shrub species in copper mine spoils. *Journal of Environmental Quality* 19, 580–586.

Salt, D. E., Pickering, I. J., Prince, R. C., Gleba, D., Dushenkov, S., Smith, R. D. and Raskin, I. 1997. Metal accumulation by aquacultured seedlings of Indian mustard. *Environmental Science and Technology* 31, 1636–1644.

Salt, D. E., Smith, R. D. and Raskin, I. 1998. Phytoremediation. *Annual Review of Plant Physiology and Plant Molecular Biology* 49, 643–68.

Schmidt, U. 2003. Enhancing phytoextraction: The effect of chemical soil manipulation on mobility, plant accumulation, and leaching of heavy metals. *Journal of Environmental Quality* 32, 1939–1954.

Sharma, S., Singh, B. and Manchanda, V. K. 2014. Phytoremediation: Role of terrestrial plants and aquatic macrophytes in the remediation of radionuclides and heavy metal contaminated soil and water. *Environmental Science and Pollution Research* 22(2), 946–962.

Shu, W. S., Ye, Z. H., Lan, C. Y., Zhang, Z. Q. and Wong, M. H. 2001. Acidification of lead/zinc mine tailings and its effect on heavy metal mobility. *Environmental International* 26, 389–394.

Singh, G., Bhati, M. and Rathod, T. 2010. Use of tree seedlings for the phytoremediation of a municipal effluent used in dry areas of north-western India: Plant growth and nutrient uptake. *Ecological Engineering* 36(10), 1299–1306.

Sivasankari, B. and Ravindran, D. 2016. A study on chemical analysis of water hyacinth (*Eichhornia crassipes*) and water lettuce (*Pistia stratiotes*). *International Journal of Innovative Research in Science and Technology* 5(10), 17566–17570.

Smith, R. A. H. and Bradshaw, A. D. 1979. The use of metal tolerant plant populations for the reclamation of metalliferous wastes. *Journal of Applied Ecology* 16, 595–612.

Snow, A. A., Andow, D. A., Gepts, P., Hallerman, E. M., Power, A., Tiedje, J. M. and Wolfenbarger, L. L. 2005. Genetically engineered organisms and the environment: Current status and recommendations. *Ecological Application* 10, 377–404.

Sood, A., Uniyal, P. L., Prasanna, R. and Ahluwalia, A. S. 2012. Phytoremediation potential of aquatic macrophyte, Azolla. *Ambio* 41, 122–137.

Stoffberg, G. H., van Rooyen, M. W., van der Linde, M. J. and Groeneveld, H. T. 2008. Predicting the growth in tree height and crown size of three street tree species in the City of Tshwane, South Africa. *Urban Forestry and Urban Greening* 7, 259–264.

Tilman, D. 1988. *Plant Strategies and the Dynamics and Structure of Plant Communities.* Princeton University Press, Princeton, NJ.

Truong, P. and Hart, B. 2001. Vetiver grass for wastewater treatment. Pacific Rim Vetiver Network Technical Bulletin No. 2001/2.

Tucker, R. K. and Shaw, J. A. 2000. Phytoremediation and public acceptance. In Raskin, I. and Ensley, B. D. (eds.), *Phytoremediation of Toxic Metals. Using Plants to Clean Up the Environment.* J. Wiley & Sons, New York, 33–42.

Turnau, K., Henriques, F. S., Anielska, T., Renker, C. and Buscot, F. 2007. Metal uptake and detoxification mechanisms in *Erica andevalensis* growing in a pyrite mine tailing. *Environmental and Experimental Botany* 61, 117–123.

United States Environmental Protection Agency (USEPA). 2000. Introduction to Phytoremediation. EPA, U.S. Environmental Protection Agency, Office of Research and Development, Cincinnati, OH.

US Congress Office of technology assessment. 1991. *Bioremediation for Marine Oil Spills Background Paper, OTA-BP-O-70.* U.S. Government Printing Office, Washington, DC.

Van Dillewijn, P., Caballero, A., Paz, J. A., Gonzalez-Perez, M. M., Oliva, J. M. and Ramos, J. L. 2007. Bioremediation of 2,4,6 trinitrotoluene under field conditions. *Environmental Science & Technology* 41, 1378–1383.

Van Rensburg, L. and Morgenthal, T. 2004. The effect of woodchip waste on vegetation establishment during platinum tailings rehabilitation. *South African Journal of Science* 100, 294–300.

Vidali, M. 2001. Bioremediation-An overview. *Pure and Applied Chemistry* 73(7), 1163–1172.

Von Willert, F. J. and Stehouwer, R. C. 2003. Compost and calcium surface treatment effects on subsoil chemistry in acidic minespoil columns. *Journal of Environmental Quality* 32, 781–788.

Wagner, S., Truong, P., Vieritz, A. and Smeal, C. 2003. Response of vetiver grass to extreme nitrogen and phosphorus supply. Proceedings of the 3rd International Conference on Vetiver and Exhibition, Guangzhou, China.

Watson, I. W., Westoby, M. and Holm, A. M. 1997. Continuous and episodic components of demographic change in arid zone shrubs: Models of two *Eremophila* species from Western Australia compared with published data on other species. *Journal of Ecology* 85, 833–846.

Windsor, D. M. and Clements, A. 2001. A germination and establishment field trial of *Themeda australis* (kangaroo grass) for mine site restoration in the Central Tablelands of New South Wales. *Restoration Ecology* 9, 104–110.

Wither, E. D. and Brooks, R. R. 1977. Hyperaccumulation of nickel by some plants of Southeast Asia. *Journal of Geochemical Exploration* 8, 579–583.

Wolfe, A. K. and Bjornstad, D. J. 2002. Why would anyone object? An exploration of social aspects of phytoremediation acceptability. *Critical Review in Plant Science* 21, 429–438.

Wolfenbarger, L. L. and Phifer, P. R. 2000. Biotechnology and ecology-the ecological risks and benefits of genetically engineered plants. Science 290, 2088–2093.

Wong, M. H. 2003. Ecological restoration of mine degraded soils, with emphasis on metal contaminated soils. *Chemosphere* 50, 775–780.

Wood, M. K., Buchanan, B. A. and Skeet, W. 1995. Shrub preference and utilization by big game on New Mexico reclaimed mine land. *Journal of Range Management* 48, 431–437.

Wu, J., Hsu, F. C. and Cunningham, S. D. 1999. Chelate-assisted Pb phytoextraction: Pb availability, uptake, and translocation constraints. *Environmental Science and Technology* 33, 1898–1904.

Ye, Z. H., Wong, J. W. C., Wong, M. H., Lan, C. Y. and Baker, A. J. M. 1999. Lime and pig manure as ameliorants for revegetating lead/zinc mine tailings: A greenhouse study. *Bioresource Technology* 69, 35–43.

Yerima, M. B., Umar, A. F., Shinkafi, S. A. and Ibrahim, M. L. 2012. Bioremediation of hydrocarbon pollution: A sustainable means of biodiversity conservation. *Journal of Environment Protection and Sustainable Development*3, 43–50.

Zhao, F. J., Lombi, E. and McGrath, S. P. 2003. Assessing the potential for zinc and cadmium phytoremediation with the hyperaccumulator *Thlaspi Caerulescens*. *Plant Soil* 249, 37–43.

Zheng, C. R., Tu, C. and Chen, H. M. 1997. Preliminary study on purification of eutrophic water with vetiver. Proceedings of the International Vetiver Workshop, Fuzhou, China.

Zhu, Y. L., Pilon-Smits, E. A. H., Jouanin, L. and Terry, N. 1999. Overexpression of glutathione synthetase in *Brassica juncea* enhances cadmium tolerance and accumulation. *Plant Physiology* 119, 73–79.

Zouboulis, A. I. and Katsoyiannis, I. A. 2005. Recent advances in the bioremediation of arsenic-contaminated groundwaters. *Environment International* 31(2), 213–219.

16 Phytoremediation
A Comprehensive Review on the Prospective Benefits and Biosafety of Vegetables

Geethu Krishna Kumar, Srija Sappa, Nikhil Kumar Ramesha, Arya Kaniyassery, Mutiu A. Alabi, Nisha Govender, and Annamalai Muthusamy

16.1 INTRODUCTION

Pollution is the entry of harmful chemicals into the soil, air, water, etc. These chemicals that cause pollution are termed *pollutants*. Human activity has an adverse effect on the environment, and the main set of organisms affected by pollution are the human population (Elliot, 2003; Aransiola et al., 2023). The industrial revolution has played a major role in introducing and raising air pollution to alarming rates. Climate changes and global warming due to pollution are causing serious problems such as food shortages, iceberg melting, deadly diseases, extinction of animals, and damage to plants. In recent times, the health effects of air pollution have raised concern and are studied intensely. The magnitude of the concentration of air pollution is much higher in rural areas of developing countries rather than the developed world due to the use of biomass fuels (McMichael et al., 1999; Smith et al., 2000). Air pollution is due to a mixture of liquid and solid particles suspended in the air, and the size of these particles can vary. In the troposphere of the atmosphere, ozone that is considered to be a strong oxidizing agent is formed via a series of complex reactions that involves the action of sunlight on hydrocarbons and nitrogen dioxide. Apart from this, burning of fossil fuels in vehicles and from heating or power generation is the major source of anthropogenic emission into the atmosphere (World Health Organization, 2000). Air pollutants including nitrogen oxide, volatile organic compounds, sulfur dioxide, carbon monoxide, ozone, and heavy metals accumulate in the air due to both natural and human activities. When exposed to living organisms over a long period of time, these can cause adverse effects. These effects in humans can be acute or chronic affecting their body system from a minor inflammation or irritation to various cancers. Pollutants like heavy metals and microplastics can contaminate food and water, which makes ingestion a route of intake of pollutants (Thron, 1996). The entry of unwanted materials into water bodies, such as discharge from industrial and domestic wastes, sewage tank leakage into the water, marine dumping, and heavy metals that changes the water quality is called *water pollution* (Alrumman et al., 2016). This is harmful to both the environment and the human health. These waste materials can accumulate in water bodies and directly damage aquatic life (Briggs, 2003).

Food safety is a concern of global scale. Environmental pollution has been reported to reduce yield production and yield quality. The entry of pollutants in the food chain is a major health issue. Biosafety methods and guidelines are necessary to ensure that the food has no more than optimal level of toxic contaminants. This chapter makes an effort to emphasis the cause and effects of the contaminants in the fruits and vegetables. It also discusses the various remedial approaches to increase the efficient biosafety methods (Figures 16.1 and 16.2).

DOI: 10.1201/9781032683751-19

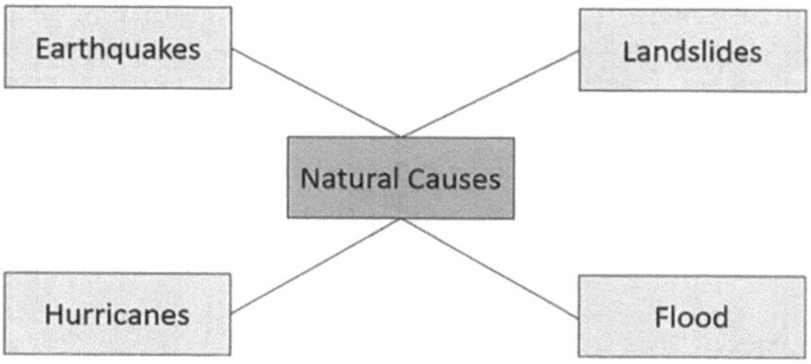

FIGURE 16.1 Sources of pollution by natural phenomena.

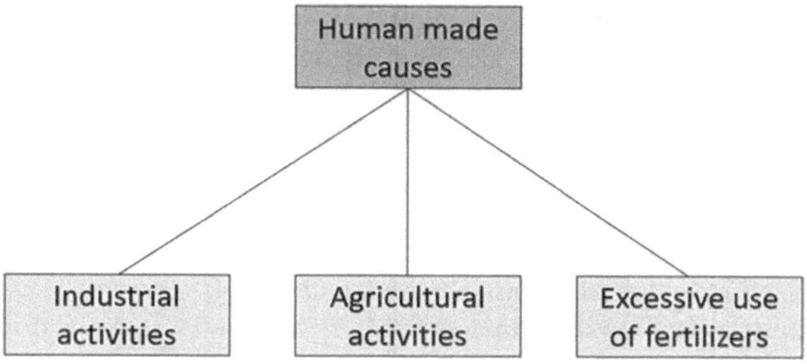

FIGURE 16.2 Sources of pollution by human activities.

16.2 ENTRY OF POLLUTANTS INTO THE FOOD CHAIN

Soil pollution is due to the intensification of agricultural practices such as excessive usage of fertilizers, salinization caused by irrigation, and accumulation of heavy metals. When the chemicals that persist in soil enter a body of water, this leads to water pollution, and when they are released into the atmosphere, this contributes to air pollution. Major chemicals found in soil are hydrocarbons, polychlorinated biphenyls, detergents, cadmium (Cd), chromium (Cr), and lead (Pb). As these chemicals are not water soluble, they tend to accumulate in the soil and enter the food chain and cause adverse effects to the organism at the top of the food chain.

Humans can obtain nutrients from plants by consuming plant-based foods such as fruits, vegetables, grains, nuts, and plant parts such as leaves and legumes. These foods provide us with essential nutrients, minerals, and fiber. Apart from nutrients, they also exhibit antioxidant properties that are important for overall health and well-being. Plants in general are a rich source of nutrients such as vitamins A, C, K, folate, potassium, and fiber. Consuming a variety of plant-based foods provide us with plenty of essential nutrients that prove to be important for maintaining good health and preventing chronic diseases. Nutrients were traditionally considered as components that the human body could not synthesize. However, recent studies show that many other plant components such as dietary fiber, phenolic acids, sterols, and flavonoids are linked with decreased disease risk (Panda et al., 2015).

16.3 PHYTOREMEDIATION

Phytoremediation can be defined as the in situ treatment of contaminated soil, water, and sediments using vegetation. Here, roots of plants access the soil containing nutrient, organic, or metal pollutants (Cunningham et al., 1996). This type of remediation is cost-effective, non-invasive, and applicable in the long term and is publicly acceptable for the elimination of pollutants from the environment (Schnoor et al., 1995; Aransiola et al., 2013). Contaminants are removed using both in situ and ex situ remediation techniques. It is possible to employ plants to gather and break down organic pollutants, and the root area helps with the microbial degradation of these compounds (Sparks et al., 2022; Aransiola et al., 2019). There are various forms of phytoremediation, including rhizodegradation and phytoextraction, phytoimmobilization, phytostabilization, and phytofiltration and rhizofiltration (Arthur et al., 2005).

16.3.1 Phytofiltration

This form of remediation uses the hydroponically cultivated plant roots to take up the contaminants from water into biomass, thus removing the pollutants. The pollutants are removed by means of precipitation, concentration, and absorption (Salt et al., 1994, 1995; Abioye et al., 2017). Phytofiltration can be helpful to recover and reuse the toxic yet valuable metals that pose serious threat. To extract and recover Pb from wastewater, Sekhar et al. (2004) presented a unique phytofiltration process that makes use of plant-based biomaterials from the bark of an Indian sarsaparilla plant.

16.3.2 Phytoextraction

The ability of plants to absorb inorganic toxins, particularly metals, is known as phytoextraction, and it is a commonly utilized remediation approach (McCutcheon and Schnoor, 2003). Plants have the ability to uptake organic as well as inorganic substances, such as soil-derived metals. While some nutrients are known to be vital, others are unknown to them. Compared to disposing of contaminated soil, these techniques are more appealing from a cost and environmental standpoint (Ernst, 1996).

16.3.3 Phytoimmobilization and Phytostabilization

Phytoimmobilization is a method that uses plants to take out pollutants; then these contaminants are released during the decomposition of plants when they are immobilized in geomat, a mineral-containing mat. Phytostabilization is the elimination of toxic metals by forming complexes with some plants' processes (Kopyra et al., 2003). This does not eliminate the contaminants from the soil; nonetheless, it does lessen the risk (Li et al., 2015) by converting contaminants into an inactive state (Cunningham et al., 1997).

16.3.4 Rhizofiltration and Phtyodegradation

Degradation of contaminants, which is the breakdown of the compounds into smaller constituents, is one of the most important stages of remediation. In the plant setup, degradation can take place at the root zone (rhizosphere) as well as within the plant. When it takes place via the roots it is termed rhizofiltration, while the latter wherein the degradation occurs when plant takes up the pollutant into the plant's tissues and enzymes works to breakdown the contaminant is known as phytodegradation (Walton and Anderson, 1990; Siciliano et al., 1998; Kaiser et al., 2021; Aransiola et al., 2022) (Figures 16.3 and 16.4).

16.4 POSSIBLE DISPOSAL OF PHYTOREMEDIATION PLANTS

Contaminant buildup in crop biomass during plant growth is a component of phytoremediation (Attinti et al., 2017). According to Shi et al. (2018), harvesting these plants produces highly contaminated biowaste, which if handled improperly could quickly turn into a secondary source of pollution. Hazardous waste is what the biowaste is. For such biowaste, proper disposal and use techniques are also necessary. Heat treatment, extraction treatment, microbiological treatment, compression landfill,

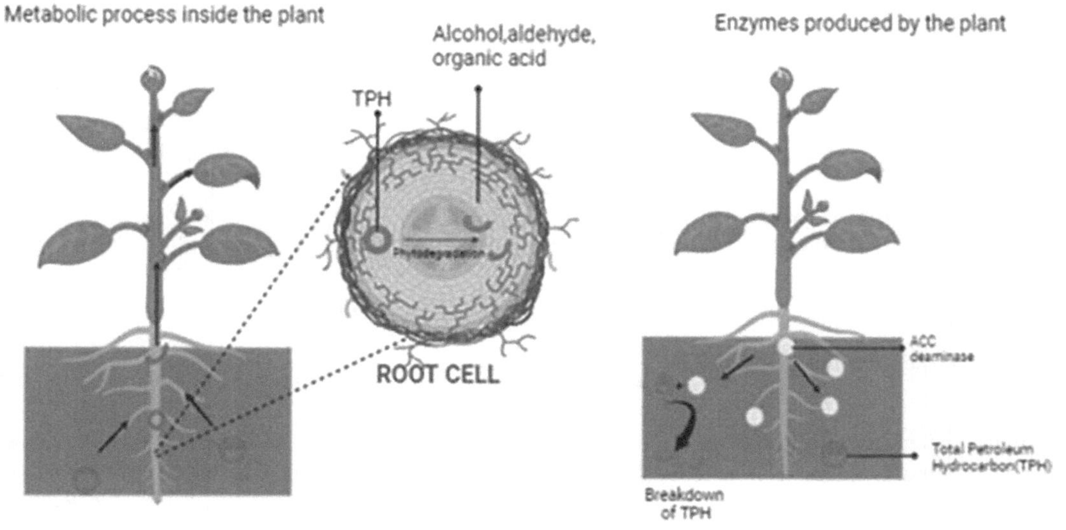

FIGURE 16.3 Mechanism of phytodegradation.

Source: Created using Biorender.com

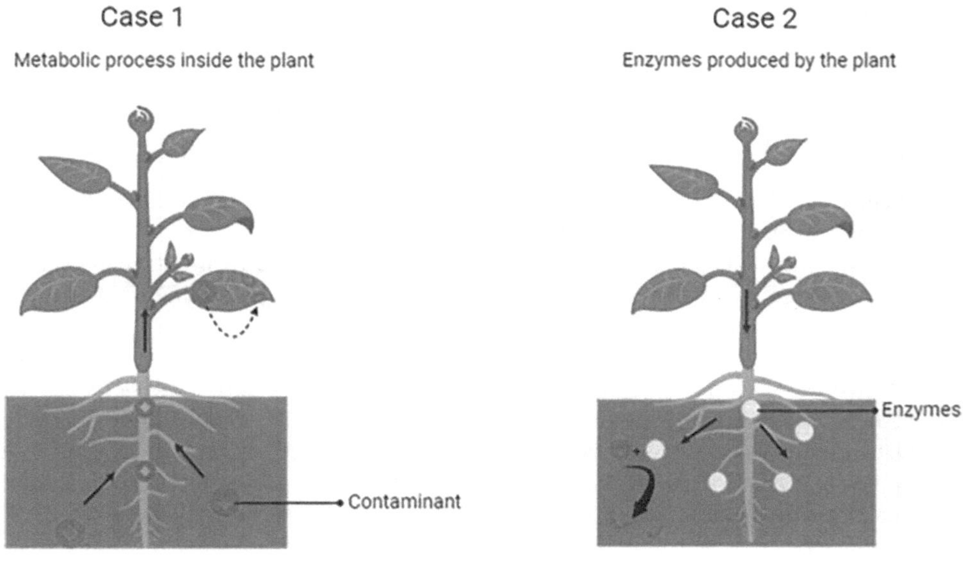

FIGURE 16.4 Phytodegradation in different cases.

and nanomaterial synthesis are the various techniques of disposal and utilization methods for phytoremediation plants with heavy metals (Liu and Tran, 2021). Most phytoremediation problems are directly related to the hyperaccumulator plants since the process of phytoremediation entails the contaminants' bioaccumulation within the plant's biomass (Mousavi et al., 2021). Therefore, harvesting these plants could produce biowastes, which are exceedingly toxic and could lead to secondary contamination if not managed properly. Development of suitable management techniques for such hazardous biowastes is therefore essential (Cai et al., 2018). In order to improve the generation of biofuel and biochemical products from various types of products made from lignocellulosic biowaste

FIGURE 16.5 Different methods for disposal of phytoremediation plants.

Source: Adopted from Tan et al. (2023).

through conversion processes, Vu et al. (2019) studied cutting-edge pre-treatment approaches that were economical and environmentally beneficial. A variety of bioprocesses, including simultaneous saccharification and fermentation and separate hydrolysis and fermentation, can be used to obtain the biochemicals, which include organic acids (e.g., lactic and acetic acid), microbial enzymes (e.g., pectinase, amylase, and cellulase), and bioactive substances (e.g., phenolic compounds and phenylpropanoids). In the meantime, the main by-products of fermentation and anaerobic digestion were liquid biofuels (e.g., fuel and ethanol) and biogas (e.g., as methane).

This section explored a number of disposal, handling, and utilization techniques, such as extraction, heat treatment, composting, compaction, synthesis of nanomaterials, and phytomining, such as heat treatment, extraction, composting, compaction, nanomaterial synthesis, and phytomining, to handle the post-phytoremediation plants containing heavy metals (Figure 16.5).

16.5 ENHANCEMENT OF PHYTOREMEDIATION POTENTIAL OF PLANTS

Physical, chemical, and microbiological techniques are among the ways that can be used to improve the effectiveness of phytoremediation. Shen et al. (2022) assert that physically and chemically enhancing the mobility and bioavailability of heavy metals can enhance their uptake by plants. The effectiveness of phytoremediation has been enhanced by physical techniques like the use of light radiation and electric fields. With the aid of blue and red light, Luo et al. (2019) were able to increase the phytoremediation efficacy of blue gum (*Eucalyptus globulus*). The phytoremediation rates of Pb, Cu, and Cd increased in their experiment, going from 28.9% to 70.6%, 71.1% to 88.7%, and 50.6% to 65.6%, respectively. In addition to applying light irradiation, they had found in a different investigation that a static magnetic field might increase blue gum's translocation factor towards these heavy metals, hence enhancing the metals' transport from roots to shoots. Conversely, chemical approaches employ biochars, acids, or artificial chelators to boost the effectiveness of phytoremediation. As an example, Jiang et al.'s (2019) study found that adding ethylene diamine tetraacetic acid as a chelator increased Pb phytoremediation within the tissues of the little bamboo (*Bambusa multiplex*) by over 40%. This suggests that the chelator improved Pb accessibility, movement, and assimilation of lead by little bamboo. Hu and colleagues (2019) reported that the addition of citric acid to plume poppy (Macleaya cordata) enhanced the uranium's bioaccumulation and bioconcentration factors. Furthermore, Zhang et al. (2019a) and Zhang et al. (2019b) applied 1.5 weight percent of biochar to lucerne (*Medicago sativa*) and were able to treble the phytoextraction rate of Cd.

Further, utilizing microorganisms from the plant rhizosphere, microbiological techniques aim to enhance the effectiveness of phytoremediation. While most living things are negatively impacted by heavy metals, many microbes are resistant to these substances and are crucial for the growth of plants. These microbes' metabolites may also contribute to a rise in the bioavailability of heavy metals, which in turn enhances the efficacy of phytoremediation (Li et al., 2020). To investigate *Pseudomonas azotoformans* ASS1's impact on the stone clover (*Trifolium arvense*) rhizosphere, Ma et al. (2017) reported that the microbe was found to be able to considerably increase the plant's chlorophyll content and growth rate, leading to an increment of the percentages of Cd, Ni, and Zn uptake by 51%, 26%, and 63%. By treating the sunflower (*Helianthus annuus*) shoots with *Trichoderma* isolated from decaying wood, Govarthanan et al. (2018) were also able to raise the arsenic (As) and Pb build up to 67% and 59%, respectively. The increasing understanding of the elements crucial to phytoremediation can serve as a foundation for the genetic engineering of plants for enhanced functionality. For years, plant attributes that are crucial to agronomy have been altered by breeders. The selection factors were frequently yield and aesthetics. A new paradigm where the value of plants is determined by their capacity to absorb, sequester, destroy, and tolerate is necessary for phytoremediation. Both conventional breeding methods and molecular biology can be used to specifically target any one of these attributes. Plant root structure could be enhanced in several ways. These consist of root density, root depth, and root penetration into anaerobic zones.

The most frequent technique is to introduce and overexpress genes involved in the absorption, translocation, and sequestration of metals. Given that metal metabolism in plants is particularly complex, there are multiple potential pathways which can be manipulated to promote metal accumulation in plant tissues, including metal mobilization and uptake via roots, chelator-ligand formation, deposition in vacuoles, and the transfer of metals roots to shoots through symplast or apoplast path (Koźmińska et al., 2018). The chloroplast transformation vectors were utilized to genetically alter *Nicotiana tabacum* plants to accumulate mercury. The same study reported that not only did the tobacco seeds germinate in media contaminated with mercury (Hg), but increased Hg accumulation was observed (Ruiz et al., 2011).

16.6 STATUS OF POLLUTANTS IN VEGETABLES AND FRUITS

Consumption of vegetables and fruits along with daily meals are good for human beings as they contain necessary components (Garba et al., 2021). Because of their high nutritional value and beneficial antioxidative effects on the human diet, eating fruits and vegetables on a regular basis is probably one of the best ways to improve your health (Atun et al., 2018). They are additionally beneficial in preventing a range of chronic diseases (Agrawal et al., 2007) as they are essential sources of micronutrients such as Cu, calcium (Ca), iron (Fe), magnesium (Mg), Zn, sodium (Na), potassium (K), antioxidants, vitamins, iodine (I), and various other metabolites (Liang et al., 2020). Fruits and vegetables are consumed in both raw and cooked forms; therefore, those containing contaminants are harmful for ingestion, and contaminants have been proven to have entered through the food chain worldwide (Ahmad et al., 2021). In terms of public health, food safety and quality are arguably the most crucial issues. Food manufacturers, state governments, and organizations that systematically monitor and control food quality should all be accountable for ensuring that food is safe and free of any chemical contaminants that could endanger the health of consumers (Igwegbe et al., 2013). The environmental conditions throughout the growing period of fruits and vegetables might serve as a cause of their heavy metal contamination (Tchounwou et al., 2012) (Table 16.1).

TABLE 16.1
Heavy Metals and Their Characteristics

Heavy Metal	Characteristics
Cadmium (Cd)	Light, ductile, bluish-white, exists in mineral form
Mercury (Hg)	Exists in the form of Hg_2^{++}, Hg^{++}
Arsenic (As)	Noxious carcinogen, exists in the form of oxides, sulfides, etc.
Lead (Pb)	Uniformly distributed trace metal, exists in ionic, oxide, hydroxide forms

Source: Manwani et al. (2022).

Trace amounts of Cu, Zn, and Ni are useful micronutrients, while others such as Pb, Hg, Cd, and As are non-essential heavy metals which when present in even trace amounts could be detrimental to human health, especially young children and pregnant women as they are more susceptible (Liang et al., 2020; Rusin et al., 2021). Organotin (Ot) is a widely used organometallic compound and is used in both the agricultural and industrial industries as a preservative in paints and a stabilizer in plastics and insecticides (Cole et al., 2015). Detections were made on 88 samples of fruits and vegetables, including peaches, tomatoes, pears, onions, and zucchini. A total of 11 Ots were found in 16 samples. Out of all of them, MPhT, a type of Ot, was found predominant with the highest detection rate of 13.6%; TcHT (5.7%) and MMT (6.8%) came next. Maximum amounts were found in a sample of papaya. These Ots were widely utilized in the production of insecticides, fungicides, bactericides, and stabilizers in agriculture and other sectors (Lespes et al., 2009). Heavy metals like As, Cd, and Ni are toxic pollutants that cause serious problems to the environment. The presence of these heavy metals in the environment can lead to bioaccumulation in the food chain, resulting in hazardous effects on human health. Even at minor concentrations, few heavy metals cause harmful effects to human health. Toxic metals like Cd, beryllium (Be), Hg, and Pb are xenobiotics as they are of no use to humans (Giller et al., 1998; Wei and Yang, 2010; Azizullah et al., 2011). Consumption of vegetables/fruits having heavy metals could lead to major human health problems. Elements like Zn, Cu, and selenium (Se)are very important for humans as they act like transcription factors, cofactors for enzymes, and antioxidants (El-Saka et al., 2020).

Globally, food security is a major concern. Contamination of vegetables and fruits with heavy metals and other pollutants raises concerns regarding food security and human health. It has been reported that approximately 20 million hectares of land is polluted and more than 50% of them are contaminated with toxic heavy metals (Hu et al., 2019). As heavy metals are non-biodegradable in nature, they accumulate in human tissues causing chronic diseases even if they are present in low concentrations (Singh et al., 2017; Rahman and Singh, 2019). A lot of medicinal plants have been reported to contain metals like Cd, As, and Cr when cultivated near industrial areas (El Hamiani et al., 2015; Bolan et al., 2017; Kim et al., 2017; Kohzadi et al., 2018). Vegetables grown in greenhouses are contaminated as well with heavy metals because of low illumination (Li et al., 2017). The main reasons of heavy metal contamination in soil are atmospheric deposition, pesticides, polluted water, and artificial fertilizers (Chary et al., 2008; Cai et al., 2009; Luo et al., 2009; Mansour et al., 2009; Gall et al., 2015; Lv et al., 2015; Elgallal et al., 2016; Woldetsadik et al., 2017; El-Kady and Abdel-Wahhab, 2018). Soil ecology is the study of natural fluctuations in soil and its environment. Soil ecosystems depend on various factors including geology, topography, climate, and human activity (Larsson et al., 1997). The toxic chemicals which are present in the irrigation water can majorly impact the growth and development of plant. Wastewater contains a good amount of trace heavy metals, which ruins the quality of soil ecosystems (Okonkwo et al., 2005). Vegetables play an important role in the human diet. Different vegetables absorb different concentrations of heavy metals (Lokeshwari and Chandrappa, 2006). There are certain plants which can accumulate large amounts of heavy metals from the environment, hence they can be used in phytoremediation (Chamley, 2003). The absorption of metal from soil depends on several factors like pH, composition, and ionic exchange (Jung, 2008). It has been reported that Cr is found to be in very high concentration in vegetables (Belay and Tizazu, 2019).

Food contamination can come from a variety of sources and entrance points. Topography and land-use interactions are two examples of the environmental risk variables that vary from farm to farm and have an effect on the occurrence and spread of food pathogens, which in turn affects the risk of contaminated products, and depending on the products' nature and the areas they are found in, different sources and pathways of contamination exist (Mostafidi et al., 2020). Alegbeleye et al. (2018) listed several ways that pathogens can spread, including oats that have precipitated; pollution of soils and groundwater by absorption; use of compost and unrefined fertilizers (or trash); being irrigated or flooded into dirty water; being bitten by insects; and animal or wildlife-caused fecal contamination. Although pathogens, parasites, and viruses are routinely identified, Pure fruits and

FIGURE 16.6 Major groups of contaminants in vegetables and fruits.

vegetables are primarily infected by yeasts and molds. The vegetables and fruits are usually contaminated prior or during harvest. In addition to harvesting instruments, transport containers, insects, rinse water, ice, and processing equipment, pre-harvest resources include soil, feces, irrigation water, fungicides, insecticides, dust, insects, and compost. Fruits and vegetables become contaminated by microorganisms when they come into contact with dirt, dust, and water, or during post-harvest procedures. They also serve as a haven for pathogenic germs (Figure 16.6; Mritunjay and Kumar, 2015).

16.7 BIOSAFETY OF VEGETABLES AND FRUITS

The agricultural sector must provide nutrient-dense plant-based diets due to the widespread deficiency of critical minerals in human populations. A viable strategy for this is the mineral biofortification of food crops (Stangoulis and Knez, 2022). Conversely, the dietary intake to heavy metal(loid)s has grown due to soil resource contamination and the resulting buildup in food crops (Anaman et al., 2022). It is crucial and extremely difficult to devise methods for making plant-based, mineral-rich diets that include only the allowed amounts of heavy metal(loid)s (Maqsood et al., 2022). Based on the high nutritional content, fruits and vegetables are essential for human health (Arora and Mathur, 2014), a great way to maintain and develop human health (Mritunjay and Kumar, 2015), and a great way to ward off diseases like cancer, heart disease, and eye problems. Food poverty, malnutrition, and lifestyle disorders like obesity, high blood pressure, cancer, and diabetes are the main global challenges that have increased demand for nutritious foods, especially fruits and vegetables (Gogo et al., 2017). Vegetables and fruits are quite market-oriented food products. It has been recently reported that a significant number of pathogens and toxic chemicals are present in vegetables and fruits. Production of organic and natural fruits and vegetables involves various activities like farming, harvesting, and post-harvest treatment and processing. Hazards can exist in any of these steps and hamper the quality and safety of vegetables and fruits. Practices can be implemented when the actual reasons for hazards are identified: microorganisms like bacteria, fungi, and protozoans. In certain cases, these microbes are introduced in the farm via pests. Sometimes, the vegetables become contaminated because of poor handling storage practices. Primary sources of microbial contamination of fresh fruits and vegetables include wastewater for irrigation, polluted soil, feces, and poor sanitary practices. Certain chemicals can pose detrimental effects to human health if they contaminate fruits and vegetables. It may cause contamination either naturally or artificially during several processes involved in agriculture production. Physical hazards can be incorporated into vegetables and fruits at various steps of production.

TABLE 16.2
Toxic Heavy Metals and Their Concentrations

Toxic Heavy Metals	Mean Concentration (mg/kg)
Zinc (Zn)	23.60e137.60
Copper (Cu)	20.30e251.60
Manganese (Mn)	10.70e112.30
Molybdenum (Mo)	5.50e213.60
Nickel (Ni)	6.50e12.00
Iron (Fe)	21.60e452.70

Source: Bhatnagar and Awasthi (2000).

Potential health risks for consumers may result from the deposition of heavy metals like Cu, Cd, Pb, and Zn in vegetables like *Beta vulgaris, Abelmoschus esculentus*, and *Brassica oleracea* that are produced in the metropolitan region of Varanasi, India (Sharma et al., 2008). Vegetables now contain higher levels of toxins, posing a health risk to consumers, because of higher Gomati River contamination with heavy metals such Cd, Cr, Ni, Pb, and Zn from industrial and agricultural rainfall (Rana, 2021). Out of the 43 most polluted towns in India, Mandi-gobindgarh ranks 17th (Gupta et al., 2013). Due to excess industrialization, the air quality of Mandi-gobindgarh is very low (CPCB, 2011). The soil of this town is also polluted because of irrigation with wastewater discharge from factories (Gupta et al., 2013; Khurana and Bansal, 2008). While root vegetables, like radishes, may have higher-than-normal U concentrations (as a radioactive danger and heavy metal) due to plants storing U in their roots, these concentrations are eliminated during vegetable washing (Evangelou et al., 2007). The fiber plant known as jute (*Corchorus capsularis*) is extensively grown and possesses several noteworthy physiological traits, such as a deep root system, biomass, and resistance to metal stress. Additionally, *Corchorus* species are native green vegetables that could remove certain heavy metals through phytoremediation. This species has been utilized in phytoremediation processes for a variety of hazardous contaminants, including Pb, Hg, Cu, and Cd (Saleem et al., 2020) (Table 16.2).

16.8 APPROACHES FOR DETECTION AND PROTECTION OF FRUITS AND VEGETABLES FROM CONTAMINANTS

Table 16.3 describes the detection of various microbes present in the fruits and vegetables and detection of pesticide residues using several spectroscopic techniques. Further, the detection of various contaminants using Raman spectroscopy and Fourier-transform infrared spectroscopy to ensure the safety of fruits and vegetables.

Various novel approaches are being used extensively in order to ensure the safety of vegetables and fruits. Several beneficial effects of edible coatings are utilized currently to enhance the acid, sugar, perfumes, ingredients and its taste and preservation of color. Furthermore, the edible coatings are very effective for reduction of storage-related infections and to maintain the quality of vegetables and fruits during transport and storage. Different plant-based compounds are being used for coating, such as dietary fibers (Alvarez et al., 2018), oil and starch (Kore et al., 2017), and chitosan (Mantilla et al., 2013). The improvement of natural antagonists which present on the surface of vegetables and fruits are being used (Sharma et al., 2009). The organic farming, beneficial microbe–directed synthesis and utilization of nanoparticles, ultraviolet-C rays, ozone, ultrasonic rays, acidic electrolyzed water, irradiation techniques using gamma and x-ray, and cold plasma were reported as efficient in controlling contaminants (Carletti et al., 2013; Dey et al., 2019; Saglam et al., 2016; Mditshwa et al., 2017; Gogo et al., 2017; Yeoh and Ali, 2017; Mridha et al., 2017).

TABLE 16.3
Approaches for Detection and Protection of Fruits and Vegetables from Contaminants

Approach	Methods	References
Microbial load	Enzyme-linked immunosorbent assay, magnetic immunity isolation and polymerase chain reaction (PCR), microarrays, and biochip technologies	Nemati et al., 2016
Cultivation and enrichment	PCR and real-time PCR	Badosa et al., 2009
Detection of pesticide residues	Liquid chromatography–mass spectrometry and gas chromatography–mass spectrometry	Zanella et al., 2012
Pesticide detection, virus detection, carotenoid and polyacetylene evaluation, carotenoid evaluation, quality evaluation, pathogen detection, pesticide penetration	Raman spectroscopy	Qin et al., 2019
Detection of adulteration, plant sugars, amino acids, and pigment content	Fourier-transform mid infrared spectroscopy	Bureau et al., 2019

16.9 CONCLUSION

Pollution of the environment has been shown to lower yield production and yield quality. Pollutants entering the food chain are a serious health concern. Despite the existence and application of numerous approaches for the remediation of contaminants such as heavy metals and radioactive minerals, phytoremediation technology is thought to be relatively simple, inexpensive, easy to monitor, environmentally benign, and less constrained. Owing to the harmful effects of pollutants and radiation, remediation needs to be handled carefully with Pollution Control Board (PCB) management. The most important consideration when choosing plants for a contaminated area is their selection. To speed up the phytoremediation process, soil amendments like biochar, humic materials, and organic acid exudates can be added to contaminated areas. Following the remediation, the use and deposit of PCBs are the main areas of concern. As a result, the authors have determined that there is a financially feasible chance to use PCBs in future research projects. Eating tainted or unhealthy food can result in serious health problems, so food safety is a major concern for any country. It is considerably more difficult to ensure food safety in a nation as diverse as India, where regional differences in dietary patterns and culinary customs are substantial. The current review has discussed the toxins in fruits and vegetables, what impacts they have, and what are the major pollutants that are present in fruits and vegetables. Several corrective measures to improve the effectiveness of biosafety procedures were also covered.

16.10 ACKNOWLEDGMENTS

We thank Manipal Academy of Higher Education (MAHE), Manipal, Karnataka, India, TIFAC-CORE and FIST, DST New Delhi, DBT BUILDER grant (BT/INF/22/SP43065/2021), DBT, New Delhi, and K-FIST, VGST, Govt. of Karnataka for the facilities. The authors (N. K. and A. K.) are thankful to MAHE for the T.M.A. Pai Ph.D. scholarship, and A. K. is thankful to the Indian Council of Medical Research (ICMR), New Delhi for financial assistance as Project Associate I through a research grant (No. 59/08/2022-TRM/BMS). We are grateful to Dr. B. S. Satish Rao, Professor and Director, Manipal School of Life Sciences, MAHE for encouragement and support.

REFERENCES

Abioye, O. P., U. J. J. Ijah, and S. A. Aransiola. "Phytoremediation of soil contaminants by biodiesel plant jatropha curcas," Chapter 4 of K. Bauddh et al. (eds.), *Phytoremediation Potential of Bioenergy Plants*. Doi: 10.1007/978-981-10-3084-0_4. Springer Nature Singapore Pte Ltd. (2017). https://link.springer.com/book/10.1007%2F978-981-10-3084-0

Agrawal, S. B., A. Singh, R. K. Sharma, and M. Agrawal. "Bioaccumulation of heavy metals in vegetables: A threat to human health." *Terrestrial and Aquatic Environmental Toxicology* 1, no. 2 (2007): 13–23.

Ahmad, Burhan, Misbah Anwar, Hammad Badar, Mubashir Mehdi, and Farooq Tanveer. "Analyzing export competitiveness of major fruits and vegetables of Pakistan: An application of revealed comparative advantage indices." *Pakistan Journal of Agricultural Sciences* 58, no. 2 (2021): 719–730.

Alegbeleye, Oluwadara Oluwaseun, Ian Singleton, and Anderson S. Sant'Ana. "Sources and contamination routes of microbial pathogens to fresh produce during field cultivation: A review." *Food Microbiology* 73 (2018): 177–208.

Alrumman, Sulaiman, Sherif Keshk, and Attallah El Kott. "Water pollution: Source & treatment." *American Journal of Environmental Engineering* (2016): 88–98.

Alvarez, María V., Alejandra G. Ponce, and María R. Moreira. "Influence of polysaccharide-based edible coatings as carriers of prebiotic fibers on quality attributes of ready-to-eat fresh blueberries." *Journal of the Science of Food and Agriculture* 98, no. 7 (2018): 2587–2597.

Anaman, Richmond, Chi Peng, Zhichao Jiang, Xu Liu, Ziruo Zhou, Zhaohui Guo, and Xiyuan Xiao. "Identifying sources and transport routes of heavy metals in soil with different land uses around a smelting site by GIS based PCA and PMF." *Science of the Total Environment* 823 (2022): 153759.

Aransiola, S. A., F. Afolabi, F. Joseph, and N.R. Maddela. "Soil enzymes: Distribution, interactions and influencing factors." In P. Jeschke and E. B. Starikov (Eds.), *Agricultural Biocatalysis: Enzymes in Agriculture and Industry* (1st ed.). Jenny Stanford Publishing, 2023, 303–333. ISBN 9789814968478. https://www.routledge.com/Agricultural-Biocatalysis-Enzymes-in-Agriculture-and-Industry/Jeschke-Starikov/p/book/9789814968478

Aransiola, S. A., U. J. J. Ijah, and O. P. Abioye. "Phytoremediation of lead polluted soil by *Glycine max* L." *Applied and Environmental Soil Science* (2013). Article ID 631619. doi: 10.1155/2013/631619. https://www.hindawi.com/journals/aess/2013/631619/abs/.

Aransiola, S. A., U. J. J. Ijah, O. P. Abioye, and J. D. Bala. "Microbial-aided phytoremediation of heavy metals contaminated soil: A review." *European Journal of Biological Research* 9, no. 2 (2019): 104–125. http://www.journals.tmkarpinski.com/index.php/ejbr/article/view/157.

Aransiola, S. A., U. J. J. Ijah, O. P. Abioye, and J. D. Bala. "Vermicompost-assisted phytoremediation of toxic trace element-contaminated soil in Madaka, Nigeria using *Melissa officinalis* L and *Sida acuta*." *International Journal of Environmental Science and Technology* (2022). http://dx.doi.org/10.1007/s13762-022-04105-y

Arora, Shreya, and Pulkit Mathur. "Nutrient profiling: An approach promoting healthy food choices." *International Journal of Health Sciences and Research* 4, no. 7 (2014): 227–239.

Arthur, Ellen L., Pamela J. Rice, Patricia J. Rice, Todd A. Anderson, Sadika M. Baladi, Keri L. D. Henderson, and Joel R. Coats. "Phytoremediation – an overview." *Critical Reviews in Plant Sciences* 24, no. 2 (2005): 109–122.

Attinti, Ramesh, Kirk R. Barrett, Rupali Datta, and Dibyendu Sarkar. "Ethylenediaminedisuccinic acid (EDDS) enhances phytoextraction of lead by vetiver grass from contaminated residential soils in a panel study in the field." *Environmental Pollution* 225 (2017): 524–533.

Atun, Sri, S. R. I. Handayani, A. N. N. A. Rakhmawati, Nur Aini Purnamaningsih, Bian Ihda An Naila, and Astuti Lestari. "Study of potential phenolic compounds from stems of dendrophthoe falcata (Loranthaceae) plant as antioxidant and antimicrobial agents." *Oriental Journal of Chemistry* 34, no. 5 (2018).

Azizullah, Azizullah, Muhammad Nasir Khan Khattak, Peter Richter, and Donat-Peter Häder. "Water pollution in Pakistan and its impact on public health – a review." *Environment International* 37, no. 2 (2011): 479–497.

Badosa, E., N. Chico, M. Pla, D. Parés, and E. Montesinos. "Evaluation of ISO enrichment real-time PCR methods with internal amplification control for detection of Listeria monocytogenes and Salmonella enterica in fresh fruit and vegetables." *Letters in Applied Microbiology* 49, no. 1 (2009): 105–111.

Belay, Tesfalem, and H. Tizazu. "Assessment of pollution status of soils and vegetables irrigated by Awash river and its selected tributaries." *International Journal of Environmental Sciences & Natural Resources* 18, no. 5 (2019): 163–168.

Bhatnagar, J. P., and S. K. Awasthi. *Prevention of food adulteration act (act no. 37 of 1954) along with central & state rules (as amended for 1999)*. Ashoka Law House, 2000.

Bolan, Shiv, Anitha Kunhikrishnan, Balaji Seshadri, Girish Choppala, Ravi Naidu, Nanthi S. Bolan, Yong Sik Ok et al. "Sources, distribution, bioavailability, toxicity, and risk assessment of heavy metal (loid) s in complementary medicines." *Environment International* 108 (2017): 103–118.

Briggs, David. "Environmental pollution and the global burden of disease." *British Medical Bulletin* 68, no. 1 (2003): 1–24.

Bureau, Sylvie, Daniel Cozzolino, and Christopher J. Clark. "Contributions of Fourier-transform mid infrared (FT-MIR) spectroscopy to the study of fruit and vegetables: A review." *Postharvest Biology and Technology* 148 (2019): 1–14.

Cai, Chen, Picheng Gong, Yue Wang, Mengmeng Wang, Bo Zhang, Bing Wang, and Huiling Liu. "Investigating the environmental risks from the use of spray-dried cephalosporin mycelial dreg (CMD) as a soil amendment." *Journal of Hazardous Materials* 359 (2018): 300–306.

Cai, Qiu, Mei-Li Long, Ming Zhu, Qing-Zhen Zhou, Ling Zhang, and Jie Liu. "Food chain transfer of cadmium and lead to cattle in a lead–zinc smelter in Guizhou, China." *Environmental Pollution* 157, no. 11 (2009): 3078–3082.

Carletti, Letizia, Rinaldo Botondi, Roberto Moscetti, Elisabetta Stella, Danilo Monarca, Massimo Cecchini, and Riccardo Massantini. "Use of ozone in sanitation and storage of fresh fruits and vegetables." *Journal of Food, Agriculture and Environment* 11, nos. 3–4 (2013): 585–589.

Chamley, Hervé, ed. *Geosciences, environment and man*. Elsevier, 2003.

Chary, N. Sridhara, C. T. Kamala, and D. Samuel Suman Raj. "Assessing risk of heavy metals from consuming food grown on sewage irrigated soils and food chain transfer." *Ecotoxicology and Environmental Safety* 69, no. 3 (2008): 513–524.

Cole, Russell F., Graham A. Mills, Ruth Parker, Thi Bolam, Andrew Birchenough, Silke Kröger, and Gary R. Fones. "Trends in the analysis and monitoring of organotins in the aquatic environment." *Trends in Environmental Analytical Chemistry* 8 (2015): 1–11.

CPCB. *Report under central pollution control board. Guidelines for the measurement of ambient air pollution*, vol. 1. Ministry of Environment and Forests, 2011.

Cunningham, Scott D., J. R. Shann, David E. Crowley, and Todd A. Anderson. *Phytoremediation of contaminated water and soil*. ACS Symposium series, vol. 664. American Chemical Society, 1997, pp. 2–17.

Cunningham, Scott D., Todd A. Anderson, A. Paul Schwab, and F. C. Hsu. "Phytoremediation of soils contaminated with organic pollutants." *Advances in Agronomy* 56, no. 1 (1996): 55–114.

Dey, Dipankar, Satadeep Singha Roy, Niharendu Saha, Anupam Datta, and Pradip Dey. "Method for estimating potentially available inorganic phosphorus under organic farming system." *International Journal of Chemical Studies* 7 (2019): 1829–1835.

Elgallal, M., L. Fletcher, and B. Evans. "Assessment of potential risks associated with chemicals in wastewater used for irrigation in arid and semiarid zones: A review." *Agricultural Water Management* 177 (2016): 419–431.

El Hamiani, Ouafae, Hicham El Khalil, Catherine Sirguey, Ahmed Ouhammou, Gabriel Bitton, Christophe Schwartz, and Ali Boularbah. "Metal concentrations in plants from mining areas in South Morocco: Health risks assessment of consumption of edible and aromatic plants." *CLEAN–Soil, Air, Water* 43, no. 3 (2015): 399–407.

El-Kady, Ahmed A., and Mosaad A. Abdel-Wahhab. "Occurrence of trace metals in foodstuffs and their health impact." *Trends in Food Science & Technology* 75 (2018): 36–45.

Elliott, Michael. "Biological pollutants and biological pollution—an increasing cause for concern." *Marine Pollution Bulletin* 46, no. 3 (2003): 275–280.

El-Saka, M. S., M. M. El-kholy, and M. A. Aiad. "The contamination status of heavy metals associated with fruits and vegetables collected from some markets at El-Gharbia Governorate." *Journal of Soil Sciences and Agricultural Engineering* 11, no. 7 (2020): 299–305.

Ernst, W. H. O. "Bioavailability of heavy metals and decontamination of soils by plants." *Applied Geochemistry* 11, no. 1–2 (1996): 163–167.

Evangelou, Michael W. H., Mathias Ebel, and Andreas Schaeffer. "Chelate assisted phytoextraction of heavy metals from soil. Effect, mechanism, toxicity, and fate of chelating agents." *Chemosphere* 68, no. 6 (2007): 989–1003.

Gall, Jillian E., Robert S. Boyd, and Nishanta Rajakaruna. "Transfer of heavy metals through terrestrial food webs: A review." *Environmental monitoring and assessment* 187 (2015): 1–21.

Garba, M., M. A. Dandago, E. C. Igwe, and K. D. Salami. "Heavy metals safety of ready-to-eat (RTE) vegetables salads (a review)." *Dutse Journal of Pure and Applied Sciences* 7, no. 2 (2021): 21–37.

Giller, Ken E., Ernst Witter, and Steve P. Mcgrath. "Toxicity of heavy metals to microorganisms and microbial processes in agricultural soils: A review." *Soil Biology and Biochemistry* 30, no. 10–11 (1998): 1389–1414.

Gogo, E. O., A. M. Opiyo, K. Hassenberg, Ch Ulrichs, and S. Huyskens-Keil. "Postharvest UV-C treatment for extending shelf life and improving nutritional quality of African indigenous leafy vegetables." *Postharvest Biology and Technology* 129 (2017): 107–117.

Govarthanan, M., R. Mythili, T. Selvankumar, S. Kamala-Kannan, and H. Kim. "Myco-phytoremediation of arsenic-and lead-contaminated soils by Helianthus annuus and wood rot fungi, Trichoderma sp. isolated from decayed wood." *Ecotoxicology and Environmental Safety* 151 (2018): 279–284.

Gupta, Shalini, D. S. Grewal, and Ajay Gupta. "Water and soil pollution in Punjab with special reference to Mandi-Gobindgarh and surrounding areas." *American Journal of Engineering Research* 2, no. 9 (2013): 251–257.

Hu, Nan, Tao Lang, Dexin Ding, Jingsong Hu, Changwu Li, Hui Zhang, and Guangyue Li. "Enhancement of repeated applications of chelates on phytoremediation of uranium contaminated soil by Macleaya cordata." *Journal of Environmental Radioactivity* 199 (2019): 58–65.

Igwegbe, Amin O., Chibugo H. Agukwe, and Charles A. Negbenebor. "A survey of heavy metal (lead, cadmium and copper) contents of selected fruit and vegetable crops from Borno State of Nigeria." *International Journal of Engineering Science* 2, no. 1 (2013): 01–05.

Jiang, Mingyan, Shiliang Liu, Yangfan Li, Xi Li, Zhenghua Luo, Huixing Song, and Qibing Chen. "EDTA-facilitated toxic tolerance, absorption and translocation and phytoremediation of lead by dwarf bamboos." *Ecotoxicology and Environmental Safety* 170 (2019): 502–512.

Jung, Myung Chae. "Heavy metal concentrations in soils and factors affecting metal uptake by plants in the vicinity of a Korean Cu-W mine." *Sensors* 8, no. 4 (2008): 2413–2423.

Kaiser, Wolfgang, Torsten Schultz-Fademrecht, Michaela Blech, Julia Buske, and Patrick Garidel. "Investigating photodegradation of antibodies governed by the light dosage." *International Journal of Pharmaceutics* 604 (2021): 120723.

Khurana, M. P. S., and R. L. Bansal. "Impact of sewage irrigation on speciation of nickel in soils and its accumulation in crops of industrial towns of Punjab." *Journal of Environmental Biology* 29, no. 5 (2008): 793–798.

Kim, Shin Woong, Yooeun Chae, Jongmin Moon, Dokyung Kim, Rongxue Cui, Gyeonghyeon An, Seung-Woo Jeong, and Youn-Joo An. "In situ evaluation of crop productivity and bioaccumulation of heavy metals in paddy soils after remediation of metal-contaminated soils." *Journal of Agricultural and Food Chemistry* 65, no. 6 (2017): 1239–1246.

Kohzadi, Shadi, Behzad Shahmoradi, Daem Raushani, and Asad Nouri. "International perspectives: Evaluation and risk assessment of heavy metals in the groundwater resources of Saqqez, Iran." *Journal of Environmental Health* 80, no. 6 (2018): E1–E9.

Kore, Vijaykumar T., Sima S. Tawade, and J. Kabir. "Application of edible coatings on fruits and vegetables." *Imperial Journal of Interdisciplinary Research* 3, no. 1 (2017): 591–603.

Kopyra, Małgorzata, and Edward A. Gwóźdź. "Nitric oxide stimulates seed germination and counteracts the inhibitory effect of heavy metals and salinity on root growth of Lupinus luteus." *Plant Physiology and Biochemistry* 41, no. 11–12 (2003): 1011–1017.

Koźmińska, Aleksandra, Alina Wiszniewska, Ewa Hanus-Fajerska, and Ewa Muszyńska. "Recent strategies of increasing metal tolerance and phytoremediation potential using genetic transformation of plants." *Plant Biotechnology Reports* 12 (2018): 1–14.

Larsson, Lisbeth, Bo Stenberg, and Lennart Torstensson. "Effects of mulching and cover cropping on soil microbial parameters in the organic growing of black currant." *Communications in Soil Science and Plant Analysis* 28, no. 11–12 (1997): 913–925.

Lespes, Gaëtane, Christophe Marcic, Julien Heroult, Isabelle Le Hecho, and Laurence Denaix. "Tributyltin and triphenyltin uptake by lettuce." *Journal of Environmental Management* 90 (2009): S60–S68.

Li, Fei-li, Wen Shi, Zan-fang Jin, Hui-mei Wu, and G. Daniel Sheng. "Excessive uptake of heavy metals by greenhouse vegetables." *Journal of Geochemical Exploration* 173 (2017): 76–84.

Li, Fei-li, Yuehua Qiu, Xinyang Xu, Feng Yang, Zhiwei Wang, Jianru Feng, and Jiade Wang. "EDTA-enhanced phytoremediation of heavy metals from sludge soil by Italian ryegrass (Lolium perenne L.)." *Ecotoxicology and Environmental Safety* 191 (2020): 110185.

Li, Xiaofang, and Longbin Huang. "Toward a new paradigm for tailings phytostabilization – nature of the substrates, amendment options, and anthropogenic pedogenesis." *Critical Reviews in Environmental Science and Technology* 45, no. 8 (2015): 813–839.

Liang, Guihong, and Zhenhua Zhang. "Reducing the nitrate content in vegetables through joint regulation of short-distance distribution and long-distance transport." *Frontiers in Plant Science* 11 (2020): 1079.

Liu, Zhongchuang, and Khanh-Quang Tran. "A review on disposal and utilization of phytoremediation plants containing heavy metals." *Ecotoxicology and Environmental Safety* 226 (2021): 112821.

Lokeshwari, H., and G. T. Chandrappa. "Impact of heavy metal contamination of Bellandur Lake on soil and cultivated vegetation." *Current Science* (2006): 622–627.

Luo, Jie, Wenxiang He, Xinli Xing, Jian Wu, and X. W. Sophie Gu. "The variation of metal fractions and potential environmental risk in phytoremediating multiple metal polluted soils using *Noccaea caerulescens* assisted by LED lights." *Chemosphere* 227 (2019): 462–469.

Luo, Y., L. Wu, L. Liu, C. Han, and Z. Li. "Heavy metal contamination and remediation in Asian agricultural land." In *MARCO symposium*, 2009.https://scholar.google.com/scholar?hl=en&as_sdt=0%2C5&q=Heavy+Metal+Contamination+and+Remediation+in+Asian+Agricultural+Land++Yongming+Luo+Longhua+Wu+Lin+Liu+Cunliang+Han+Zhu+Li&btnG= accessed on 15/1/2024

Lv, Jianshu, Yang Liu, Zulu Zhang, Jierui Dai, Bin Dai, and Yuchi Zhu. "Identifying the origins and spatial distributions of heavy metals in soils of Ju country (Eastern China) using multivariate and geostatistical approach." *Journal of Soils and Sediments* 15 (2015): 163–178.

Ma, Ying, Mani Rajkumar, António Moreno, Chang Zhang, and Helena Freitas. "Serpentine endophytic bacterium Pseudomonas azotoformans ASS1 accelerates phytoremediation of soil metals under drought stress." *Chemosphere* 185 (2017): 75–85.

Mansour, Sameeh A., Mohamed H. Belal, Asem A. K. Abou-Arab, and Marwa F. Gad. "Monitoring of pesticides and heavy metals in cucumber fruits produced from different farming systems." *Chemosphere* 75, no. 5 (2009): 601–609.

Mantilla, N., M. E. Castell-Perez, C. Gomes, and Rosana G. Moreira. "Multilayered antimicrobial edible coating and its effect on quality and shelf-life of fresh-cut pineapple (Ananas comosus)." *LWT-Food Science and Technology* 51, no. 1 (2013): 37–43.

Manwani, Seema, C. R. Vanisree, Vibha Jaiman, Kumud Kant Awasthi, Chandra Shekhar Yadav, Mahipal Singh Sankhla, Pritam P. Pandit, and Garima Awasthi. "Heavy metal contamination in vegetables and their toxic effects on human health." *Sustainable Crop Production: Recent Advances* 181–197 (2022).

Maqsood, Quratulain, Nazim Hussain, Mehvish Mumtaz, Muhammad Bilal, and Hafiz M. N. Iqbal. "Novel strategies and advancement in reducing heavy metals from the contaminated environment." *Archives of Microbiology* 204, no. 8 (2022): 478.

McCutcheon, S. C., and J. L. Schnoor. "Overview of phytotransformation and control of wastes." *Phytoremediation: Transformation and Control of Contaminants* (2003): 1–58.

McMichael, Anthony J., and Kirk R. Smith. "Seeking a global perspective on air pollution and health." *Epidemiology* (1999): 1–4.

Mditshwa, Asanda, Lembe Samukelo Magwaza, Samson Zeray Tesfay, and Nokwazi Carol Mbili. "Effect of ultraviolet irradiation on postharvest quality and composition of tomatoes: A review." *Journal of Food Science and Technology* 54 (2017): 3025–3035.

Mostafidi, Mahdieh, Mohammad Reza Sanjabi, Faezeh Shirkhan, and Maryam Tamaskani Zahedi. "A review of recent trends in the development of the microbial safety of fruits and vegetables." *Trends in Food Science & Technology* 103 (2020): 321–332.

Mousavi, Afsaneh, Latifeh Pourakbar, Sina Siavash Moghaddam, and Jelena Popović-Djordjević. "The effect of the exogenous application of EDTA and maleic acid on tolerance, phenolic compounds, and cadmium phytoremediation by okra (Abelmoschus esculentus L.) exposed to Cd stress." *Journal of Environmental Chemical Engineering* 9, no. 4 (2021): 105456.

Mridha, F. A., R. O. Huque, A. F. Khatun, M. A. Islam, A. R. Hossain, M. A. Hossain, and M. S. Kabir. Effects of gamma irradiation on antioxidant markers, microbial population an d sensory attributes of strawberry (Fragaria× ananassa Duch.) cv. festival. *Plant Cell Biotechnology and Molecular Biology* 18, nos. 5–6 (2017): 208–218.

Mritunjay, Sujeet Kumar, and Vipin Kumar. "Fresh farm produce as a source of pathogens: A review." *Research Journal of Environmental Toxicology* 9, no. 2 (2015): 59–70.

Nemati Josheghani, Reza. "Novel methods of chromatography and mass spectrometry for quantitative investigation of bacteria-derived lipopeptides and their relationship with human disease." (2016).

Okonkwo, Jonathan O., and M. Mothiba. "Physico-chemical characteristics and pollution levels of heavy metals in the rivers in Thohoyandou, South Africa." *Journal of Hydrology* 308, no. 1–4 (2005): 122–127.

Pais, Istvan, and JonesJr JB. *The handbook of trace elements*. CRC Press, 1997. 1–27.

Panda, T., N. Mishra, B. K. Pradhan, and R. B. Mohanty. "Diversity of leafy vegetables and its significance to rural households of Bhadrak district, Odisha, India." *Scientia Agriculturae* 11, no. 3 (2015): 114–123.

Qin, Jianwei, Moon S. Kim, Kuanglin Chao, Sagar Dhakal, Byoung-Kwan Cho, Santosh Lohumi, Changyeun Mo, Yankun Peng, and Min Huang. "Advances in Raman spectroscopy and imaging techniques for quality and safety inspection of horticultural products." *Postharvest Biology and Technology* 149 (2019): 101–117.

Rahman, Zeeshanur, and Ved Pal Singh. "The relative impact of toxic heavy metals (THMs)(arsenic (As), cadmium (Cd), chromium (Cr)(VI), mercury (Hg), and lead (Pb)) on the total environment: An overview." *Environmental Monitoring and Assessment* 191 (2019): 1–21.

Rana, Vivek. "Assessment of surface water quality of river Kali-east: A tributary of river Ganga in Uttar Pradesh, India." (2021).[https://doi.org/10.21203/rs.3.rs-591475/v1].

Ruiz, Oscar N., Derry Alvarez, Cesar Torres, Laura Roman, and Henry Daniell. "Metallothionein expression in chloroplasts enhances mercury accumulation and phytoremediation capability." *Plant Biotechnology Journal* 9, no. 5 (2011): 609–617.

Rusin, Monika, Joanna Domagalska, Danuta Rogala, Mehdi Razzaghi, and Iwona Szymala. "Concentration of cadmium and lead in vegetables and fruits." *Scientific Reports* 11, no. 1 (2021): 11913.

Saleem, Muhammad Hamzah, Shah Fahad, Shahid Ullah Khan, Mairaj Din, Abid Ullah, Ayman EL Sabagh, Akbar Hossain, Analía Llanes, and Lijun Liu. "Copper-induced oxidative stress, initiation of antioxidants and phytoremediation potential of flax (Linum usitatissimum L.) seedlings grown under the mixing of two different soils of China." *Environmental Science and Pollution Research* 27 (2020): 5211–5221.

Salt, David E., Michael Blaylock, Nanda P. B. A. Kumar, Viatcheslav Dushenkov, Burt D. Ensley, Ilan Chet, and Ilya Raskin. "Phytoremediation: A novel strategy for the removal of toxic metals from the environment using plants." *Biotechnology* 13, no. 5 (1995): 468–474.

Salt, D. E., P. B. A. Nanda Kumar, S. Dushenkov, and I. Raskin. "Phytoremediation: A new technology for the environmental cleanup of toxic metals." *Proceedings of the International Symposium on Resource Conservation and Environmental Technologies in Metallurgical Industries* (1994): 381–384.

Saglam, Necdet, Ozfer Yesilada, Ahmet Cabuk, Mesut Sam, Semran Saglam, Sedef Ilk, Ezgi Emul, Pınar Aytar Celik, and Ekrem Gurel. "Innovation of strategies and challenges for fungal nanobiotechnology." *Advances and Applications Through Fungal Nanobiotechnology* (2016): 25–46.

Schnoor, Jerald L., Louis A. Light, Steven C. McCutcheon, N. Lee Wolfe, and Laura H. Carreia. "Phytoremediation of organic and nutrient contaminants." *Environmental Science & Technology* 29, no. 7 (1995): 318A–323A.

Sekhar, K. Chandra, C. T. Kamala, N. S. Chary, A. R. K. Sastry, T. Nageswara Rao, and M. J. J. O. H. M. Vairamani. "Removal of lead from aqueous solutions using an immobilized biomaterial derived from a plant biomass." *Journal of Hazardous Materials* 108, no. 1–2 (2004): 111–117.

Sharma, Rajesh Kumar, Madhoolika Agrawal, and Fiona M. Marshall. "Heavy metal (Cu, Zn, Cd and Pb) contamination of vegetables in urban India: A case study in Varanasi." *Environmental Pollution* 154, no. 2 (2008): 254–263.

Sharma, R. R., Dinesh Singh, and Rajbir Singh. "Biological control of postharvest diseases of fruits and vegetables by microbial antagonists: A review." *Biological Control* 50, no. 3 (2009): 205–221.

Shen, Xing, Min Dai, Jiawei Yang, Lin Sun, Xiao Tan, Changsheng Peng, Imran Ali, and Iffat Naz. "A critical review on the phytoremediation of heavy metals from environment: Performance and challenges." *Chemosphere* 291 (2022): 132979.

Shi, Taoran, Jin Ma, Xiao Wu, Tienan Ju, Xianglong Lin, Yunyun Zhang, Xinhong Li, Yiwei Gong, Hong Hou, Long Zhou, Fuyong Wu "Inventories of heavy metal inputs and outputs to and from agricultural soils: A review." *Ecotoxicology and Environmental Safety* 164 (2018): 118–124.

Siciliano, S. D., H. Goldie, and J. J. Germida. "Enzymatic activity in root exudates of Dahurian wild rye (Elymus dauricus) that degrades 2-chlorobenzoic acid." *Journal of Agricultural and Food Chemistry* 46, no. 1 (1998): 5–7.

Singh, Nitika, Vivek Kumar Gupta, Abhishek Kumar, and Bechan Sharma. "Synergistic effects of heavy metals and pesticides in living systems." *Frontiers in Chemistry* 5 (2017): 70.

Smith, Allan H., Elena O. Lingas, and Mahfuzar Rahman. "Contamination of drinking-water by arsenic in Bangladesh: A public health emergency." *Bulletin of the World Health Organization* 78, no. 9 (2000): 1093–1103.

Sparks, Donald L., Balwant Singh, and Matthew G. Siebecker. *Environmental Soil Chemistry* (edited by D. L. Sparks, B. Singh, and M. G. Siebecker). Elsevier, 2022.

Stangoulis, James C. R., and Marija Knez. "Biofortification of major crop plants with iron and zinc-achievements and future directions." *Plant and Soil* 474, no. 1 (2022): 57–76.

Tan, Hui Wun, Yean Ling Pang, Steven Lim, and Woon Chan Chong. "A state-of-the-art of phytoremediation approach for sustainable management of heavy metals recovery." *Environmental Technology & Innovation* 30 (2023): 103043.

Tchounwou, Paul B., Clement G. Yedjou, Anita K. Patlolla, and Dwayne J. Sutton. "Heavy metal toxicity and the environment." *Molecular, Clinical and Environmental Toxicology: Volume 3: Environmental Toxicology* (2012): 133–164.

Thron, Raymond W. "Direct and indirect exposure to air pollution." *Otolaryngology – Head and Neck Surgery* 114, no. 2 (1996): 281–285.

Vu, Minh Thai, Thi Minh Hao Dong, and Huu Cuong Le. "The applicability of using biofuel for diesel engines." *World Wide Journal of Multidisciplinary Research and Development* no. 5 (2019): 85–90.

Walton, Barbara T., and Todd A. Anderson. "Microbial degradation of trichloroethylene in the rhizosphere: Potential application to biological remediation of waste sites." *Applied and Environmental Microbiology* 56, no. 4 (1990): 1012–1016.

Wei, Binggan, and Linsheng Yang. "A review of heavy metal contaminations in urban soils, urban road dusts and agricultural soils from China." *Microchemical Journal* 94, no. 2 (2010): 99–107.

Woldetsadik, Desta, Pay Drechsel, Bernard Keraita, Fisseha Itanna, and Heluf Gebrekidan. "Heavy metal accumulation and health risk assessment in wastewater-irrigated urban vegetable farming sites of Addis Ababa, Ethiopia." *International Journal of Food Contamination* 4 (2017): 1–13.

World Health Organization. *Air quality guidelines for Europe*. World Health Organization. Regional Office for Europe, 2000.

Yeoh, Wei Keat, and Asgar Ali. "Ultrasound treatment on phenolic metabolism and antioxidant capacity of fresh-cut pineapple during cold storage." *Food Chemistry* 216 (2017): 247–253.

Zanella, R., O. Damian, C. D. Amaral Friggi, M. Leonardo, and M. Bohrer. (2012). *An overview about recent advances in sample preparation techniques for pesticide residues analysis in cereals and feedstuffs*. InTech. http://dx.doi.org/10.5772/48752.

Zhang, Cuimei, Shangli Shi, Zhen Liu, Fan Yang, and Guoli Yin. "Drought tolerance in alfalfa (Medicago sativa L.) varieties is associated with enhanced antioxidative protection and declined lipid peroxidation." *Journal of Plant Physiology* 232 (2019a): 226–240.

Zhang, Chen, Zemin Li, Chong-Yu Zhang, Mengmeng Li, Yunkyoung Lee, and Gui-Guo Zhang. "Extract methods, molecular characteristics, and bioactivities of polysaccharide from alfalfa (Medicago sativa L.)." *Nutrients* 11, no. 5 (2019b): 1181.

17 Soil Assessment Policy and Permissible Levels of Contaminants in Plant Food Products

Harshitha Shettigar, Haya Fathima, Manoj Kumar, Harsha K. Chandrashekar, Nisha Govender, and Annamalai Muthusamy

17.1 INTRODUCTION

The combination of organic materials and rock fragments present on the surface of the Earth is called soil. Rocks that have been raised above sea level weather and turn into soils throughout geologic time. The Latin word *solum*, which means earthy material required for plant growth, is where the word "soil" originates. A mixture of organic and weathered rock materials found in the earth's uppermost layer is referred to as soil. Numerous slow-moving, long-term chemical, physical, and biological processes combine to form the soil. In addition to giving terrestrial ecosystems water and nutrients, they also store and exchange gases with the atmosphere and have an impact on surface and groundwater flow (Montgomery et al. 2000). Soil is a multiphase system that is made up of solids, liquids, and gases. Solids, liquids, and gases make up roughly 50%, 20%–30%, and 20%–30% of the total volume of a typical soil, respectively. The solid phase includes minerals and organic matter with sizes ranging from submicroscopic to visually distinguishable. The liquid phase consists of various salts dissolved in water making up a solution. The gaseous phase has a high concentration of water vapor and resembles air in composition (Young 2012). The two components of the solid phase are inorganic and organic matter, the latter comprising 1% to 5% of the soil. Soil organic matter (SOM) is an essential resource that serves as the foundation for soil fertility. In general, the term SOM refers to all organic materials in soil containing carbon (C; Senesi and Loffredo 2018). It can be separated into fractions that are living and non-living. The non-living fraction is not significant, but aids in the formation of stable aggregates, soil structure, and fertility, as well as the soil's water and nutrient holding capacity (Skjemstad et al. 1998). Most of the soil is composed primarily of inorganic material. It is mainly made up of mineral fragments, each of which possesses distinct physical and chemical properties that vary according to the parent material and environmental variables at the time the soil was formed. The inorganic part of soils determines their physical features, including texture. Density, structure, and water retention are all heavily influenced by this. The inorganic component of soil is mostly composed of four elements – oxygen (O), silicon (Si), aluminum (Al), and iron (Fe) – in addition to magnesium (Mg), calcium (Ca), sodium (Na), and potassium (K; Knoepp et al. 2005; Aransiola et al. 2019).

Soil is very important for agriculture as it affects the quality and quantity of food production. Food scarcity is of global concern, and maintenance of soil health is vital. Due to natural and human-made activities, contaminants have been reported in the soil. These contaminants can also enter food and cause major health risks in humans and animals. Therefore, the present review aims to compile the information available on soil and contaminants and the assessment methods. The

DOI: 10.1201/9781032683751-20

chapter also emphasizes the interventions of governments and civil bodies in soil sustainability to improve plant productivity.

17.2 SOIL: TYPES AND PROFILING

Soil classification refers to the method of categorizing soils based on shared characteristics. Since the earth is covered in a vast array of soil types, it is preferable to systematize or group the soils according to their general characteristics. In general, soils can be divided into cohesive as well as coarse- and fine-grained categories. A soil's classification includes terms like boulder clay, laterite, loam, caliche, hardpan, loess, topsoil, peat, marl, moorum, China clay, fill, varved clay, bentonite, and black cotton soil. India is a diverse nation with a wide range of vegetation types, climatic zones, relief features, and landforms. Indian soils are divided into several groups according to the predominant particle size in the soil or the area where the soil is found. Soil is categorized as alluvial, laterite, red, black, desert, or mountain soil depending on its geographical location. Soil is divided into three categories based on the predominant particle sizes: sand, peat, and clay (Siddiqui and Fatima 2017; Chandan and Thakur 2018).

17.2.1 Alluvial Soil

The majority of India's arable land is composed of alluvial soils, which are primarily located along the Ganges, Kosi, Mahananda, and their tributaries river plains. Approximately 15,000 km^2, or 45.6% of the nation's total land area, is covered by them, and they make significant contributions to the national food basket (Reza et al. 2017). Alluvial soils' morphological, physical, chemical, and mineralogical properties are highly impacted by the features of the alluvial parent material from which they evolved, especially when the soils are young (Boettinger et al. 2010). The alluvium is material that has been deposited by rivers, winds, glaciers, and sea waves; soils that contain alluvium are known as alluvial soils. The older alluviums are found to be on slightly elevated terrain, are clayey and sticky, and have a darker color with nodules of lime concretions. The flood plains and deltas are home to the new, lighter-colored alluviums (Siddiqui and Fatima 2017).

17.2.2 Red Soil

Sesquioxide concentrations are high in red soils, which are highly leached soils found in humid tropical regions. Red soils are found in most of South America, South and Southeast Asia, Central Africa, Japan, and Australia, as well as India. They are found in the Deccan Plateau in the east of India, as well as the Chota Nagpur Plateau (Jharkhand) states from the south such as Tamil Nadu, Kerala, and Karnataka. Approximately 35,000 km^2, or 10.6% of the country's total land area, is covered by red soil. Despite their low water-holding capacity, these soils often provide good physical conditions for plant development. Low natural fertility is the main barrier to excellent yields of crops on these soils; they are often acidic and lacking in all critical nutrients required for the crop plants (Siddiqui and Fatima 2017; Baligar et al. 2004).

17.2.3 Laterite Soil

The by-product of laterization is called "leached residue" from laterite soil, and it is now found all over the world. Lateralization is an uncontrollably occurring natural process that has been shaped by the interplay of geological and atmospheric factors (Santha Kumar et al. 2022). The highly weathered material known as laterite is rich in either or both secondary forms of Fe and Al. After being exposed to periodic wetting and drying, it becomes either hard or susceptible to hardening. Laterite soils are naturally granular in structure. They have little plasticity and excellent drainage.

When remolded in water, they can turn clayey and plastic depending on the degree of disturbance (Winterkorn and Chandrasekharan 1951). The soils lack humus and are mostly devoid of bases, silica, and weatherable primary minerals, as well as silicate clays. Andhra Pradesh, Karnataka, Kerala, Tamil Nadu, Pondicherry, Maharashtra, Odisha, and West Bengal are among the states with laterite soils (Mishra 2016).

17.2.4 Black Soil

Black soils have a high content of organic matter in the top layer and a deep humus layer that makes them soft. Black soils are classified as having a high organic C content and black or dark surface horizons by the international network of black soils (INBS), the thickness of the horizons' very dark to black surfaces. Because they are good for producing yield, they are important for addressing issues like land degradation, food security, and climate change (Sulaeman et al. 2023). Indian black soils are also referred to as *regur* or black cotton soils; they are very difficult to work with and manage because they swell when wet and shrink when dry (Bhattacharyya et al. 2014). Black soil covers 16.6% of India's land area, with the majority of this land located in Madhya Pradesh, Maharashtra, Pondicherry, Tamil Nadu, Uttar Pradesh, Bihar, Odisha, Andhra Pradesh, and Gujarat (Mishra 2016).

17.2.5 Desert Soil

Desert soil is typically found in regions with little rainfall and is 90% to 95% sand. Its low levels of organic matter and nitrogen (N), combined with high levels of phosphate and calcium carbonate, make it infertile. Ten times as much Ca is found in the lowest layer as in the topsoil (Dwevedi et al. 2017). Texture, pH, organic matter concentration, and landscape orientation are the main properties of desert soil that affect water and nutrient availability. Desert soils typically exhibit restricted parent material growth (Holzapfel 2009). In dryland ecosystems, water availability is determined by the balance of precipitation and evaporation, as well as salt levels (Jansson and Hofmockel 2020). Desert soil is found in Rajasthan, Haryana, and South Punjab, and it covers 4.32% of the area in total (Siddiqui and Fatima 2017).

17.2.6 Mountain Soil

Soils found in forests and mountains can be found not only at higher altitudes but also at lower altitudes that receive sufficient rainfall. Their composition varies depending on parent rocks, ground structure, and climate. They are produced by the deposition of organic matter from forest growth (Dwevedi et al. 2017). India's mountainous regions, including the states of Sikkim, Assam, Arunachal Pradesh, Uttarakhand, and Kashmir, are home to the majority of the country's mountain soils. Together with these areas, there are also significant amounts of mountain soils in the Peninsula, Eastern Ghats, and the summits of the Western Ghats, collectively referred to as Sahyadri (Sharma et al. 2018).

17.2.7 Soil Profiling

A vertical section of mature soil reveals multiple layers, referred to as soil horizons or horizons, each with unique physical and chemical characteristics. The soil profile is made up of all of these horizons or layers, arranged from top to bottom. The fundamental building blocks of morphological research and soil classification schemes are vertical cross sections of soil profiles (Hartemink et al. 2020). To learn more about the composition, functions, and characteristics of soil, soil profiles are utilized. Every horizon has a unique porosity, texture, color, structure, thickness, and other characteristics. These soil horizons are also distinguished by variations in chemical characteristics like pH and cation exchange capacity (Montgomery et al. 2000).

Soil horizons reflect soil processes and provide information about previous and current soil conditions. In some soils, however, they are oriented differently because of pedogenesis (Owens 2023). Five horizons can be used to roughly categorize the soil profile. These can be referred to as the O-, A-, B-, C-, and R-horizons, descending from the surface. The true soil, or solum, is made up of the A and B zones combined. The O-horizon, which consists of organic materials in varying states of breakdown, is mostly found in forests because that's where debris from fallen leaves and twigs accumulates. The A horizon makes up the surface layer of many agricultural lands, in which all or most of the original structure of the underlying rocks or geological deposits has been destroyed. The E horizon is a mineral horizon where most of the original rock has been destroyed. It is characterized by the eluvial loss of silicate clay, Fe, Al, Si, or a combination of these. This leaves a residual concentration of sand and silt particles. E-horizon has undergone significant leaching, which is the process by which soluble nutrients are removed from the soil as a result of irrigation or precipitation. The horizon is usually bright in color and is most frequently found beneath the O horizon. Under an A, E, or O horizon, the B horizon is a mineral horizon where most or all the original parent material structures have been destroyed. Numerous pedogenic characteristics arising from in situ processes, soil material translocation, or both may be present in the B horizon. The composition, color, and structure of the B horizon might be different from that of the original parent material. Various pedogenic characteristics may arise in the B horizon due to in situ processes, such as material translocation. The composition, color, and structure of the B horizon might be different from that of the original parent material. It is an underground horizon created by the layer or layers above. It is where minerals that have leached from the layer or layers above have been deposited. The C horizon is subsurface. Except for hard bedrock that is firmly cemented, it is the least worn mineral horizon. By definition, it is devoid of the characteristics of the O, A, E, and B horizons and is not significantly impacted by pedogenic processes. It is a loose, unconsolidated parent material that is occasionally referred to as saprolite (Hartemink et al. 2020; Siddiqui and Fatima 2017).

17.3 SOIL CONTAMINANTS

Soils are formed through the long-term breakdown of organic matter and rock over many years. The characteristics of soils differ from location to location due to variations in climate, bedrock composition, and other factors. Some chemical elements that are naturally present in soils as components of minerals may occasionally be toxic at levels above those advised for the well-being of humans, animals, or plants (Balali-Mood et al. 2021). A contaminant is a substance that is not present in the environment but is incorporated into the soil by human activity and natural processes, changing the soil's composition. It can be classified as biological, chemical, physical, or radioactive substances that can negatively impact living things employing soil, water, air, and food when present in sufficient concentrations. Both human-made and natural sources discharge contaminants into the soil. The distribution of various soil contaminants in diverse ecosystems is significantly influenced by human-associated factors. This underscores a crucial connection between anthropogenic activities and soil contamination extending beyond the natural sources of specific contaminants, such as metals and metalloids (Liu et al. 2023). Naturally, deterioration of soil parent material and volcanic eruptions introduce contaminants into the soil. The primary source for soil and water contamination is anthropogenic activity. The overuse of fertilizers and biocides in agriculture, heavy mechanical industries, mining and smelting, and the tanning of leather are some of the major human activities (Saeed et al. 2023).

These activities contaminate the soil with various organic and inorganic substances, making it dangerous for people, domesticated animals, plants, and microorganisms. Point sources and diffuse sources are the two categories into which sources can be separated based on the point of origin and mode of contamination. Untreated wastewater from industrial sewage plants, agricultural farm building effluents, solid waste disposal sites, plant growth regulator application sites, and pesticide manufacturing sites are the main sources for contaminants released into soils (Valentín et al. 2013). A few types of forestry and agriculture, transportation, and poor waste and wastewater management are linked to diffuse contamination. Low-volatile compound atmospheric deposition is a different diffuse source of contamination. Heavy metals and persistent organic pollutants (POPs) are prominent examples of contaminants from diffuse sources (Fabietti et al. 2009). Diffuse contamination can cause soils to lose essential functions and lower the overall quality of the environment.

17.4 CLASSIFICATION OF CONTAMINANTS

17.4.1 Organic Contaminants

Halogenated aliphatic and aromatic hydrocarbons, which come from a variety of point and non-point sources, constitute the main group of chemicals. These include petroleum hydrocarbons, benzenes, organochlorine pesticides, phthalate esters, polynuclear hydrocarbons (PAHs), polychlorinated biphenyls, grease, and oil (Valentín et al. 2013). The physicochemical characteristics of specific organic compounds, the various local point sources, the operating parameters of wastewater treatment plants, and wastewater quality all have a substantial impact on the presence and number of organic contaminants. Generally speaking, industrial sewage has a greater number of organic contaminants than home effluent (Rajasulochana and Preethy 2016). Persistent organic pollutants (POPs), known for their high molecular masses and lipid affinity, accumulate in the adipose tissues of living organisms. Despite their low concentrations, chronic exposure raises concerns for potential adverse effects, especially as some POPs act as endocrine disruptors, harming the reproductive system and wildlife development (Rhind 2005).

17.4.2 Inorganic Contaminants

Inorganic pollutants, including toxic metals, nutrients, and salts, exist as dissolved anions and cations in soil, affecting soil quality, fertility, and the ecosystem (Kumar 2020). Heavy metals are the main category of inorganic pollutants. Contaminants containing heavy metals lower soil quality, negatively impact biomass production quality, and harm aquatic environments (Makovnikova et al. 2006). Toxic metals, various nutrients, and salts are examples of inorganic contaminants. These pollutants typically manifest as dissolved cations and anions. While some inorganic pollutants break down or change quickly, others remain in the soil for an infinite amount of time. The physical, biological, and fertility levels of the soil are all negatively impacted by inorganic contaminants, which ultimately leads to a decrease in productivity. Of all the inorganic pollutants, heavy metals have received the greatest attention and have had the greatest impact on the system. An extensive range of heavy metals are utilized in industry, transportation, agriculture, and other sectors of the economy. These include cadmium (Cd), arsenic (As), chromium (Cr), zinc (Zn), nickel (Ni), copper (Cu), mercury (Hg), and lead (Pb; Kumar 2020).

The incorporation of heavy metals into the trophic chain may impact the health of animals and people. Soil contamination by heavy metals from wastewater irrigation raises health concerns associated with consuming crops grown in contaminated areas (Sharma et al. 2006). Even at low concentrations, heavy metals have detrimental effects on plants, soil, humans, and animals due to their toxic characteristics (Rajendran et al. 2022).

17.5 SOIL ASSESSMENT

Along with water and air quality, soil quality is one of the three elements of environmental quality. The fundamental determinant of water and air quality is the quantity of pollution, which directly impacts the health and consumption of humans, animals, and natural systems. On the other hand, a soil's "capacity to function within ecological systems and land utilization boundaries to preserve biological productivity, keep environmental quality, and encourage plant and animal health" is a much more widely used definition of soil quality, which is not restricted to the level of soil pollution

(Bünemann et al. 2018). The composition of soil includes solid, liquid, and gaseous phases, which makes it more complex compared to the quality of air or water. Additionally, the range of uses for soil is greater (Nortcliff 2002). The quality of the soil can be evaluated for natural ecosystems, where the primary goals are to preserve biodiversity and environmental quality, as well as for agricultural ecosystems, where efficiency is the primary, though not exclusive, ecosystem service (Blum 2005).

The various uses of the soil, such as agriculture, forestry, rangeland, preservation of nature, recreation, or urban development, have to be taken into account in any assessment of the quality of the soil. But when it comes to soil quality, agro-ecosystems are the most globally significant and commonly accepted concept. The two primary questions that require to be addressed in any soil-quality assessment are the way the soil functions and what methods are suitable for conducting the assessment. Following the completion of those inquiries, a variety of parameter values or indexes that signify a soil's optimal performance can be computed through the application of pedogenesis knowledge, landscape features, and a deeper comprehension of the changing events taking place in the soil. To assess how sustainable soil management techniques are, soil-quality assessment places a strong emphasis on dynamic elements; however, it has to be grounded in the intrinsic soil factors.

17.6 APPROACHES FOR SOIL ASSESSMENT

17.6.1 Visual Soil Assessment

The use of visual soil assessment (VSA) methods for research, consulting, and teaching in agriculture and the field of soil science has grown in popularity (Ball et al. 2017). The potential utility of VSA techniques is highlighted by their low equipment requirements, quick results, and strong correlation with quantitative evaluations of physical and biochemical properties (Mutuku et al. 2021). Examining the soil visually is a crucial step in determining the quality of agricultural soil. One important component of the soil's condition that is vulnerable to soil deterioration is soil structure. Combining meteorological data with particular hazard indicators, like a high risk of drought, flooding, or contamination, with visually identifiable "core" soil indicators, like soil structure, rooting depth, wetness, and slope, can yield an internationally applicable overall soil quality rating (Mueller et al. 2012).

The primary techniques for assessing soil structure visually center on characterizing soil aggregates, porosity, and rooting as they pertain to the movement and storage of water, the growth of roots, and the absorption of nutrients. Soil type may have an impact on the estimate of quality, but it has little bearing on how the estimate is applied in terms of soil function – for instance, in severely degraded soils. Soil structure is a universal indicator of soil quality (Ball et al. 2017). Visual soil examination and evaluation (VSEE) techniques are inexpensive, rapid, low-tech, and allow for direct field evaluation of soil structure. In addition, these methods simplify the process of making immediate judgments regarding soil management for non-soil professionals such as farmers and extension staff by comparing soil quality scores with acceptable thresholds (Guimarães et al. 2017). The two categories of VSEE techniques are "profile" and "spade." The foundation of spade methods is the removal of a soil block with a spade, enabling up to 0.5 m of structural examination. Profile techniques, on the other hand, require mechanical excavation of soil profiles in order to conduct structural assessments down to a depth of 1.5 m. As a result, whereas profile tests take longer and can only review a limited number of points, spade tests may be done in the field more often and analyze a greater number of blocks. As a consequence, distinct soil profiles are assessed using profile methodologies. These approaches provide detailed vertical and lateral data for specific sample spots. In contrast, spade techniques analyze up to 50 cm below the surface, need a spade for soil exposure, and offer structural quality information across a greater geographic region more swiftly (Emmet-Booth et al. 2016).

17.6.2 Qualitative Soil Assessment

Soil quality evaluation usually employs analytical procedures that leverage biological, physical, and chemical indicators, as well as data sets translated into soil quality indexes (Andrews et al. 2004). These studies are costly and time-consuming, however, and they contradict the notion that indices should be easy to measure, generally available, and suited for field situations. When resources, whether financial or human, are limited, soil quality qualitative indicators may be the solution. Furthermore, qualitative indicators enable continuous evaluation of soil quality and, as a result, continuous monitoring of the condition of soil functions in agricultural systems (Sonneveld et al. 2014; Van Dang 2007).

Within the framework of agroecology, where farmers are contributors to knowledge about production systems, qualitative assessments of soil quality are highly prevalent (Wezel et al. 2020). The objective of the quantitative indicators was to assess the buffering and structural support functions of soil using soil bulk density, aggregate weighted mean diameter, and soil texture. We measured the pH of water, exchangeable Ca^{+2}, Al^{+3}, Mg^{+2}, and available phosphorus (P) and potassium (K) to analyze the buffering and productivity, and soil functions. The soil nutrient cycling functions were assessed using total nitrogen (TN), total organic carbon (TOC), and soil organic matter in size fractions (Comin et al. 2024). In the laboratory, soil samples can be analyzed using standard and conventional analytical techniques such as atomic absorption spectroscopy (AAS), atomic fluorescence spectroscopy (AFS), high-resolution gas chromatography and mass spectrometry (HRGC/HRMS), and ion-coupled plasma spectroscopy (ICP MS/ICP OES). Using these techniques, qualitative and quantitative analysis of metals and metalloids – like Cr, Ni, Cu, Zn, manganese (Mn), silver (Ag), Cd, Pb, Fe, Hg and As – can be achieved (Pan et al. 2020; Bonelli et al. 2017). The soil can be assessed in situ and ex situ by using field portable x-ray fluorescence (FP-XRF) spectroscopy. On-site evaluation of organic contaminants and heavy metals associated with particulate organic carbon can be easily made through this technique as it is faster and more economical (Hu et al. 2014; Bonelli et al. 2017)

17.6.3 Biological Soil Assessment

To evaluate a variety of soil chemical, physical, and biological properties, laboratory soil tests are available. Additionally, farmers can keep an eye on the condition of their soil based on how it generally looks and reacts to disturbances. Significant contributions to the sustainability of agriculture are made by soil organisms. On-farm tests for soil mesofauna, microfauna, and microorganisms are not frequently utilized, although they are available for evaluating the diversity and abundance of larger soil fauna (e.g., earthworms and other macrofauna). Farmers may be able to evaluate soil mesofauna and arbuscular mycorrhizal (AM) fungi on their farms by modifying laboratory techniques for these measurements (Mahdi et al. 2017; Lee and Pankhurst 1992). Root health assessment, organic matter content, populations of beneficial and parasitic nematodes, potentially mineralizable nitrogen, organic matter decomposition rate, particulate organic matter, active carbon test of the weed seed bank, microbial respiration rate, and glomalin content are the factors used to evaluate biological soil quality (Karlen et al. 2003).

17.7 GOVERNMENT POLICIES IN INDIA FOR SUSTAINABLE SOIL HEALTH

Several environmentally friendly and site-specific technologies have been developed by the Indian Council of Agricultural Research (ICAR) institutes, State Agricultural Universities, and All India Co-ordinated Research Projects (AICRP) to enhance and preserve the physical, chemical, and biological health of soils and produce sustainability. Additionally, the Indian government has started a number of initiatives to support farmers, including the Soil Health Card program, the sale of urea coated in neem oil, the National Mission on Sustainable Agriculture (NMSA), the National Action

Plan for Climate Change (NAPCC), the National Water Mission, the National Mission for a Green India under the NAPCC, the Mission for Integrated Development of Horticulture, and the National Mission on Oilseeds & Oil Palm NAAS (2018); ICAR (2023). With effect from April 2010, the Indian government is putting into practice the nutrient-based subsidy policy for P and K fertilizers. A set rate of subsidy (in rupees per kilogram) is announced annually on nutrients under the same policy. Urea is excluded from the policy, however, for many reasons. To ensure efficient use of N and prevent diversion to non-agricultural uses, all of the urea is currently marketed as neem oil–coated urea. The government's policy, which limits the amount of urea that can be sold in India to only that which is coated in neem, has undoubtedly increased crop yields by 6%–17% while decreasing consumption by 10%. The benefits must be measured in terms of actual use reduction, yield enhancement, nitrate leaching, and greenhouse gas reduction. Only urea is still subject to statutory price controls, and the government sets the maximum retail price for this fertilizer. Subsidies up to the difference between the concessional price that is administered and the cost of production are permitted. The Indian government encourages the sustainable use of natural resources through the NMSA and the NAPCC for India. In addition, the NAPCC's National Water Mission and National Mission for a Green India are directly related to the management of rainfed areas and natural resources. Saline, alkali, and acid soil reclamation are some of the components of problem soils that the government has included under the NMSA. The government is supporting balanced and integrated nutrient management based on soil tests under the Soil Health Management (SHM) component of the NMSA. This is accomplished through the establishment of compost and biofertilizer units, the use of micronutrients, training segments, and demonstrations on the balanced use of fertilizers, and the setup or strengthening of soil testing laboratories. A National Mission on Soil Health Card has been initiated, whereby each farmer receives a Soil Health Card from the soil testing laboratory situated in each state district. This has been done to give all of the nation's farmers recommendations for fertilizer based on soil tests. Rashtriya Krishi Vikas Yojana, Mission for Integrated Development of Horticulture, National Mission on Oilseeds and Oil Palm, and National Programme for Organic Production of the Agricultural and Processed Food Products Export Development Authority are some of the schemes and programs that support organic farming. The Indian government launched the Paramparagat Krishi Vikas Yojana in 2015 as part of SHM to encourage organic farming in the nation. This program is being implemented using a cluster approach, with each cluster made up of 50 farmers who have 1 acre to convert to organic farming. Over 3 years, 10,000 of these clusters were to be established nationwide. National regulations have been implemented to ensure that crop residues are properly recycled and do not burn, thus enhancing soil health.

17.8 REGULATORY BODIES OF SOIL ASSESSMENT IN INDIA

National Bureau of Soil Survey (NBSS) and Land Use Planning (LUP): Within the ICAR, the NBSS & LUP is a well-known organization with expertise in land use planning and national soil surveys.

State soil survey organizations: In India, each state normally has its own state soil survey organization that conducting soil surveys and assessments at the state level. An example is the Madhya Pradesh State Soil Conservation Board (Dharumarajan et al. 2019).

Soil testing laboratories (STLs): STLs, both at the state and district levels, are critical in soil assessment because they analyze soil samples for nutrient content and make suggestions to farmers. As an example, consider the National Portal of India Soil Testing Laboratories.

Ministry of Agriculture and Farmers Welfare: The Ministry of Agriculture and Farmers Welfare, through its many agencies, plays a significant role in developing soil health and evaluation policies.

State agriculture departments: State agriculture departments take an active role in soil evaluation, implementing recommendations, and promoting sustainable soil management techniques. An example is the Government of Karnataka Department of Agriculture (Nandi and Swamikannu 2019).

Indian Institute of Soil Science (IISS): IISS, which is part of the ICAR, conducts advanced soil research and supports soil management methods (Dharumarajan et al. 2019). Soil quality and environmental protection have achieved international policy recognition. Because soil is considered a living body, there is an urgent need to increase our knowledge and take action to safeguard soil quality. Effective soil assessment requires collaboration between government agencies, research institutions, and agricultural stakeholders (Meena and Kumar 2023).

17.9 PERMISSIBLE LEVELS OF CONTAMINANTS IN PLANT FOOD PRODUCTS

Environmental chemical contamination is a major global food safety concern that poses a substantial risk to human health. These molecules fall into a variety of categories, such as radioactive elements, plastics, nanoparticles, pharmaceuticals and personal care products, perfluorinated compounds, metals/metalloids, PAHs, POPs, and nanoparticles. While some of these come from artificial sources, others happen naturally in the surroundings. They might poison our drinking water and contaminate our food, including shellfish, cattle, and crops, which would be harmful to our health. It is critical to evaluate any possible risks that may be involved. To ensure that food is safer everywhere, monitoring pollution levels, putting control measures in place like cleanup, and considering the sociopolitical fallout are crucial. It is essential to monitor contamination levels, implement control methods such as remediation, and take sociopolitical consequences into account (Thompson and Darwish 2019). The two most significant public health concerns are food safety and quality. Food producers should be held accountable for ensuring that the food they sell is safe, as should state governments and organizations that tally and oversee food quality. Food should not contain any chemicals that could endanger the health of consumers (Igwegbe et al. 2013). Foods derived from plants are valuable resources that provide essential biological elements needed for human survival on Earth. Over 70% of human energy and nutritional needs are met by plant-based diets. Pollution of arable soils, linked to human activities like metal ore mining and processing, is what causes contamination of food crops. When vegetables are grown at home, the buildup of metals in the edible plant parts can have a negative impact on the health of the residents (Dziubanek et al. 2015). Plant-based diets include a large number of critical trace elements for humans, such as Fe, Cu, Zn, Cr, selenium (Se), Mn, and molybdenum (Mo). The essentiality of As and Ni for their biological roles in humans has not been established, even though these elements may be beneficial to humans at ultra-trace levels. Conversely, toxic elements (e.g., Cd, Pb, and Hg) have no biological significance in the human body (Liu et al. 2013).

In total diet research from the United Kingdom and France, the As content in various grains was less than the maximum permitted level (Leblanc et al. 2005). Safe levels of Cd were found in cereal grains sourced from regional markets in India, Ethiopia, Australia, Nigeria, the United Kingdom, and France. A study on Swedish rye found lead concentrations below the detection limit (Jorhem et al. 2001). A study by Rahman et al. (2014) reported that Pb concentrations in most nations' market surveys were safe. Food grains from marketplaces in many nations have acceptable amounts of Ni. Concentrations in carrot, cabbage, and potato from Croatia and Cambodia, as well as in cauliflower, carrot, and potato from local markets in Spain, varied from undetectable to safe levels (Stancic et al. 2016; Martorell et al. 2011). Cadmium levels in fruit samples from markets and production areas in most nations were under safe limits. Mulberry cultivated on roadsides in Turkey had a Cd concentration under the permissible range; however, the Cd concentration in apples (Hamurcu et al. 2010) and plums (Pehluvan et al. 2012) was higher than the safe limit Vegetables purchased from marketplaces in Nigeria, Yemen, Serbia, Saudi Arabia, and Tanzania had Pb amounts lower than permitted. Similarly, selected vegetables produced in agricultural settings in Kyrgyzstan have up to 0.18 mg Pb/kg (Usubalieva et al. 2013).

A large number of fruits found in local markets and farms in the United States, Nigeria, India, Turkey, Pakistan, Finland, Brazil, and the Slovak Republic have safe quantities of Ni. Tomatoes and cauliflower both have average concentrations of Cr, which are within acceptable ranges.

The Standardization Administration of China (SAC) has set a limit on Cr levels in vegetables of 0.5 mg/kg (Alegría et al. 1991). Permissible Pb levels for leafy vegetables are defined by Alimentarius (2011) at 0.2 mg/kg and for fruity vegetables as 0.05 mg/kg (Real et al. 2017).

17.10 CONCLUSION

This chapter describes the types of soil found in the Indian subcontinent. Soil classification is based on its geographical distribution, composition, particle size, and so on. Black, red, alluvial, and lateritic soils are present in significant amounts in India. This soil can grow a variety of food and commercial crops. Soil health is vital for agriculture, as it directly affects the yield production and nutrition profile of the yield. Hence soil assessment is important. Contaminants like organic compounds and inorganic heavy metals enter the soil for various reasons. Agricultural activity and industrial wastes are the main reasons for the presence of soil contaminants. To ensure the safety of agricultural crops grown in contaminated regions, soil assessment is a necessary step. There are many approaches to soil assessments. Visual soil assessment, where the soil texture, color, rooting depth, and topography are evaluated, and risk assessment are made. Qualitative and quantitative assessment methods, including analytical techniques such as AAS, IC PMS, AFS, and FP-XRF spectroscopy of soil samples, will be conducted and thorough elemental profiling will be done. The chapter emphasizes the Indian involvement of governments and civil bodies in soil sustainability. The policies and subsidies to help with the maintenance of soil health have been mentioned in the chapter. Regulatory bodies in India help make such policies and subsidies to help the farming community with soil health sustenance.

17.11 ACKNOWLEDGMENTS

We thank Manipal Academy of Higher Education (MAHE), Manipal, Karnataka, India, TIFAC-CORE and FIST, DST New Delhi, DBT BUILDER grant (BT/INF/22/SP43065/2021), DBT, New Delhi, and K-FIST, VGST, Govt. of Karnataka for the facilities. The authors (M. K. and H.K.C.) are thankful to MAHE for the T.M.A. Pai Ph.D. scholarship. We are grateful to Dr. B. S. Satish Rao, Professor and Director, Manipal School of Life Sciences, MAHE for encouragement and support.

REFERENCES

Alegría, A., R. Barberá, R. Boluda, F. Errecalde, R. Farré, and M. J. Lagarda. 1991. "Environmental cadmium, lead and nickel contamination: Possible relationship between soil and vegetable content." *Fresenius' Journal of Analytical Chemistry* 339 (9):654–657. doi: 10.1007/BF00325553.

Alimentarius, C. 2011. "Codex standard." *Rome, Food Agriculture Organization*:33–198.

Andrews, Susan S., Douglas L. Karlen, and Cynthia A. Cambardella. 2004. "The soil management assessment framework." *Soil Science Society of America Journal* 68 (6):1945–1962. doi: 10.2136/sssaj2004.1945.

Aransiola, S.A., U.J.J. Ijah, O.P. Abioye, and J.D. Bala. 2019. "Microbial-aided phytoremediation of heavy metals contaminated soil: A review." *European Journal of Biological Research* 9 (2):104–125. http://www.journals.tmkarpinski.com/index.php/ejbr/article/view/157.

Balali-Mood, M., K. Naseri, Z. Tahergorabi, M. R. Khazdair, and M. Sadeghi. 2021. "Toxic mechanisms of five heavy metals: Mercury, lead, chromium, cadmium, and arsenic." *Frontiers in Pharmacology* 12:643972. doi: 10.3389/fphar.2021.643972.

Baligar, V.C., N.K. Fageria, H. Eswaran, M.J. Wilson, and Zhenli He. 2004. "Nature and properties of red soils of the world." *The Red Soils of China: Their Nature, Management and Utilization*:7–27.

Ball, Bruce C., Rachel M. L. Guimarães, Joanna M. Cloy, Paul R. Hargreaves, T. Graham Shepherd, and Blair M. McKenzie. 2017. "Visual soil evaluation: A summary of some applications and potential developments for agriculture." *Soil and Tillage Research* 173:114–124. doi: 10.1016/j.still.2016.07.006.

Bhattacharyya, Tapas, D. Sarkar, S.K. Ray, P. Chandran, D.K. Pal, D.K. Mandal, J. Prasad, G.S. Sidhu, K.M. Nair, and A.K. Sahoo. 2014. "Soil information system: Use and potentials in humid and semi-arid tropics." *Current Science*:1550–1564.

Blum, Winfried E. H. 2005. "Functions of soil for society and the environment." *Reviews in Environmental Science and Bio/Technology* 4 (3):75–79. doi: 10.1007/s11157-005-2236-x.

Boettinger, J., G. Busch, B. Fonnesbeck, J. Lawley, A. Croft, T. Edwards, and J. Macmahon. 2010. "Soil-landscape relationships and soil properties associated with rare plants in the eastern Mojave Desert near Las Vegas, Nevada, USA." *Proceedings of the 19th World Congress of Soil Science*, Brisbane, Australia.

Bonelli, Maria Grazia, Mauro Ferrini, and Andrea Manni. 2017. "Artificial neural networks to evaluate organic and inorganic contamination in agricultural soils." *Chemosphere* 186:124–131.

Bünemann, Else K., Giulia Bongiorno, Zhanguo Bai, Rachel E. Creamer, Gerlinde De Deyn, Ron de Goede, Luuk Fleskens, Violette Geissen, Thom W. Kuyper, Paul Mäder, Mirjam Pulleman, Wijnand Sukkel, Jan Willem van Groenigen, and Lijbert Brussaard. 2018. "Soil quality – A critical review." *Soil Biology and Biochemistry* 120:105–125. doi: 10.1016/j.soilbio.2018.01.030.

Chandan, Ritula Thakur, and R. Thakur. 2018. "An intelligent model for Indian soil classification using various machine learning techniques." *International Journal of Computational Engineering Research (IJCER)* 33 (2250):3005.

Comin, J.J., C.A. Bourscheid, L.D. Giumbelli, M. Souza Teixeira, R.D.R. Couto, P.E. Lovato, G. Brunetto, F.M. Vezzani, C.R. Lourenzi, and A. Loss. 2024. "Qualitative soil quality assessment is efficient in a grazing system with or without trees." *Journal of the Science of Food and Agriculture* 104 (4):1962–1970. doi: 10.1002/jsfa.13084.

Dharumarajan, S., Rajendra Hegde, N. Janani, and S. K. Singh. 2019. "The need for digital soil mapping in India." *Geoderma Regional* 16. doi: 10.1016/j.geodrs.2019.e00204.

Dwevedi, Alka, Promod Kumar, Pravita Kumar, Yogendra Kumar, Yogesh K. Sharma, and Arvind M. Kayastha. 2017. "Soil sensors: Detailed insight into research updates, significance, and future prospects." *New Pesticides and Soil Sensors*:561–594.

Dziubanek, G., A. Piekut, M. Rusin, R. Baranowska, and I. Hajok. 2015. "Contamination of food crops grown on soils with elevated heavy metals content." *Ecotoxicology and Environmental Safety* 118:183–189. doi: 10.1016/j.ecoenv.2015.04.032.

Emmet-Booth, J.P., P.D. Forristal, O. Fenton, B.C. Ball, N.M. Holden, and Michael Goss. 2016. "A review of visual soil evaluation techniques for soil structure." *Soil Use and Management* 32 (4):623–634. doi: 10.1111/sum.12300.

Fabietti, Gabriele, Mattia Biasioli, Renzo Barberis, and Franco Ajmone-Marsan. 2009. "Soil contamination by organic and inorganic pollutants at the regional scale: The case of Piedmont, Italy." *Journal of Soils and Sediments* 10 (2):290–300. doi: 10.1007/s11368-009-0114-9.

Guimarães, Rachel M. L., Afrânio F. Neves Junior, Wellington G. Silva, Craig D. Rogers, Bruce C. Ball, Célia R. Montes, and Bruno F. F. Pereira. 2017. "The merits of the visual evaluation of soil structure method (VESS) for assessing soil physical quality in the remote, undeveloped regions of the Amazon basin." *Soil and Tillage Research* 173:75–82. doi: 10.1016/j.still.2016.10.014.

Hamurcu, M., M.M. Ozcan, N. Dursun, and S. Gezgin. 2010. "Mineral and heavy metal levels of some fruits grown at the roadsides." *Food and Chemical Toxicology* 48 (6):1767–70. doi: 10.1016/j.fct.2010.03.031.

Hartemink, A.E., Y. Zhang, J.G. Bockheim, N. Curi, S.H.G. Silva, J. Grauer-Gray, D.J. Lowe, and P. Krasilnikov. 2020. "Soil horizon variation: A review." 125–185. doi: 10.1016/bs.agron.2019.10.003.

Holzapfel, C. 2009. "These deserts lie within the subtropical high pressure belt where the descending part of the Hadley's cell air•• circulation causes general aridity. Rain shadow deserts. They are found on the landward side of coastal mountain ranges. Coastal deserts. Found along coasts bordering very cold." *Ecosystem Ecology*:222.

Hu, W., B. Huang, D.C. Weindorf, and Y. Chen. 2014. "Metals analysis of agricultural soils via portable X-ray fluorescence spectrometry." *Bulletin of Environmental Contamination and Toxicology* 92 (4):420–426. doi: 10.1007/s00128-014-1236-3.

Igwegbe, A. O., C. H. Agukwe, and C. A. Negbenebor. 2013. "A survey of heavy metal (lead, cadmium and copper) contents of selected fruit and vegetable crops from Borno State of Nigeria." *International Journal of Engineering Science* 2 (1):01–05.

Jansson, J.K., and K.S. Hofmockel. 2020. "Soil microbiomes and climate change." *Nature Reviews Microbiology* 18 (1):35–46. doi: 10.1038/s41579-019-0265-7.

Jorhem, Lars, Birgitta Sundström, and Joakim Engman. 2001. "Cadmium and other metals in Swedish wheat and rye flours: Longitudinal study, 1983–1997." *Journal of AOAC International* 84 (6):1984–1992. doi: 10.1093/jaoac/84.6.1984.

Karlen, Douglas L., Craig A. Ditzler, and Susan S. Andrews. 2003. "Soil quality: Why and how?" *Geoderma* 114 (3–4):145–156. doi: 10.1016/s0016-7061(03)00039-9.

Knoepp, Jennifer D., Leonard F. DeBano, and Daniel G. Neary. 2005. "Soil chemistry." *Wildland Fire in Ecosystem*:42–44.

Kumar, Anil. 2020. "Inorganic soil contaminants and their biological remediation." *Plant Responses to Soil Pollution*:133–153.

Leblanc, J.C., A. Tard, J.L. Volatier, and P. Verger. 2005. "Estimated dietary exposure to principal food mycotoxins from the first French total diet study." *Food Additives & Contaminants* 22 (7):652–672. doi: 10.1080/02652030500159938.

Lee, K.E., and C.E. Pankhurst. 1992. "Soil organisms and sustainable productivity." *Soil Research* 30 (6):855–892.

Liu, X., Q. Song, Y. Tang, W. Li, J. Xu, J. Wu, F. Wang, and P.C. Brookes. 2013. "Human health risk assessment of heavy metals in soil-vegetable system: A multi-medium analysis." *Science of the Total Environment* 463–464:530–540. doi: 10.1016/j.scitotenv.2013.06.064.

Liu, Y.R., L. Guo, Z. Yang, Z. Xu, J. Zhao, S.H. Wen, M. Delgado-Baquerizo, and L. Chen. 2023. "Multidimensional drivers of mercury distribution in global surface soils: Insights from a global standardized field survey." *Environmental Science & Technology* 57 (33):12442–12452. doi: 10.1021/acs.est.3c04313.

Mahdi, Jwad E., Lynette K. Abbott, Natasha Pauli, and Zakaria M. Solaiman. 2017. "Biological Indicators for Soil Health: Potential for Development and Use of On-Farm Tests." In *Modern Tools and Techniques to Understand Microbes*, 123–134.

Makovnikova, Jarmila, Gabriela Barančíková, Pavel Dlapa, and Katarina Dercova. 2006. "Inorganic contaminants in soil ecosystems." *Chemicke listy* 100 (6).

Martorell, I., G. Perelló, R. Marti-Cid, J.M. Llobet, V. Castell, and J.L. Domingo. 2011. "Human exposure to arsenic, cadmium, mercury, and lead from foods in Catalonia, Spain: Temporal trend." *Biological Trace Element Research* 142 (3):309–322. doi: 10.1007/s12011-010-8787-x.

Meena, Ram Swaroop, and Sandeep Kumar. 2023. "Soil quality protection policies and plans to ensure sustainability." In *Encyclopedia of Soils in the Environment*, 457–472.

Mishra, B.B. 2016. "Indian system of soil classification: A way forward." *Agricultural Research & Technology: Open Access Journal* 3 (1). doi: 10.19080/artoaj.2016.03.555606.

Montgomery, David R., Darlene Zabowski, Fiorenzo C. Ugolini, Rolf O. Hallberg, and Henri Spaltenstein. 2000. "Soils, watershed processes, and marine." *Earth System Science: From Biogeochemical Cycles to Global Changes* 72:159.

Mueller, Lothar, Uwe Schindler, T. Graham Shepherd, Bruce C. Ball, Elena Smolentseva, Chunsheng Hu, Volker Hennings, Peter Schad, Jutta Rogasik, Jutta Zeitz, Sandro L. Schlindwein, Axel Behrendt, Katharina Helming, and Frank Eulenstein. 2012. "A framework for assessing agricultural soil quality on a global scale." *Archives of Agronomy and Soil Science* 58 (sup1):S76–S82. doi: 10.1080/03650340.2012.692877.

Mutuku, Eunice A., Bernard Vanlauwe, Dries Roobroeck, Pascal Boeckx, and Wim M. Cornelis. 2021. "Visual soil examination and evaluation in the sub-humid and semi-arid regions of Kenya." *Soil and Tillage Research* 213. doi: 10.1016/j.still.2021.105135.

Nandi, Ravi, and Nedumaran Swamikannu. 2019. "Agriculture extension system in India: A meta-analysis." *Agricultural Science Research Journal* 10:473–479.

Nortcliff, Stephen. 2002. "Standardisation of soil quality attributes." *Agriculture, Ecosystems & Environment* 88 (2):161–168.

Owens, A. 2023. *Soil and Landscape Assessment of the Hydrological Capacity of a~ 45-Year-Old Spray Irrigation Wastewater Treatment System*. The Pennsylvania State University.

Pan, F., Y. Yu, L. Yu, H. Lin, Y. Wang, L. Zhang, D. Pan, and R. Zhu. 2020. "Quantitative assessment on soil concentration of heavy metal-contaminated soil with various sample pretreatment techniques and detection methods." *Environmental Monitoring and Assessment* 192 (12):800. doi: 10.1007/s10661-020-08775-4.

Pehluvan, Mücahit, Huseyin Karlidag, and Metin Turan. 2012. "Heavy metal levels of mulberry (MORUS ALBA L.) grown at different distances from the roadsides." *Journal of Animal and Plant Sciences* 22:665–670.

Rahman, M.A., M.M. Rahman, S.M. Reichman, R.P. Lim, and R. Naidu. 2014. "Heavy metals in Australian grown and imported rice and vegetables on sale in Australia: Health hazard." *Ecotoxicology and Environmental Safety* 100:53–60. doi: 10.1016/j.ecoenv.2013.11.024.

Rajasulochana, P., and V. Preethy. 2016. "Comparison on efficiency of various techniques in treatment of waste and sewage water – A comprehensive review." *Resource-Efficient Technologies* 2 (4):175–184. doi: 10.1016/j.reffit.2016.09.004.

Rajendran, S., A.K. Priya, P. Senthil Kumar, T.K.A. Hoang, K. Sekar, K.Y. Chong, K.S. Khoo, H.S. Ng, and P.L. Show. 2022. "A critical and recent developments on adsorption technique for removal of heavy metals from wastewater-A review." *Chemosphere* 303 (Pt 2):135146. doi: 10.1016/j.chemosphere.2022.135146.

Real, M.I.H., H.M. Azam, and N. Majed. 2017. "Consumption of heavy metal contaminated foods and associated risks in Bangladesh." *Environmental Monitoring and Assessment* 189 (12):651. doi: 10.1007/s10661-017-6362-z.

Reza, S.K., D.C. Nayak, S. Mukhopadhyay, T. Chattopadhyay, and S.K. Singh. 2017. "Characterizing spatial variability of soil properties in alluvial soils of India using geostatistics and geographical information system." *Archives of Agronomy and Soil Science* 63 (11):1489–1498. doi: 10.1080/03650340.2017.1296134.

Rhind, S.M. 2005. "Are endocrine disrupting compounds a threat to farm animal health, welfare and productivity?" *Reproduction in Domestic Animals* 40 (4):282–290.

Saeed, Maimona, Noshin Ilyas, Fatima Bibi, Sumera Shabir, Sabiha Mehmood, Nosheen Akhtar, Iftikhar Ali, Sami Bawazeer, Abdel Rahman Al Tawaha, and Sayed M. Eldin. 2023. "Nanoremediation approaches for the mitigation of heavy metal contamination in vegetables: An overview." *Nanotechnology Reviews* 12 (1). doi: 10.1515/ntrev-2023-0156.

Santha Kumar, G., P.K. Saini, Rajesh Deoliya, Aman Kumar Mishra, and S.K. Negi. 2022. "Characterization of laterite soil and its use in construction applications: A review." *Resources, Conservation & Recycling Advances* 16. doi: 10.1016/j.rcradv.2022.200120.

Senesi, Nicola, and Elisabetta Loffredo. 2018. "The chemistry of soil organic matter." In *Soil Physical Chemistry*, 239–370. CRC Press.

Sharma, L.K., N.N. Sirdesai, K.M. Sharma, and T.N. Singh. 2018. "Experimental study to examine the independent roles of lime and cement on the stabilization of a mountain soil: A comparative study." *Applied Clay Science* 152:183–195. doi: 10.1016/j.clay.2017.11.012.

Sharma, R.K., M. Agrawal, and F. Marshall. 2006. "Heavy metal contamination in vegetables grown in wastewater irrigated areas of Varanasi, India." *Bulletin of Environmental Contamination & Toxicology* 77 (2).

Siddiqui, Shahid Ali, and Naseem Fatima. 2017. "Indian soils: Identification and classification." *Earth Sci. India* 10 (Iii):1–14.

Skjemstad, J.O., Leslie J. Janik, and J.A. Taylor. 1998. "Non-living soil organic matter: What do we know about it?" *Australian Journal of Experimental Agriculture* 38 (7):667–680.

Sonneveld, M.P.W., G.B.M. Heuvelink, and S.W. Moolenaar. 2014. "Application of a visual soil examination and evaluation technique at site and farm level." *Soil Use and Management* 30 (2):263–271. doi: 10.1111/sum.12117.

Stancic, Z., D. Vujevic, A. Gomaz, S. Bogdan, and D. Vincek. 2016. "Detection of heavy metals in common vegetables at Varazdin City Market, Croatia." *Arh Hig Rada Toksikol* 67 (4):340–350. doi: 10.1515/aiht-2016-67-2823.

Sulaeman, Yiyi, Sukarman Sukarman, Risma Neswati, Nurdin Nurdin, and Tony Basuki. 2023. "Characteristics and utilization of black soils in Indonesia." *SAINS TANAH – Journal of Soil Science and Agroclimatology* 20 (1). doi: 10.20961/stjssa.v20i1.70343.

Thompson, L.A., and W.S. Darwish. 2019. "Environmental chemical contaminants in food: Review of a global problem." *Journal of Toxicology* 2019:2345283. doi: 10.1155/2019/2345283.

Usubalieva, Aigul Z., M.B. Batkibekova, Holger Hintelmann, and Ramzan Judge. 2013. "The content of zinc, copper, lead and cadmium in some vegetables of Kyrgyzstan." *Pakistan Journal of Food Sciences* 23 (4):189–193.

Valentin, L., A. Nousiainen, and A. Mikkonen. 2013. "Introduction to organic contaminants in soil: Concepts and risks." *Emerging Organic Contaminants in Sludges: Analysis, Fate and Biological Treatment*:1–29.

Van Dang, Minh. 2007. "Quantitative and qualitative soil quality assessments of tea enterprises in Northern Vietnam." *African Journal of Agricultural Research* 2 (9):455–462.

Wezel, Alexander, Barbara Gemmill Herren, Rachel Bezner Kerr, Edmundo Barrios, André Luiz Rodrigues Gonçalves, and Fergus Sinclair. 2020. "Agroecological principles and elements and their implications for transitioning to sustainable food systems. A review." *Agronomy for Sustainable Development* 40 (6). doi: 10.1007/s13593-020-00646-z.

Winterkorn, Hans F., and E.C. Chandrasekharan. 1951. *Laterite Soils and Their Stabilization*. Winterkorn Road Research Institute.

Young, R. 2012. *Soil Properties and Behaviour*. Vol. 5. Elsevier.

18 Plant Uptake of Emerging Contaminants in Foods, Regulatory Considerations, and Future Directions

Gundarapu Vidya Sagar Reddy and Bellamkonda Ramesh

18.1 INTRODUCTION

The growing concern about the prevalence of novel contaminants in our food supply arises from their potential to harm human health. A noteworthy component catching interest is the absorption of these pollutants by plants, culminating in their buildup in the crops we rely on for nourishment. This chapter attempts to examine the scale of the problems associated with the absorption of these newly found pollutants in the context of food production.

Generally, not tracked down or controlled, emerging contaminants are a broad category of substances found in our food supply and environment. These materials are categorized by Marmugi et al. (2011) as industrial chemicals, pesticides, flame retardants, pharmaceuticals, and personal hygiene items. Epoxide resins and polycarbonate plastics are frequently made with the chemical bisphenol A (BPA). Marmugi et al. have raised concerns regarding the potential influence of BPA, which is known to disrupt hormones, on human health due to its presence in multiple food products (Marmugi et al., 2011).

Numerous factors influence the complex process of new toxins being absorbed by plants. The properties of soil, such as pH, organic matter concentration, and composition, have an impact on the availability and movement of pollutants in the soil (Aransiola et al., 2022). Meanwhile, aspects specific to plants, like the physiology, morphology, and existence of transporters, influence how pollutants are absorbed and moved around the plant. The complex dynamics of pollutant uptake and accumulation in plants are further influenced by environmental factors such as temperature, humidity, and rainfall (Shahid et al., 2017).

Understanding the implications of the uptake of newly developing toxins by plants is crucial since it provides the means of assessing any health concerns to humans. A direct exposure pathway to consumers is established in the event that these pollutants build up in edible plant parts, such as fruits, vegetables, and grains. In addition, the process of bioaccumulation of pollutants in the food chain promotes their migration to trophic levels at which humans are more likely to be exposed (Li et al., 2005).

In an attempt to shed light on the various elements that influence this complex process, this chapter delves into the current understanding of plant uptake of developing toxins in food. The possible effects on food safety and public health will be emphasized. Moreover, evaluating the degree of impacts resulting from plant absorption will be emphasized in order to establish a foundation for the creation of tactics intended to reduce the hazards connected with new pollutants entering the food chain.

18.2 ACCUMULATION OF CONTAMINANTS IN FOOD CROPS

The possible effects of pollutant accumulation in agricultural crops on human health and food safety make it a serious concern. Applying sewage sludge, irrigating with treated municipal wastewater,

DOI: 10.1201/9781032683751-21

and adding manure and biosolids as soil amendments are some of the ways that emerging contaminants, like triclosan, triclocarban, and heavy metals, can find their way into agricultural fields (Christou et al., 2019; Macherius et al., 2012).

Based on research, it has been demonstrated that these pollutants can be assimilated by crop plants and subsequently accumulate in edible plant parts. When food crops are irrigated or have soil amended with known concentrations of individual or mixes of contaminants of emerging concern (CECs), the uptake and bioaccumulation of these contaminants have been shown in controlled settings. Low micrograms to low milligrams per kilogram are possible ranges for CEC uptake by plants (Hyland et al., 2015).

There are several factors that affect how much contamination is absorbed and accumulates in food crops. Organic matter content, pH, and composition of the soil can all have an impact on the mobility and availability of pollutants in the soil, which in turn can have an impact on how readily plants can absorb them. According to Macherius et al., plants' metabolic capacities also influence how pollutants are transformed and stored inside their tissues.

When pollutants are found in food crops, there is cause for concern over their possible introduction into the food chain and eventual exposure to humans. The possibility of human exposure can be further increased by the transmission of contaminants to higher trophic levels via animal ingestion of contaminated crops (Macherius et al., 2012). This emphasizes how critical it is to evaluate the level of pollutant uptake by plants in order to determine the possible dangers to human health (Onakpa et al., 2018).

18.2.1 Soil Factors

Soil factors play a crucial role in the accumulation of contaminants in food crops. Several studies have investigated the influence of soil characteristics on the uptake and accumulation of contaminants by plants.

Soil carbon sequestration (SCS) and the addition of biochar to land have the potential to produce negative emissions, according to a study by Smith (2016). Comparing SCS and biochar to other negative emissions technologies showed that they have more potential for meaningful negative emissions and may have less of an effect on land, water use, nutrients, albedo, energy requirements, and cost (Smith, 2016).

The application of biosolids and manures as soil supplements, as well as the reuse of treated wastewater (TWW) for irrigation, are two significant pathways by which CECs may enter the agricultural environment, according to Christou et al. (2019). Surface and groundwater systems, sediments receiving TWW, soils altered with manure and biosolids, and agricultural soils receiving TWW irrigation have all been reported to contain CECs. It has been shown that crop plants grown in contaminated agricultural contexts absorb and hold on to CECs in their tissues, which could cause these pollutants to enter the food chain. Crop species have varying capacities for absorbing and accumulating CECs in edible tissues; leafy vegetables are more capable than root vegetables, cereals, and fruit vegetables.

Antimony (Sb) is another important pollutant that can build up in plants. According to studies, terrestrial plants typically take up Sb in proportion to the amount of soluble Sb present in the soil. Recent years have seen a rise in the amount of Sb contamination of soil brought on by human activity, underscoring the need for more study on the behavior and uptake of Sb in the soil-plant system (Tschan et al., 2009).

Understanding the soil conditions that influence plant absorption is critical for minimizing the danger of pollutants entering the food chain. Contaminant buildup and loss, as well as bioavailability and translocation across the food chain, can all be influenced by soil conditions. Soil pH, for example, can be controlled to reduce the likelihood of pollutants entering the food chain in large numbers (McGrath & Zhao, 2013).

18.2.2 Plant Factors

In food crops, plant factors are important because they influence how pollutants are absorbed and build up. The mechanisms and procedures involved in the uptake of pollutants by plants have been the subject of research.

According to Jones and Finlay (2009), the movement of carbon (C) and nitrogen (N) in the rhizosphere, or soil-root interface, is extremely complicated and influenced by a variety of plant and environmental factors. One of the main causes of the significant rhizosphere alterations is the loss of organic C and N from roots into the soil. Mechanism of bidirectional flow of C and N at the soil-root interface occurs where these compounds are simultaneously taken up from the soil and lost from the roots. The active cycling of high and low molecular weight organic compounds in the rhizosphere has been explained by a variety of theories. These include interroot and root-microbial signal exchange, direct re-uptake to increase plant efficiency, indirect root exudate recapture, and direct uptake to recapture organic nutrients (Jones & Finlay, 2009).

Plants and their environmental context determine the type and quantity of rhizodeposits, including mucilage, border cells, and exudates. Rhizodeposition's quantity and its function in ecosystem processes like carbon sequestration and nutrient uptake are difficult to measure and study because of this fluctuation. More investigation is required to gain a deeper comprehension of the quantitative significance of various carbon and nitrogen fluxes in the rhizosphere, as well as the fluctuations in these fluxes over time and space (Jones & Finlay, 2009).

According to Onakpa et al. (2018), plants have the ability to absorb heavy metals like arsenic (As) and accumulate them in their edible sections. Understanding and managing As pollution in soil-plant systems requires an understanding of the absorption, transport, and detoxifying mechanisms of As in plants, particularly crops. According to Tang and Zhao (2020), membrane transporters are essential for the absorption and translocation of As in plants (Onakpa et al., 2018; Tang & Zhao, 2020).

18.2.3 Environmental Factors

Contaminants in food crops build up due in large part to environmental causes. The impact of environmental factors on the uptake and accumulation of pollutants by plants has been the subject of numerous investigations.

The relationship between environmental sustainability, human health, and global diets has been emphasized by Tilman and Clark (2014). Diets heavy in meats, fats, oils, and refined sugars have become popular, which has led to a rise in land clearance and greenhouse gas emissions from agriculture. In addition to having an impact on the environment, this dietary change is associated with an increased risk of chronic noncommunicable diseases such as type 2 diabetes and coronary heart disease. Adopting better alternative diets appears to be one way to address these issues; it promises to cut greenhouse gas emissions, reduce land clearance, and enhance public health in general (Tilman & Clark, 2014).

According to Vitousek et al. (1997), the proliferation of adventive plants in forest parks is influenced by factors such as urbanization, habitat fragmentation, and forest structure. While the fraction of native species did not decline, the study indicated that urbanization enhanced the richness and abundance of adventive plant species. Another factor affecting the extent of adventive plant colonization was the existence of recently drawn forest boundaries. The condition of the understory vegetation was not shown to be affected by edge effects, according to the study (Vitousek et al., 1997).

Public health problems are raised by heavy metal poisoning of the food chain. Plants have the ability to take up heavy metals from the soil, water, and air and store them in their edible sections. Because of bioaccumulation, the quantities of heavy metals in soil and food crops can vary greatly and pose different health risks to the local community. The environment becomes contaminated with heavy metals due to a number of reasons, such as industrial activities, agricultural practices, and vehicle emissions (Rehman et al., 2018).

A number of variables, such as plant species and environmental parameters like moisture and temperature, can lead to mold infection of food crops and the consequent formation of mycotoxins. Mold contamination can also be influenced by agricultural activities. Controlling mold infestation and guaranteeing food safety need an understanding of these variables (Kyei et al., 2021).

Food crops may absorb nitrofuran antibiotics, which are carcinogenic and mutagenic, after they are released into the environment. Because of the possibility of drug-resistant bacteria and human exposure through food intake, nitrofuran-induced environmental pollution poses a health risk to the general population. Research has indicated that the edible portions of food crops have a high rate of nitrofuran absorption and accumulation, underscoring the necessity for mitigation and monitoring measures (Wang et al., 2017).

18.3 CONTAMINANTS OF EMERGING CONCERN IN AGRICULTURAL WATER REUSE

A heterogeneous group of bioactive chemicals known as pharmaceuticals and personal care products (PPCPs) can enter aquatic ecosystems via both treated and untreated effluent. These PPCPs can be introduced continuously to surface waters, even at low concentrations, exposing aquatic organisms to them throughout their entire life cycles. The continuous introduction of PPCPs into aquatic environments gives rise to concerns regarding their potential effects on aquatic life and the environment.

In agricultural water reuse, water quality is a crucial factor. The buildup of heavy metals in vegetation and soil, including cadmium (Cd), chromium (Cr), nickel (Ni), lead (Pb), copper (Cu), and zinc (Zn), can significantly compromise agricultural output and food safety. Likewise, the existence of emerging contaminants in effluent, such as pharmaceuticals, organic solvents, and pesticides, may present hazards to the well-being of humans, the environment, and water (Shahid et al., 2017).

Antibiotics are an additional category of pollutants that may be detected in animal and municipal effluent. Antibiotics' environmental fate is determined by the mechanisms of sorption and transformation. There is increased likelihood that specific antibiotics, including fluoroquinolones and sulfonamides, will be detected in agricultural runoff and wastewater effluent, which raises concerns regarding their potential environmental and water quality impacts.

Therefore, while utilizing recycled farm water, external factors greatly affect the development and ultimate placement of contaminants that should be taken into consideration. Antibiotics, prescription drugs, heavy metals, and novel pollutants may pose a threat to human health, water quality, and soil quality in wastewater. Reducing the potential harm that contaminants may do to the environment and public health requires maintaining the safe use of wastewater in agriculture via proper management, treatment, and legal compliance (Manyi-Loh et al., 2018).

18.4 PLANT UPTAKE OF EMERGING CONTAMINANTS

The absorption and accumulation of CECs in agricultural plants, as well as their subsequent entrance into the food chain, is an increasingly important issue in environmental study. CECs are introduced into the agricultural environment by reusing treated wastewater for irrigation and using biosolids and manures as soil additions. Numerous studies have demonstrated that plants exposed to CECs via irrigation with treated wastewater or CEC-spiked irrigation water, or cultivated in soil altered with manure and biosolids, may absorb and deposit these pollutants in their tissues. However, there is currently a shortage of reliable data on CEC absorption by agricultural plants under real-world farming circumstances (Christou et al., 2019).

Root absorption is thought to be the primary route for CEC buildup in plants. Studies have revealed that the concentration of CECs in root tissue of crops such as strawberries and lettuce is linked to the aqueous concentration of these pollutants. Furthermore, a recent study related the distribution of CECs in plant tissues to their accumulation in root tissue. Understanding root absorption of CECs is critical because it influences the buildup of these pollutants in plants (Hyland et al., 2015).

Christou et al. graded crop plants on their ability to absorb and accumulate CECs in edible tissue. The capacity of crop plants to absorb and store CECs declines in the following order: leafy vegetables > root vegetables > cereals and fodder crops > fruit vegetables (Christou et al., 2019).

The presence of CECs in the food chain poses a significant public health risk. However, the degree of CECs' potential consequences on public health as a result of contaminated food intake remains unknown. More study is needed to determine the possible consequences of consuming CEC-contaminated food.

18.4.1 Uptake of Heavy Metals

Lead accumulation in Indian mustard (*Brassica juncea*) is an important topic in the study of plant uptake of specific contaminants. Indian mustard has been widely studied for its ability to accumulate high concentrations of Pb in its tissues when grown in contaminated soil (Lim et al., 2012). Cobb et al. investigated the accumulation of heavy metals, including Pb, by vegetables grown in mine wastes. The study found that Indian mustard effectively accumulated and translocated Pb, along with other metals such as Cd, As, and Zn. The accumulation pattern varied among different plant species, with lettuce leaves and radish roots showing higher concentrations of metals compared to bean and tomato fruits. This suggests that consumption of lettuce leaves or radish roots from plants grown in mine wastes may pose greater risks to humans and wildlife than consumption of beans or tomatoes from the same area (Cobb et al., 2000).

In addition to mine wastes, other sources of Pb contamination in agricultural soils include atmospheric deposition, livestock manures, fertilizers, agrochemicals, sewage irrigation, and sewage sludge. Luo et al. conducted an inventory of trace element inputs to agricultural soils in China and found that atmospheric deposition accounted for a significant portion of Pb inputs, along with other trace elements (Luo et al., 2009).

18.4.2 Uptake of Flame Retardants, Triclocarban, Pesticides, and Herbicides

Polybrominated diphenyl ethers and organophosphate flame retardants (OPFRs) are widely used. These are used in textiles, polyurethane foams, electrical appliances, upholstered furniture, building insulation, automobiles, airplanes, and so on. These fire retardants have a deleterious effect on fertility and also function as neurotoxins and endocrine disruptors.

Flame retardants enter the environment through many ways such as sludge application, irrigation with recovered water, and recycling of electronic debris. Once these are in the surroundings, plants have the ability to absorb them through their roots and branches. Certain crop characteristics and different soil conditions have a significant impact on the absorption and intrinsic transport behavior of these chemicals in plants (Blum et al., 2019).

It has been established by studies that plants may gather OPFRs and BFRs from the air, water, and soil. These substances have the ability to move about and change inside plants, as well as create new harmful metabolites that may be discharged into the environment or come into contact with people through the food chain (Zhang et al., 2021).

Wu et al. investigated the uptake of PPCPs by soybean plants from soils applied with biosolids and irrigated with contaminated water. The study focused on the uptake of three pharmaceuticals (carbamazepine, diphenhydramine, and fluoxetine) and two personal care products (triclosan and triclocarban) by soybean plants. The results showed that these PPCPs can be taken up by soybean plants and accumulate in their tissues. This highlights the potential for these contaminants to enter the food chain through the consumption of crops grown in contaminated soils or irrigated with contaminated water (Wu et al., 2010).

Pesticides and herbicides are commonly used in agriculture to control pests and weeds. These chemicals can enter the environment through various pathways, including soil application and spray drift. The uptake of pesticides and herbicides by plants depends on factors such as their physicochemical properties, application methods, and plant species. Some plants have been found to have the ability to take up and accumulate pesticides and herbicides in their tissues, while others may have lower uptake rates or exhibit detoxification mechanisms to minimize their accumulation (Tudi et al., 2021).

18.5 EMERGING CONTAMINANTS IN WASTEWATER TREATMENT PLANTS

Wastewater treatment plants (WWTPs) play a significant role in the presence of emerging contaminants in the environment. PPCPs are considered emerging contaminants and are found ubiquitously in wastewater and surface waters worldwide). These compounds enter WWTPs through incomplete metabolism in humans and subsequent excretion in human waste, leading to their discharge into surface waters through WWTP effluent. One specific pharmaceutical that has been found in high abundance in WWTP effluent and surface water studies is metformin, which is widely prescribed as an antidiabetic drug (Morin-Crini et al., 2022; Niemuth et al., 2014).

The presence of emerging contaminants in WWTPs is a growing concern due to their potential impacts on aquatic organisms and ecosystems. Studies have shown that exposure to environmentally relevant concentrations of metformin can cause potential endocrine disruption in adult male fish. Metformin exposure has been found to upregulate the expression of vitellogenin, an egg-protein messenger ribonucleic acid, in male fish, indicating endocrine disruption. Additionally, interactions between insulin signaling and steroidogenesis suggest potential endocrine-disrupting effects of metformin in the aquatic environment (Unuofin et al., 2019).

Efforts to remove emerging contaminants from wastewater have been explored. Heterogeneous photocatalytic treatment using supported photocatalysts has been proposed as an advanced oxidation treatment for the removal of emerging contaminants from wastewater effluents. This process utilizes photocatalysis to degrade and remove contaminants, such as pharmaceuticals, from wastewater (Morin-Crini et al., 2022; Petrie, 2021).

Bioaccumulation of human pharmaceuticals, including emerging contaminants, in fish has been observed in various habitats, including urban bayous influenced by tidal activity. The bioaccumulation of these contaminants in fish highlights the importance of understanding the risks associated with their presence in aquatic systems (Ortuzar et al., 2022).

One potential solution to address the presence of emerging contaminants in wastewater is the use of oxidative enzymes, such as laccase, for treatment. Laccase, a polyphenol oxidase, has shown promise in oxidizing aromatic contaminants in wastewater. The use of enzymatic treatment, including laccase, has been explored as a means to enhance wastewater treatment and promote environmental sustainability (Unuofin et al., 2019).

18.6 REGULATORY CONSIDERATIONS

Regulatory considerations for emerging contaminants in food are crucial to ensure food safety and protect consumer health. Here are some key aspects:

Global harmonization of regulations: Harmonized global regulatory systems can be very important for making sure that rules and standards are followed to protect customer health, support fair trade, and maintain people's trust in the global food control system (van der Meulen et al., 2022).

Risk mitigation strategies: Food regulators need to adopt key strategies to mitigate various risks associated with foods. This includes addressing emerging challenges in the area of food safety.

Balancing innovation and safety: Modern food regulations need to find and implement a balance between encouraging innovation in the food industry and ensuring food safety.

Consumer education: The US Food and Drug Administration and Food Safety and Standards Authority of India, for example, always consider a variety of rules and steps to decrease exposure, including educating consumers about how they may reduce the dangers caused by certain pollutants.

Industry responsibility: Regulatory bodies may require or encourage the industry to take steps to reduce the presence of these contaminants in products.

It is important to note that these regulations can be different depending on the type of new pollutant, the food product, and the rules in the area or country (Sorbo et al., 2022).

18.7 FUTURE DIRECTIONS

These plans for the future are meant to make sure that food production and usage will be better and last longer.

Technological advancements: New technologies have created great opportunities to make the food processing business safer and more environmentally friendly. For example, work is being done to create new mass spectrometry methods and electrochemical aptamer–based biosensors that can quickly and cheaply find new contaminant residues (Amit et al., 2017).

Sustainability and food safety: As the world's population grows, more people are turning to both animal and plant products to meet their energy needs. It's crucial to implement stringent food safety measures to safeguard consumer health. Food safety and sustainability go hand in hand. To keep up with demand and ease the strain on the earth, we need to cut down on food waste.

Complex food supply chain: Strong tracking systems will be needed as the food supply chain gets bigger and more complicated because of globalization and population growth (Konfo et al., 2023).

Consumer concerns over food safety: Strict food safety measures and labeling will be necessary to reassure shoppers that food products are safe.

Emerging research topics: More study needs to be done on new contaminants (e.g., microplastics), fast ways to find contaminants (e.g., biosensors), and eco-friendly materials for food packing (e.g., biodegradable polymers; Bradu et al., 2023).

Regulatory evolution: Regulations will need to evolve to keep pace with these changes, ensuring that they adequately address the risks posed by emerging contaminants while also promoting innovation in the food industry (FAO, 2022).

18.8 CONCLUSIONS

The increasing concern about novel contaminants in our food supply stems from their potential to harm human health. These pollutants are absorbed by plants, leading to their buildup in crops we rely on for nourishment. Emerging contaminants include industrial chemicals, pesticides, flame retardants, pharmaceuticals, and personal hygiene items. The presence of BPA in various food products has raised concerns about its impact on human health. Soil properties, plant characteristics, and environmental factors such as temperature, humidity, and rainfall also influence the uptake and accumulation of pollutants in plants.

Agricultural hazards, such as sewage sludge, treated municipal wastewater, and manure and biosolids, can lead to potential health concerns for humans. Factors such as pH, organic matter content, soil composition, and metabolic capacities affect the mobility and availability of pollutants in the soil, which in turn affect how readily plants can absorb them. Plant species have varying capacities for absorbing and accumulating CECs, and understanding soil conditions that influence plant absorption is critical for minimizing the danger of pollutants entering the food chain.

Environmental factors, such as diets high in meats, fats, oils, and refined sugars, can also contribute to the uptake and accumulation of pollutants in food crops. Mitigation and monitoring measures are necessary to address these issues.

PPCPs can enter aquatic ecosystems through treated and untreated effluent, exposing aquatic organisms to heavy metals and emerging contaminants. Wastewater treatment plants play a significant role in the presence of emerging contaminants in the environment. Future directions include technological advancements, sustainability, food safety, complex food supply chains, consumer concerns, emerging research topics, and regulatory evolution.

REFERENCES

Amit, S. K., Uddin, M. M., Rahman, R., Islam, S. M. R., & Khan, M. S. (2017). A review on mechanisms and commercial aspects of food preservation and processing. *Agriculture & Food Security*, *6*(1), 51. https://doi.org/10.1186/s40066-017-0130-8

Aransiola, S. A., Ijah, U. J. J., Abioye, O. P., & Bala, J. D. (2022). Vermicompost-assisted phytoremediation of toxic trace element-contaminated soil in Madaka, Nigeria using Melissa officinalis L and Sida acuta. *International Journal of Environmental Science and Technology*. http://dx.doi.org/10.1007/s13762-022-04105-y

Blum, A., Behl, M., Birnbaum, L., Diamond, M. L., Phillips, A., Singla, V., Sipes, N. S., Stapleton, H. M., & Venier, M. (2019). Organophosphate ester flame retardants: Are they a regrettable substitution for polybrominated diphenyl ethers? *Environmental Science & Technology Letters*, *6*(11), 638–649. https://doi.org/10.1021/acs.estlett.9b00582

Bradu, P., Biswas, A., Nair, C., Sreevalsakumar, S., Patil, M., Kannampuzha, S., Mukherjee, A. G., Wanjari, U. R., Renu, K., Vellingiri, B., & Gopalakrishnan, A. V. (2023). Recent advances in green technology and Industrial Revolution 4.0 for a sustainable future. *Environmental Science and Pollution Research*, *30*(60), 124488–124519. https://doi.org/10.1007/s11356-022-20024-4

Christou, A., Papadavid, G., Dalias, P., Fotopoulos, V., Michael, C., Bayona, J. M., Piña, B., & Fatta-Kassinos, D. (2019). Ranking of crop plants according to their potential to uptake and accumulate contaminants of emerging concern. *Environmental Research*. https://doi.org/10.1016/j.envres.2018.12.048

Cobb, G. P., Sands, K., Waters, M. S., Wixson, B. G., & Dorward-King, E. J. (2000). Accumulation of heavy metals by vegetables grown in mine wastes. *Environmental Toxicology and Chemistry*. https://doi.org/10.1002/etc.5620190311

FAO. (2022). *Thinking about the future of food safety*. FAO. https://doi.org/10.4060/cb8667en

Hyland, K., Blaine, A. C., & Higgins, C. P. (2015). Accumulation of contaminants of emerging concern in food crops-part 2: Plant distribution. *Environmental Toxicology and Chemistry*. https://doi.org/10.1002/etc.3068

Jones, D. R., & Finlay, R. D. (2009). Carbon flow in the rhizosphere: Carbon trading at the soil–root interface. *Plant and Soil*. https://doi.org/10.1007/s11104-009-9925-0

Konfo, T. R. C., Djouhou, F. M. C., Hounhouigan, M. H., Dahouenon-Ahoussi, E., Avlessi, F., & Sohounhloue, C. K. D. (2023). Recent advances in the use of digital technologies in agri-food processing: A short review. *Applied Food Research*, *3*(2), 100329. https://doi.org/10.1016/j.afres.2023.100329

Kyei, N. N. A., Waid, J. L., Ali, N., & Gabrysch, S. (2021). Awareness, experience, and knowledge of farming households in rural Bangladesh regarding mold contamination of food crops: A cross-sectional study. *International Journal of Environmental Research and Public Health*. https://doi.org/10.3390/ijerph181910335

Li, H., Sheng, G., Chiou, C. T., & Xu, O. (2005). Relation of organic contaminant equilibrium sorption and kinetic uptake in plants. *Environmental Science & Technology*. https://doi.org/10.1021/es050424z

Lim, J.-M., Jin, B., & Butcher, D. J. (2012). A comparison of electrical stimulation for electrodic and EDTA-enhanced phytoremediation of lead using Indian Mustard (Brassica juncea). *Bulletin of the Korean Chemical Society*. https://doi.org/10.5012/bkcs.2012.33.8.2737

Luo, L., Ma, Y., Zhang, S., Wei, D., & Zhu, Y.-G. (2009). An inventory of trace element inputs to agricultural soils in China. *Journal of Environmental Management*. https://doi.org/10.1016/j.jenvman.2009.01.011

Macherius, A., Eggen, T., Lorenz, W., Moeder, M., Ondruschka, J., & Reemtsma, T. (2012). Metabolization of the bacteriostatic agent triclosan in edible plants and its consequences for plant uptake assessment. *Environmental Science & Technology*. https://doi.org/10.1021/es3028378

Manyi-Loh, C., Mamphweli, S., Meyer, E., & Okoh, A. (2018). Antibiotic use in agriculture and its consequential resistance in environmental sources: Potential public health implications. *Molecules: A Journal of Synthetic Chemistry and Natural Product Chemistry, 23*(4), 795. https://doi.org/10.3390/molecules23040795

Marmugi, A., Ducheix, S., Lasserre, F., Polizzi, A., Paris, A., Priymenko, N., Bertrand-Michel, J., Pineau, T., Guillou, H., Martin, P. G. P., & Mselli-Lakhal, L. (2011). Low doses of bisphenol a induce gene expression related to lipid synthesis and trigger triglyceride accumulation in adult mouse liver. *Hepatology*. https://doi.org/10.1002/hep.24685

McGrath, S. P., & Zhao, F.-J. (2013). Concentrations of metals and metalloids in soils that have the potential to lead to exceedance of maximum limit concentrations of contaminants in food and feed. *Soil Use and Management*. https://doi.org/10.1111/sum.12080

Morin-Crini, N., Sharma, V. K., Zheng, M., Balaram, V., Ribeiro, A. C. F., Lu, Z., Stock, F., Carmona, E., Teixeira, M. R., Picos-Corrales, L. A., Moreno-Piraján, J. C., Giraldo, L., Li, C., Pandey, A., Hocquet, D., Torri, G., & Crini, G. (2022). Worldwide cases of water pollution by emerging contaminants: A review. *Environmental Chemistry Letters*. https://doi.org/10.1007/s10311-022-01447-4

Niemuth, N. J., Jordan, R., Crago, J., Blanksma, C. A., Johnson, R. D., & Klaper, R. D. (2014). Metformin exposure at environmentally relevant concentrations causes potential endocrine disruption in adult male fish. *Environmental Toxicology and Chemistry*. https://doi.org/10.1002/etc.2793

Onakpa, M. M., Njan, A. A., & Kalu, O. C. (2018). A review of heavy metal contamination of food crops in Nigeria. *Annals of Global Health*. https://doi.org/10.29024/aogh.2314

Ortuzar, M., Esterhuizen, M., Olicón-Hernández, D. R., González-López, J., & Aranda, E. (2022). Pharmaceutical pollution in aquatic environments: A concise review of environmental impacts and bioremediation systems. *Frontiers in Microbiology, 13*. https://www.frontiersin.org/articles/10.3389/fmicb.2022.869332

Petrie, B. (2021). A review of combined sewer overflows as a source of wastewater-derived emerging contaminants in the environment and their management. *Environmental Science and Pollution Research*. https://doi.org/10.1007/s11356-021-14103-1

Rehman, Z. U., Khan, S., Shah, M. M., Brusseau, M. L., Khan, S. G., & Mainhagu, J. (2018). Transfer of heavy metals from soils to vegetables and associated human health risks at selected sites in Pakistan. *Pedosphere*. https://doi.org/10.1016/s1002-0160(17)60440-5

Shahid, M., Dumat, C., Khalid, S., Schreck, E., Xiong, T., & Niazi, N. K. (2017). Foliar heavy metal uptake, toxicity and detoxification in plants: A comparison of foliar and root metal uptake. *Journal of Hazardous Materials*. https://doi.org/10.1016/j.jhazmat.2016.11.063

Smith, P. (2016). Soil carbon sequestration and biochar as negative emission technologies. *Global Change Biology*. https://doi.org/10.1111/gcb.13178

Sorbo, A., Pucci, E., Nobili, C., Taglieri, I., Passeri, D., & Zoani, C. (2022). Food safety assessment: Overview of metrological issues and regulatory aspects in the European Union. *Separations, 9*(2), Article 2. https://doi.org/10.3390/separations9020053

Tang, Z., & Zhao, F.-J. (2020). The roles of membrane transporters in arsenic uptake, translocation and detoxification in plants. *Critical Reviews in Environmental Science and Technology*. https://doi.org/10.1080/10643389.2020.1795053

Tilman, D., & Clark, M. R. (2014). Global diets link environmental sustainability and human health. *Nature*. https://doi.org/10.1038/nature13959

Tschan, M., Robinson, B., & Schulin, R. (2009). Antimony in the soil – plant system – a review. *Environmental Chemistry*. https://doi.org/10.1071/en08111

Tudi, M., Daniel Ruan, H., Wang, L., Lyu, J., Sadler, R., Connell, D., Chu, C., & Phung, D. T. (2021). Agriculture development, pesticide application and its impact on the environment. *International Journal of Environmental Research and Public Health, 18*(3), 1112. https://doi.org/10.3390/ijerph18031112

Unuofin, J. O., Okoh, A. I., & Nwodo, U. U. (2019). Aptitude of oxidative enzymes for treatment of wastewater pollutants: A laccase perspective. *Molecules*. https://doi.org/10.3390/molecules24112064

van der Meulen, B., Card, M. M., Din, A., Fortin, N. D., Mahmudova, A., Maister, B., Türkoğlu, H. G., Bilgin, F. K., Lederman, J., Poto, M. P., Sattigeri, V. D., Sohn, M., Sun, J., Urazbaeva, A., & Vasiliev, Y. (2022). Chapter 3 – Food regulation around the world. In A. Martinović, S. Oh, & H. Lelieveld (Eds.), *Ensuring global food safety* (2nd ed., pp. 11–137). Academic Press. https://doi.org/10.1016/B978-0-12-816011-4.00001-X

Vitousek, P. M., Mooney, H. A., Lubchenco, J., & Melillo, J. M. (1997). Human domination of earth's ecosystems. *Science*. https://doi.org/10.1126/science.277.5325.494

Wang, Y.-N., Chan, K. M., & Chan, W. (2017). Plant uptake and metabolism of nitrofuran antibiotics in spring onion grown in nitrofuran-contaminated soil. *Journal of Agricultural and Food Chemistry*. https://doi.org/10.1021/acs.jafc.7b01050

Wu, C., Spongberg, A. L., Witter, J. D., Fang, M., & Czajkowski, K. (2010). Uptake of pharmaceutical and personal care products by soybean plants from soils applied with biosolids and irrigated with contaminated water. *Environmental Science & Technology*. https://doi.org/10.1021/es1011115

Zhang, Q., Yao, Y., Wang, Y., Zhang, Q., Cheng, Z., Li, Y., Yang, X., Wang, L., & Sun, H. (2021). Plant accumulation and transformation of brominated and organophosphate flame retardants: A review. *Environmental Pollution*, *288*, 117742. https://doi.org/10.1016/j.envpol.2021.117742

19 Toxic Metals and Emerging Contaminants in Food

A Global Perspective

Deborah O. Aderibigbe, Peter O. Oladoye, Adedayo O. Adeboye, and Abdur-Rahim A. Giwa

19.1 INTRODUCTION

All foods contain chemical substances that are expected to perform specific functions in the body after ingestion. Naturally, these chemical substances are expected to be safe for consumption, but when taken above the recommended limits, some of these substances become hazardous or toxic. Also, some of these naturally occurring chemicals in food originally exist within the food material as a form of defense mechanism (Urugo & Tringo, 2023) for the crop, but they have begun to emerge as having negative implications on the health of consumers.

The presence of harmful chemicals in food poses a significant threat to human health, particularly for children who are more vulnerable to their effects. This is because children consume a larger proportion of food in relation to their body weight, have developing immune systems, and have limited dietary options. The level of risk from food contamination depends on the amount of food consumed and the concentration of chemical contaminants in the food. These contaminants have been linked to various health issues such as cancer and developmental and neurological problems. To reduce the health risks associated with food contamination, it is crucial to take appropriate measures to minimize exposure to these chemicals by monitoring and controlling the sources of contamination. Some of the most commonly reported chemical contaminants in food include heavy metals, pesticides, and mycotoxins (Onyeaka et al., 2024). Therefore there is a need to identify these potentially harmful chemical constituents and their food sources so that efforts can be made to address the safety concerns or hazards that they present.

19.2 NATURALLY OCCURRING TOXICANTS IN FOODS

Due to the nature of food especially the cultivation, there are toxicants that naturally occur with them. Table 19.1 shows some of the toxicants and the foods that may likely contain them.

19.3 HEAVY METALS IN RAW AND PROCESSED FOODS

Soil, which is the source of life for food crops, can be a primary carrier for heavy metals from various sources such as industrial waste, sewage effluents, vehicle exhaust, agricultural runoff, and storage systems (Gadd, 2010; Gall et al., 2015; Aransiola et al., 2013) These contaminated soils, through soil-food crop interactions, lead to a transfer of the heavy metals into the food crops. According to Rai et al. (2019), these heavy metal contaminations vary across locations in terms of concentrations and the nature of food crops being contaminated. In Asia, sewage effluent and industrial waste are reportedly the major sources of heavy metal contamination in foods, and industrial or modern intensive agriculture has been implicated as a source of contamination in South America

DOI: 10.1201/9781032683751-22

TABLE 19.1
Naturally Occurring Toxicants and Their Sources

Toxicant	Food Source(s)	Reference(s)
Estrogens	Carrots, soya beans, potatoes, wheat, rice, barley, apples, plums, cherries, corn, olives, peanut and cottonseed oils	(Bejarano et al., 2000; Urugo & Tringo, 2023)
Goitrogens	Broccoli, kale, cabbage, cauliflower, spinach, legumes, nuts, milk	(Urugo & Tringo, 2023;Bejarano et al., 2000)
Antivitamins	Raw fish, eggs	(Beier & Nigg, 1992)
Enzyme inhibitors	Legumes, cereals, potatoes, eggs, broccoli, apples, oranges	(Beier & Nigg, 1992)
Hemagglutinins or lectins	Legumes, potatoes, cereals	(Alves et al., 2017)
Cyanogenic glycosides	Cassava, flaxseed, sorghum, cocoyam	(Bejarano et al., 2000;Urugo & Tringo, 2023)
Glycoalkaloids	Potatoes, tomatoes, eggplant	(Agarwal, 1988)
Glucosinolates	Broccoli, cauliflower, cabbage	(Schilter et al., 2014)
Lectins	Cereals, beans, wheat	(Madariaga-Mazón et al., 2019)
Trace mineral elements such as mercury, cadmium, lead, and arsenic	Seafood	(Barbosa et al., 2018)
Nitrate, nitrite, and nitrosamines	Water, leafy vegetables	(Schrenk et al., 2020)
Oxalate	Spinach, cocoa, tea, almonds, cashew nuts	(Giera et al., 2022)
Alkaloids	Potatoes	(Urugo & Tringo, 2023)

(Corguinha et al., 2015). In Australia, according to Rahman et al. (2014), arsenic (As) and metal-contaminated groundwater are the principal causes of contamination.

Cooking and processing are expected to affect the extent of heavy metal contamination in processed foods (Oyekunle et al., 2020). According to J. G. Lee et al., 2019, there was significant reduction in heavy metal concentration when oil seeds were subjected to extraction using pressing and solvent extraction methods. The study also reported that boiling did not significantly reduce heavy metal concentration for most of the metals except for aluminum (Al), which significantly decreased as boiling time increased. Studies on heavy metal contents of rice in Iran (Naseri et al., 2014) and fish (Atta et al., 1997) also confirmed this trend. Grilling reduced zinc (Zn), cadmium (Cd), copper (Cu), and manganese (Mn) content in meat but increased iron (Fe) content (Perelló et al., 2008). In an assessment of heavy metals in crops grown in Ghana, Adjei-Mensah et al. (2021) reported that irrespective of location, boiling, frying, and roasting methods significantly reduced Mn, lead (Pb), Cd, As, and mercury (Hg) found in cassava, cocoyam, and yam. The Cd levels were below detectable limits (Table 19.2).

19.4 GLOBAL TRENDS IN FOOD CONTAMINATION (AFRICA, AMERICA, ASIA, EUROPE)

One prominent avenue through which food is contaminated is the "feed-to-food" pathway, whereby humans as ultimate consumers are exposed to emerging and legacy contaminants. Among other possible exposure routes, the feed-to-food pathway alone accounts for 90%, thereby making it a critical pathway of concern (Ding et al., 2020). Emerging contaminants (ECs) are chemicals that produce adverse or toxic effects in human body, even when ingested at low concentrations (Li et al., 2023). As a result of the presence of ECs in different environmental compartments, it is necessary to say that pollution of animal- and plant-derived items is highly feasible. The bio-accumulative and persistence nature of ECs generally also warrant their tendency to enrich the food chain via the feed

TABLE 19.2
Heavy Metals in Different Types of Food across the Globe

Heavy Metal	Food Type	Country	Concentration (mg/kg)	References
Lead	Fruits	South Africa	1.52–3.45	(Nkhebenyane & Tywabi-Ngeva, n.d.)
	Vegetable		1.57–4.52	
	Chicken, pork, beef, goat (raw)		0.006–0.054	
	Chicken, pork, beef, goat (processed)		0.146–2.290	
	Vegetables (Amaranthus, tomatoes, cabbage, eggplant)	Uganda	3.9–12.9	(Kasozi et al., 2021)
	Apple	Ukraine	1.15–3.89	(Medvedeva et al., 2021)
	Full fat milk	North Cyprus	0.00–8.00	(Dağcilar & Gezer, 2021)
	Full fat yogurt		0.00–7.00	
	Halloumi cheese		10.00–81.00	
	Kashar cheese		N.D.	
	White cheese		N.D.	
	Lor cheese		N.D.	
	Cassava (raw, boiled, roasted, fried)	Ghana	0.07–2.11	(Adjei-Mensah et al., 2021)
	Cocoyam (raw, boiled, roasted, fried)		0.20–0.95	
	Plantain (raw, boiled, roasted, fried)		0.04–1.11	
	Yam (raw, boiled, roasted, fried)		0.01–1.00	
	Raw milk	Egypt	0.1016	
	Kariesh cheese		0.2924	(Elafify et al., 2023)
	Processed cheese		0.1256	
	Milk powder		0.3352	
	Vegetables (Amaranthus, tomatoes, cabbage, eggplant)	Uganda	0.6–2.1	(Kasozi et al., 2021)
	Apple	Ukraine	0.09–0.28	(Medvedeva et al., 2021)
	Full fat milk	North Cyprus	0.00–15.00	(Dağcilar & Gezer, 2021)
	Full fat yogurt		23.00–55.00	
	Halloumi cheese		31.00–58.00	
	Kashar cheese		0.00–1.00	
	White cheese		0.00–2.00	
	Lor cheese		N.D.	
	Raw milk	Egypt	0.07	
	Kariesh cheese		0.1172	(Elafify et al., 2023)
	Processed cheese		0.0576	
	Milk powder		0.1252	
Arsenic	Eggs	Iran	0.52–1.07	(Abedi et al., 2023)
	Full fat milk	North Cyprus	0.00–5.00	(Dağcilar & Gezer, 2021)
	Full fat yogurt		0.00–12.00	
	Halloumi cheese		3.00–25.00	
	Kashar cheese		0.00–1.00	
	White cheese		5.00–7.00	
	Lor cheese		0.00–1.00	
	Cassava (raw, boiled, roasted, fried)	Ghana	0.01–0.02	(Adjei-Mensah et al., 2021)
	Cocoyam (raw, boiled, roasted, fried)		0.02–0.07	
	Plantain (raw, boiled, roasted, fried)		0.01–0.09	
	Yam (raw, boiled, roasted, fried)		0.01–0.02	
	Raw milk	Egypt	0.012	
	Kariesh cheese		0.0764	(Elafify et al., 2023)
	Processed cheese		0.052	
	Milk powder		0.12	

(Continued)

TABLE 19.2 ***(Continued)***
Heavy Metals in Different Types of Food across the Globe

Heavy Metal	Food Type	Country	Concentration (mg/kg)	References
Mercury	Eggs	Iran	0.1–0.26	(Abedi et al., 2023)
	Full fat milk	North Cyprus	0.00–6.00	(Dağcilar & Gezer, 2021)
	Full fat yoghurt		2.00–10.00	
	Halloumi cheese		6.00–24.00	
	Kashar cheese		0.00–5.00	
	White cheese		0.00–3.00	
	Lor cheese		N.D.	
	Cassava (raw, boiled, roasted, fried)	Ghana	0.02–0.16	(Adjei-Mensah et al., 2021)
	Cocoyam (raw, boiled, roasted, fried)		0.14–0.19	
	Plantain (raw, boiled, roasted, fried)		0.04–0.13	
	Yam (raw, boiled, roasted, fried)		0.02–1.34	
	Raw milk	Egypt	0.0014	(Elafify et al., 2023)
	Kariesh cheese		0.0378	
	Processed cheese		0.0012	
	Milk powder		0.0204	
Nickel	Vegetables (Amaranthus, tomatoes, cabbage, eggplant)	Uganda	0.9–4.6	(Kasozi et al., 2021)
Cobalt	Vegetables (Amaranthus, tomatoes, cabbage, eggplant)	Uganda	0.6–3.6	(Kasozi et al., 2021)
Copper	Vegetables (Amaranthus, tomatoes, cabbage, eggplant)	Uganda	0.9–8.0	(Kasozi et al., 2021)
	Apple	Ukraine	0.77–1.26	(Medvedeva et al., 2021)
	Full fat milk	North Cyprus	N.D.	(Dağcilar & Gezer, 2021)
	Full fat yogurt		0.00–33.00	
	Halloumi cheese		466.00–738.00	
	Kashar cheese		235.00–289.00	
	White cheese		182.00–443.00	
	Lor cheese		212.00–300.00	
	Raw milk	Egypt	0.0656	
	Kariesh cheese		0.188	(Elafify et al., 2023)
	Processed cheese		0.1272	
	Milk powder		0.2496	
Chromium	Vegetables (Amaranthus, tomatoes, cabbage, eggplant)	Uganda	0.5–6.9 0.02–0.31	(Kasozi et al., 2021)
	Apple	Ukraine	0.1044	(Medvedeva et al., 2021)
	Raw milk	Egypt	0.174	(Elafify et al., 2023)
	Kariesh cheese		0.1016	
	Processed cheese		0.208	
	Milk powder			
Zinc	Vegetables (Amaranthus, tomatoes, cabbage, eggplant)	Uganda	17.9–61.9	(Kasozi et al., 2021)
	Apple	Ukraine	0.96–4.19	(Medvedeva et al., 2021)
	Fufu (cassava and plantain meal)	Ghana	0.09–7.53	(Nkansah et al., 2021)

to food. Globally, the annual demand for protein is estimated at 202 million with animal-derived products occupying 43% of the total demand (Food and Agriculture Organization of the United Nations, 2012). Examples of ECs that have been investigated and reported to be present in feed and food (products) are per- and polyfluoroalkyl substances (PFASs), short-chain chlorinated paraffins (SCCPs), novel brominated flame retardants (NBFRs), organophosphate esters (OPEs), and polybrominated diphenyl ethers (PBDEs).

OPEs are applied as flame retardants and as plasticizers in the manufacturing of electronics, textiles, plastics, antifoaming agents, and all sort of polymers. Although OPEs were developed several years ago, only recently have they been manufactured in large amount due to the implementation of the regulation of brominated fire retardants (BFRs; Fu et al., 2021). In the environment, the persistence nature of chlorinated OPEs and the bioaccumulation tendency of aryl and alkyl OPEs in aquatic food chain have been reported (Fu et al., 2021). As revealed in Table 19.3, food and feed contamination with OPEs is possible. According to Zhao et al. the trend of OPEs in protein feed, such as blood meal, plant-derived meal, feather meal, and meat meal, are 28.0 ± 12.0, 41.9 ± 16.0, 54.6 ± 30.0, and 117 ± 75.6 ng/g dw, respectively (Zhao et al., 2020). In terms of biomagnification, Fu et al. reported trophic magnification factors of 2.92, 5.20, and 2.74 for tris(2-chloroisopropyl) phosphate, tris(2-chloroethyl) phosphate, and triphenyl phosphate, respectively. These serve as indicator to the presence of OPEs in feed, food, and biomagnification potential across the food chain (Fu et al., 2020).

PFASs are unconventionally referred to as "forever chemicals" with a general formula of $F(CF_2)_xR$. The nature *R* substituent determines the classification that any PFASs belong to, such as fluorotelomer alcohols, perfluoroalkyl sulfonates, perfluorophosphonates, perfluoroalkyl carboxylates, polyfluoroalkyl phosphate esters, and perfluoroalkane sulfonyl fluoride, respectively. Due to environmental contamination, food and feed may be contaminated with PFASs and thus they are ubiquitous, and have been quantified in a variety of animal protein feed, such as fish meal, meat meal, and blood meal, with average concentrations of 6.74, 18.2, and 22.9 ng/g dw, respectively (Li, Dong, et al., 2019; Li, Gao, et al., 2019). As indicated in Table 19.3, seafood and fish, representing aquatic food products, possessed the highest concentration of PFASs when compared to other food types. This is an eye-opener to consumers of seafood as there is need to be conscious of intake of PFASs via animal-derived and fish-derived products.

The occurrence and levels of ECs in food are presented in Table 19.3. Relevant information regarding the locations of study is also provided in order to have a global overview of ECs presence and levels in food, such as PFASs, SCCPs, NBFRs, OPEs, and PBDEs.

19.5 REGULATORY STANDARDS FOR FOOD CONTAMINATION

To ensure the health of consumers of food items are protected, government authorities and agencies have put in place public health regulatory standards for a variety of food contaminants and additives in the form of maximum residue/risk limit (MRL) or maximum permissible level (MPL; Shaw & Vannoort, 2001).

The main consideration in setting these limits (MRL and MPL) depends on the health impact (chronic or acute toxicity) for each chemical substance upon human consumption. Furthermore, several other factors are considered at the national level in setting these norms, limits, and/or standards for food contaminant. For instance, the tendency to comply by food consumers, societal nutritional requirements, and global population diet are among several other factors in reaching any decision on standard setting. In other words, whatever is feasible and appropriate for the society are trade-offs for health effects related to any food contaminant during regulatory standard setting.

Globally, the Food and Agricultural Organization of the United Nations (FAO) and the World Health Organization (WHO) have been given the responsibility to set regulatory standards for various toxic chemical substances in food, including assessing risk linked to pesticides, drugs, herbicides, toxic heavy metals, persistent organic pollutants, food additives, and so on. In achieving

TABLE 19.3
Levels of Emerging Contaminants in Food

Matrix	Location	Food Type	Year of Sampling	Concentration	Unit	Number of Samples Taken	References
PFASs	Netherland	Fish (herring)	2012–2018	0.24(<MDL–1.10)	ng/g ww	N/A	(Zafeiraki et al., 2019)
	China	Meat	2011–2013	0.2	ng/g ww	Pooled sample	(Wang et al., 2019)
	Netherland	Beef	2009	94	pg/g product	Pooled sample	(Zafeiraki et al., 2019)
	China	All types	2012	97.3 (<MDL–1523)	ng/g ww	121	(Zhang et al., 2017)
	Netherland	Milk	2009	55	pg/g product	Pooled sample	(Zafeiraki et al., 2019)
	Sweden	All types	2010	32	pg/g	Pooled sample	(Wang et al., 2019)
	Netherland	Fish	2009	180	pg/g product	Pooled sample	(Noorlander et al., 2011)
	Sweden	All types	1999	20	pg/g	Pooled sample	(Gebbink et al., 2015)
	Norway	Fish	2012–2018	0.11 (<MDL–0.50)	ng/g ww	N/A	(Zafeiraki et al., 2019)
	South America	Fishmeal	2012–2014	12.4 (1.06–42.1)	ng/g dw	45	(Li, Dong, et al., 2019)
	United States	Fishmeal	2012–2014	28.3 (0.65–85.5)	ng/g dw	9	(Li, Dong, et al., 2019)
OPEs	Australia	Meat	2018	1	ng/g ww	12	(He et al., 2018)
	United States	Fish	N/A	8.7	ng/g ww	22	(Han et al., 2019)
	Sweden	Fruits and vegetables	2015	0.23–1.92	ng/g ww	10	(Ding et al., 2018)
	China	Chicken	N/A	4.1 ± 1.5	ng/g ww	4	(Ding et al., 2018)
	Australia	Seafood and fish	2018	1.8	ng/g ww	9	(He et al., 2018)
	Sweden	Meat	2015	<LOQ–2.76	ng/g ww	5	(Poma et al., 2017)
SCCPs	China	Cereals	2017–2018	17	ng/g ww	Pooled sample	(Cui et al., 2020)
	Korea	Egg	2016–2018	286	ng/g ww	14	(Lee et al., 2020)
	China	Meat (products)	2011	129 ± 4.1 (40.6–469)	ng/g ww	20	(Huang et al., 2018)
	Korea	Meat	2016–2019	89.8	ng/g ww	35	(Lee et al., 2020)
	Korea	Grain	2016–2019	45.3	ng/g ww	7	(Lee et al., 2020)
	China	Milk	2017–2018	38	ng/g ww	Pooled sample	(Cui et al., 2020)
PBDEs	United Kingdom	Meat	2015	13.7–40.7	ng/g ww	12	(Tao et al., 2017)
	Netherlands	Fish (herring)	2009–2014	1323 ± 959	ng/g ww	8	(Gebbink et al., 2019)
	United States	Meat	2006	383 ± 190 (39–1426)	pg/g ww	18	(Schecter et al., 2006)
	United States	Fish	2015–2016	438–3680	pg/g ww	7	(Cade et al., 2018)
	Netherlands	Meat (beef)	2009–2014	113 ± 143 (<10–900)	pg/g ww	83	(Gebbink et al., 2019)
	Poland	Milk	2015–2018	0.23 ± 0.09 (0.10–0.50)	pg/g ww fat	30	(Pietron et al., 2021)

FIGURE 19.1 Internationally recognized food regulatory bodies.

these, the WHO and FAO have put in place the Joint Expert Committee on Food Additives and Contaminants and the Joint Meeting on Pesticide Residues to conduct the above chemical contaminant risk assessment. Sequel to this assessment, the WHO/FAO Codex Alimentarius Commission has established international regulatory standards for food (Figure 19.1; Cheng, 2018).

19.6 CONCLUSION

Safeguarding health, food security, and responsible consumption begins with understanding food quality and potential contaminants. This chapter delves into the critical issue of heavy metals and emerging contaminants found in various food sources worldwide. By investigating data from international regulatory bodies like the WHO and the FAO, we gain valuable insights for setting safe limits and conducting risk assessments. This is particularly crucial for determining the cancer-causing potential of specific contaminants in different food groups. Ultimately, ensuring robust food quality standards is key to achieving Zero Hunger (SDG 2), promoting Good Health and Well-Being (SDG 3), and fostering Responsible Consumption and Production (SDG 12). By prioritizing food safety, we protect both human health and the planet's resources, paving the way for a more sustainable future.

REFERENCES

Abedi, A. S., Hoseini, H., Mohammadi-Nasrabadi, F., Rostami, N., & Esfarjani, F. (2023). Consumer health risk assessment of Arsenic and Mercury in hen eggs through Monte Carlo simulations. *BMC Public Health*, *23*(1). https://doi.org/10.1186/s12889-023-16223-4

Adjei-Mensah, R., Ofori, H., Tortoe, C., Torgbor Johnson, P. N., Aryee, D., & Kofi Frimpong, S. (2021). Effect of home processing methods on the levels of heavy metal contaminants in four food crops grown in and around two mining towns in Ghana. *Toxicology Reports*, *8*, 1830–1838. https://doi.org/10.1016/j.toxrep.2021.11.001

Agarwal, S. K. (1988). The importance of naturally occurring toxicants in foods. *Biochemical Education*, *16*(4), 212–213. https://doi.org/10.1016/0307-4412(88)90124-0

Alves, R. N., Maulvault, A. L., Barbosa, V. L., Cunha, S., Kwadijk, C. J. A. F., Álvarez-Muñoz, D., Rodríguez-Mozaz, S., Aznar-Alemany, Ò., Eljarrat, E., Barceló, D., Fernandez-Tejedor, M., Tediosi, A., & Marques,

A. (2017). Preliminary assessment on the bioaccessibility of contaminants of emerging concern in raw and cooked seafood. *Food and Chemical Toxicology*, *104*, 69–78. https://doi.org/10.1016/j.fct.2017.01.029

Aransiola, S. A., Ijah, U. J. J., & Abioye, O. P. (2013). Phytoremediation of lead polluted soil by *Glycine max* L. *Applied and Environmental Soil Science*, *2013*. Article ID 631619. doi:10.1155/2013/631619. https://www.hindawi.com/journals/aess/2013/631619/abs/

Atta, M. B., El-Sebaie, L. A., Noaman, M. A., & Kassab, H. E. (1997). The effect of cooking on the content of heavy metals in fish (Tilapia nilotica). *Food Chemistry*, *58*(1–2), 1–4. https://doi.org/10.1016/0308-8146(95)00205-7

Barbosa, V., Maulvault, A. L., Alves, R. N., Kwadijk, C., Kotterman, M., Tediosi, A., Fernández-Tejedor, M., Sloth, J. J., Granby, K., Rasmussen, R. R., Robbens, J., De Witte, B., Trabalón, L., Fernandes, J. O., Cunha, S. C., & Marques, A. (2018). Effects of steaming on contaminants of emerging concern levels in seafood. *Food and Chemical Toxicology*, *118*, 490–504. https://doi.org/10.1016/j.fct.2018.05.047

Beier, R. C., & Nigg, H. N. (1992). Natural Toxicants in Foods. In *Phytochemical Resources for Medicine and Agriculture* (pp. 247–367). Springer US. https://doi.org/10.1007/978-1-4899-2584-8_12

Bejarano, L., Mignolet, E., Devaux, A., Espinola, N., Carrasco, E., & Larondelle, Y. (2000). Glycoalkaloids in potato tubers: The effect of variety and drought stress on the ?-solanine and ?-chaconine contents of potatoes. *Journal of the Science of Food and Agriculture*, *80*(14), 2096–2100. https://doi.org/10.1002/1097-0010(200011)80:14<2096::AID-JSFA757>3.0.CO;2-6

Cade, S. E., Kuo, L.-J., & Schultz, I. R. (2018). Polybrominated diphenyl ethers and their hydroxylated and methoxylated derivatives in seafood obtained from Puget Sound, WA. *Science of the Total Environment*, *630*, 1149–1154. https://doi.org/10.1016/j.scitotenv.2018.02.301

Cheng, C. (2018). Codex alimentarius commission. In *Encyclopedia of Food Security and Sustainability*. https://doi.org/10.1016/B978-0-08-100596-5.22376-7

Corguinha, A. P. B., Souza, G. A. de, Gonçalves, V. C., Carvalho, C. de A., Lima, W. E. A. de, Martins, F. A. D., Yamanaka, C. H., Francisco, E. A. B., & Guilherme, L. R. G. (2015). Assessing arsenic, cadmium, and lead contents in major crops in Brazil for food safety purposes. *Journal of Food Composition and Analysis*, *37*, 143–150. https://doi.org/10.1016/j.jfca.2014.08.004

Cui, L., Gao, L., Zheng, M., Li, J., Zhang, L., Wu, Y., Qiao, L., Xu, C., Wang, K., & Huang, D. (2020). Short- and medium-chain chlorinated paraffins in foods from the sixth Chinese total diet study: Occurrences and estimates of dietary intakes in South China. *Journal of Agricultural and Food Chemistry*, *68*(34), 9043–9051. https://doi.org/10.1021/acs.jafc.0c03491

Dağcilar, K., & Gezer, C. (2021). Heavy metal residues in milk and dairy products produced in northern Cyprus. *Progress in Nutrition*, *23*(1). https://doi.org/10.23751/pn.v23i1.9384

Ding, J., Deng, T., Xu, M., Wang, S., & Yang, F. (2018). Residuals of organophosphate esters in foodstuffs and implication for human exposure. *Environmental Pollution*, *233*, 986–991. https://doi.org/10.1016/j.envpol.2017.09.092

Ding, Y., Han, M., Wu, Z., Zhang, R., Li, A., Yu, K., Wang, Y., Huang, W., Zheng, X., & Mai, B. (2020). Bioaccumulation and trophic transfer of organophosphate esters in tropical marine food web, South China Sea. *Environment International*, *143*, 105919. https://doi.org/10.1016/j.envint.2020.105919

Elafify, M., EL-Toukhy, M., Sallam, K. I., Sadoma, N. M., Mohammed Abd-Elghany, S., Abdelkhalek, A., & El-Baz, A. H. (2023). Heavy metal residues in milk and some dairy products with insight into their health risk assessment and the role of Lactobacillus rhamnosus in reducing the lead and cadmium load in cheese. *Food Chemistry Advances*, *2*. https://doi.org/10.1016/j.focha.2023.100261

Food and Agriculture Organization of the United Nations. (2012). World agriculture towards 2030/2050. *Science*, *376* (6589), 133–134. https://doi.org/10.1126/science.abo7429

Fu, J., Fu, K., Chen, Y., Li, X., Ye, T., Gao, K., Pan, W., Zhang, A., & Fu, J. (2021). Long-range transport, trophic transfer, and ecological risks of organophosphate esters in remote areas. *Environmental Science & Technology*, *55*(15), 10192–10209. https://doi.org/10.1021/acs.est.0c08822

Fu, J., Fu, K., Gao, K., Li, H., Xue, Q., Chen, Y., Wang, L., Shi, J., Fu, J., Zhang, Q., Zhang, A., & Jiang, G. (2020). Occurrence and trophic magnification of organophosphate esters in an Antarctic ecosystem: Insights into the shift from legacy to emerging pollutants. *Journal of Hazardous Materials*, *396*, 122742. https://doi.org/10.1016/j.jhazmat.2020.122742

Gadd, G. M. (2010). Metals, minerals and microbes: Geomicrobiology and bioremediation. *Microbiology*, *156*(3), 609–643. https://doi.org/10.1099/mic.0.037143-0

Gall, J. E., Boyd, R. S., & Rajakaruna, N. (2015). Transfer of heavy metals through terrestrial food webs: A review. *Environmental Monitoring and Assessment*, *187*(4), 201. https://doi.org/10.1007/s10661-015-4436-3

Gebbink, W. A., Glynn, A., Darnerud, P. O., & Berger, U. (2015). Perfluoroalkyl acids and their precursors in Swedish food: The relative importance of direct and indirect dietary exposure. *Environmental Pollution*, *198*, 108–115. https://doi.org/10.1016/j.envpol.2014.12.022

Gebbink, W. A., van der Lee, M. K., Peters, R. J. B., Traag, W. A., Dam, G. ten, Hoogenboom, R. L. A. P., & van Leeuwen, S. P. J. (2019). Brominated flame retardants in animal derived foods in the Netherlands between 2009 and 2014. *Chemosphere*, *234*, 171–178. https://doi.org/10.1016/j.chemosphere.2019.06.046

Giera, D. S., Preisitsch, M., Brevard, H., & Nemetz, J. (2022). Quantitative removal of pyrrolizidine alkaloids from essential oils by the hydrodistillation step in their manufacturing process. *Planta Medica*, *88*(7), 538–547. https://doi.org/10.1055/a-1534-6928

Han, L., Sapozhnikova, Y., & Nuñez, A. (2019). Analysis and occurrence of organophosphate esters in meats and fish consumed in the United States. *Journal of Agricultural and Food Chemistry*, *67*(46), 12652–12662. https://doi.org/10.1021/acs.jafc.9b01548

He, C., Wang, X., Tang, S., Thai, P., Li, Z., Baduel, C., & Mueller, J. F. (2018). Concentrations of organophosphate esters and their specific metabolites in food in Southeast Queensland, Australia: Is dietary exposure an important pathway of organophosphate esters and their metabolites? *Environmental Science & Technology*, *52*(21), 12765–12773. https://doi.org/10.1021/acs.est.8b03043

Huang, H., Gao, L., Zheng, M., Li, J., Zhang, L., Wu, Y., Wang, R., Xia, D., Qiao, L., Cui, L., Su, G., Liu, W., & Liu, G. (2018). Dietary exposure to short- and medium-chain chlorinated paraffins in meat and meat products from 20 provinces of China. *Environmental Pollution*, *233*, 439–445. https://doi.org/10.1016/j.envpol.2017.10.022

Kasozi, K. I., Otim, E. O., Ninsiima, H. I., Zirintunda, G., Tamale, A., Ekou, J., Musoke, G. H., Muyinda, R., Matama, K., Mujinya, R., Matovu, H., Ssempijja, F., Eze, E. D., Atino, M., Udechukwu, B., Kayima, R., Etiang, P., Ayikobua, E. T., Kembabazi, S., . . . Otim, O. (2021). An analysis of heavy metals contamination and estimating the daily intakes of vegetables from Uganda. *Toxicology Research and Application*, *5*, 239784732098525. https://doi.org/10.1177/2397847320985255

Lee, J. G., Hwang, J. Y., Lee, H. E., Kim, T. H., Choi, J. D., & Gang, G. J. (2019). Effects of food processing methods on migration of heavy metals to food. *Applied Biological Chemistry*, *62*(1). https://doi.org/10.1186/s13765-019-0470-0

Lee, S., Choo, G., Ekpe, O. D., Kim, J., & Oh, J.-E. (2020). Short-chain chlorinated paraffins in various foods from Republic of Korea: Levels, congener patterns, and human dietary exposure. *Environmental Pollution*, *263*, 114520. https://doi.org/10.1016/j.envpol.2020.114520

Li, X., Dong, S., Zhang, W., Fan, X., Wang, R., Wang, P., & Su, X. (2019). The occurrence of perfluoroalkyl acids in an important feed material (fishmeal) and its potential risk through the farm-to-fork pathway to humans. *Journal of Hazardous Materials*, *367*, 559–567. https://doi.org/10.1016/j.jhazmat.2018.12.103

Li, X., Gao, K., Dong, S., Liu, X., Fu, K., Wang, P., Zhang, A., Su, X., & Fu, J. (2019). Length-specific occurrence and profile of perfluoroalkyl acids (PFAAs) in animal protein feeds. *Journal of Hazardous Materials*, *373*, 224–231. https://doi.org/10.1016/j.jhazmat.2019.03.074

Li, X., Liu, Y., Yin, Y., Wang, P., & Su, X. (2023). Occurrence of some legacy and emerging contaminants in feed and food and their ranking priorities for human exposure. *Chemosphere*, *321*, 138117. https://doi.org/10.1016/j.chemosphere.2023.138117

Madariaga-Mazón, A., Hernández-Alvarado, R. B., Noriega-Colima, K. O., Osnaya-Hernández, A., & Martinez-Mayorga, K. (2019). Toxicity of secondary metabolites. *Physical Sciences Reviews*, *4*(12). https://doi.org/10.1515/psr-2018-0116

Medvedeva, Y., Kucher, A., Lipsa, J., & Hełdak, M. (2021). Human health risk assessment on the consumption of apples growing in urbanized areas: Case of Kharkiv, Ukraine. *International Journal of Environmental Research and Public Health*, *18*(4), 1–14. https://doi.org/10.3390/ijerph18041504

Naseri, M., Rahmanikhah, Z., Beiygloo, V., & Ranjbar, S. (2014). Effects of two cooking methods on the concentrations of some heavy metals (cadmium, lead, chromium, nickel and cobalt) in some rice brands available in Iranian market. *Journal of Chemical Health Risks*, *4*(2).

Nkansah, M. A., Agorsor, P. I., & Opoku, F. (2021). Heavy metal contamination and health risk assessment of mechanically milled delicacy called fufu. *International Journal of Food Contamination*, *8*(1). https://doi.org/10.1186/s40550-021-00085-y

Nkhebenyane, J., & Tywabi-Ngeva, Z. (n.d.). *Heavy Metal Contamination in Food: The Perspective of the Sub-Saharan Informal Food Trade*. www.intechopen.com

Noorlander, C. W., van Leeuwen, S. P. J., te Biesebeek, J. D., Mengelers, M. J. B., & Zeilmaker, M. J. (2011). Levels of perfluorinated compounds in food and dietary intake of PFOS and PFOA in The Netherlands. *Journal of Agricultural and Food Chemistry*, *59*(13), 7496–7505. https://doi.org/10.1021/jf104943p

Onyeaka, H., Ghosh, S., Obileke, K., Miri, T., Odeyemi, O. A., Nwaiwu, O., & Tamasiga, P. (2024). Preventing chemical contaminants in food: Challenges and prospects for safe and sustainable food production. *Food Control*, *155*, 110040. https://doi.org/10.1016/j.foodcont.2023.110040

Oyekunle, J. A. O., Ore, O. T., Durodola, S. S., Oyinloye, J. A., Oyebode, B. A., & Ajanaku, O. L. (2020). Heavy metal levels and changes in trimethylamine content of smoked fish and meat under different storage conditions. *SN Applied Sciences*, *2*(6), 1004. https://doi.org/10.1007/s42452-020-2844-7

Perelló, G., Martí-Cid, R., Llobet, J. M., & Domingo, J. L. (2008). Effects of various cooking processes on the concentrations of arsenic, cadmium, mercury, and lead in foods. *Journal of Agricultural and Food Chemistry*, *56*(23), 11262–11269. https://doi.org/10.1021/jf802411q

Pietron, W. J., Warenik-Bany, M., & Wozniak, B. (2021). Polybrominated diphenyl ethers (PBDEs) in raw milk from different animal species and in infant formula. Occurrence and risk assessment. *Chemosphere*, *278*, 130479. https://doi.org/10.1016/j.chemosphere.2021.130479

Poma, G., Glynn, A., Malarvannan, G., Covaci, A., & Darnerud, P. O. (2017). Dietary intake of phosphorus flame retardants (PFRs) using Swedish food market basket estimations. *Food and Chemical Toxicology*, *100*, 1–7. https://doi.org/10.1016/j.fct.2016.12.011

Rahman, M. A., Rahman, M. M., Reichman, S. M., Lim, R. P., & Naidu, R. (2014). Heavy metals in Australian grown and imported rice and vegetables on sale in Australia: Health hazard. *Ecotoxicology and Environmental Safety*, *100*, 53–60. https://doi.org/10.1016/j.ecoenv.2013.11.024

Rai, P. K., Lee, S. S., Zhang, M., Tsang, Y. F., & Kim, K. H. (2019). Heavy metals in food crops: Health risks, fate, mechanisms, and management. In *Environment International* (Vol. 125, pp. 365–385). Elsevier Ltd. https://doi.org/10.1016/j.envint.2019.01.067

Schecter, A., Päpke, O., Harris, T. R., Tung, K. C., Musumba, A., Olson, J., & Birnbaum, L. (2006). Polybrominated diphenyl ether (PBDE) levels in an expanded market basket survey of U.S. food and estimated PBDE dietary intake by age and sex. *Environmental Health Perspectives*, *114*(10), 1515–1520. https://doi.org/10.1289/ehp.9121

Schilter, B., Constable, A., & Perrin, I. (2014). Naturally occurring toxicants of plant origin. In *Food Safety Management* (pp. 45–57). Elsevier. https://doi.org/10.1016/B978-0-12-381504-0.00003-2

Schrenk, D., Bignami, M., Bodin, L., Chipman, J. K., del Mazo, J., Grasl-Kraupp, B., Hogstrand, C., Hoogenboom, L. (Ron), Leblanc, J., Nebbia, C. S., Nielsen, E., Ntzani, E., Petersen, A., Sand, S., Schwerdtle, T., Vleminckx, C., Marko, D., Oswald, I. P., Piersma, A., . . . Wallace, H. (2020). Risk assessment of aflatoxins in food. *EFSA Journal*, *18*(3). https://doi.org/10.2903/j.efsa.2020.6040

Shaw, I., & Vannoort, R. (2001). Pesticides. In *Food Chemical Safety* (pp. 218–237). Elsevier. https://doi.org/10.1533/9781855736320.2.218

Tao, F., Abou-Elwafa Abdallah, M., Ashworth, D. C., Douglas, P., Toledano, M. B., & Harrad, S. (2017). Emerging and legacy flame retardants in UK human milk and food suggest slow response to restrictions on use of PBDEs and HBCDD. *Environment International*, *105*, 95–104. https://doi.org/10.1016/j.envint.2017.05.010

Urugo, M. M., & Tringo, T. T. (2023). Naturally occurring plant food toxicants and the role of food processing methods in their detoxification. In *International Journal of Food Science* (Vol. 2023). Hindawi Limited. https://doi.org/10.1155/2023/9947841

Wang, J.-X., Bao, L.-J., Shi, L., Liu, L.-Y., & Zeng, E. Y. (2019). Characterizing PBDEs in fish, poultry, and pig feeds manufactured in China. *Environmental Science and Pollution Research*, *26*(6), 6014–6022. https://doi.org/10.1007/s11356-018-04057-2

Zafeiraki, E., Gebbink, W. A., Hoogenboom, R. L. A. P., Kotterman, M., Kwadijk, C., Dassenakis, E., & van Leeuwen, S. P. J. (2019). Occurrence of perfluoroalkyl substances (PFASs) in a large number of wild and farmed aquatic animals collected in the Netherlands. *Chemosphere*, *232*, 415–423. https://doi.org/10.1016/j.chemosphere.2019.05.200

Zhang, H., Vestergren, R., Wang, T., Yu, J., Jiang, G., & Herzke, D. (2017). Geographical differences in dietary exposure to perfluoroalkyl acids between manufacturing and application regions in China. *Environmental Science & Technology*, *51*(10), 5747–5755. https://doi.org/10.1021/acs.est.7b00246

Zhao, N., Fu, J., Liu, Y., Wang, P., Su, X., & Li, X. (2020). Animal-derived and plant-derived protein supplement feeds are important sources of organophosphate esters in the food supply. *Journal of Agricultural and Food Chemistry*, *68*(42), 11694–11701. https://doi.org/10.1021/acs.jafc.0c04235

Index

Note: Page numbers in **bold** indicate a table on the corresponding page.

F

G

H

I

Q

R

X

Y

Z